AA002967

2nd International Symposium on Fundamental Aspects of Rare-earth Elements Mining and Separation and Modern Materials Engineering (REES-2015)

IOP Conference Series: Materials Science and Engineering
Volume 112

Belokuriha, Russia
7 – 15 September 2015

Editors:

Levon Tavady
Viktor Sachkov
Anna Godymchuk
Anna Bogdan

ISBN: 978-1-5108-2087-6
ISSN: 1757-8981

2nd International Symposium on Fundamental Aspects of Rare-earth Elements Mining and Separation and Modern Materials Engineering (REES-2015)

IOP Conference Series: Materials Science and Engineering
Volume 112

Belokuriha, Russia
7 – 15 September 2015

IOP
Institute of Physics

Printed from e-media with permission by:

Curran Associates, Inc.
57 Morehouse Lane
Red Hook, NY 12571

Some format issues inherent in the e-media version may also appear in this print version.

Copyright© (2015) by the Institute of Physics
All rights reserved. The material featured in this book is subject to IOP copyright protection, unless otherwise indicated.

Printed by Curran Associates, Inc. (2016)

For permission requests, please contact the Institute of Physics at the address below.

Institute of Physics
Dirac House, Temple Back
Bristol BS1 6BE UK

Phone: 44 1 17 929 7481
Fax: 44 1 17 920 0979

techtracking@iop.org

Additional copies of this publication are available from:

Curran Associates, Inc.
57 Morehouse Lane
Red Hook, NY 12571 USA
Phone: 845-758-0400
Fax: 845-758-2634
Email: curran@proceedings.com
Web: www.proceedings.com

Table of contents

Volume 112

2nd International Symposium on Fundamental Aspects of Rare-earth Elements Mining and Separation and Modern Materials Engineering (REES-2015)
7-15 September 2015, Altay, Russia

Accepted papers received: 29 December 2015
Published online: 18 February 2016

Preface
011001

2nd International Symposium on Fundamental Aspects of Rare-earth Elements Mining and Separation and Modern Materials Engineering (REES-2015) OPEN ACCESS Levon Tavadyan Prof, Viktor Sachkov Prof, Dr. Anna Godymchuk and Anna Bogdan

011002

Peer review statement OPEN ACCESS

Papers

012001

Extraction of rare earth elements from hydrate-phosphate precipitates of apatite processing OPEN ACCESS M O Andropov, A V Anufrieva, A S Buynovskiy, Y N Makaseev, I N Mazov, R A Nefedov, V I Sachkov, O B Stepanova and AV Valkov pg. 1

012002

Rare earth elements materials production from apatite ores OPEN ACCESS A V Anufrieva, O S Andrienko, A S Buynovskiy, Y N Makaseev, I N Mazov, R A Nefedov, V I Sachkov, O B Stepanova and A V Valkov pg. 5

012003

Acoustoplasma Technique for Obtaining of Metal-Based Nanostructures OPEN ACCESS N A Bulychev and M A Kazaryan pg. 11

012004

The effects of liquid-phase oxidation of multiwall carbon nanotubes on their surface characteristics OPEN ACCESS I N Burmistrov, D S Muratov, I A Ilinykh, E A Kolesnikov, A Yu Godymchuk and D V Kuznetsov pg. 18

012005

Study of sulfur isotopes by vibrational spectroscopy and quantum chemistry OPEN ACCESS O Kh Poleshchuk, N B Egorov, D V Akimov, I I Zherin and N A Zhuravlev pg. 22

012006

Simulation of hydrogen adsorption on carbon nanotubes with different chirality parameters OPEN ACCESS S V Gromov, I N Burmistrov, I A Ilinykh and D V Kuznetsov pg. 30

012007

Spectrophotometric Procedure for Fast Reactor Advanced Coolant Manufacture Control OPEN ACCESS O S Andrienko, N B Egorov, I I Zherin and D V Indyk pg. 38

012008

The efficiency of rare elements used for hydrogen production from polymer waste OPEN ACCESS S A Garelina, A L Gysev, R A Zakharyan, M A Kazaryan, G Martoyan, V I Sachkov and I N Feofanov pg. 43

012009

Purification of rare-earth metals as the approach to improving properties of hard magnetic $Nd_2Fe_{14}B$-based materials OPEN ACCESS N B Kolchugina, G S. Burkhanov, A G Dormidontov, A A Lukin, Yu S Koshkid'ko, K Skotnicova, H Drulis and B Smetana pg. 49

012010

Corundum ceramic materials modified with silica nanopowders: structure and mechanical properties OPEN ACCESS M. A. Kostytsyn, D. S. Muratov, D. V. Lysov, K. O. Chuprunov, A. G. Yudin and D. V. Leybo pg. 56

012011

Energy effective approach for activation of metallurgical slag OPEN ACCESS I N Mazov, B B Khaydarov, S L Mamulat, D S Suvorov, Y S Saltikova, A G Yudin and D V Kuznetsov pg. 61

012012

Ultrasound-assisted synthesis of nanosized zero-valent iron for metal cations extraction and wastewater treatment applications OPEN ACCESS I Yu Mikhailov, D V Lysov, V V Levina, I N Mazov, A A Gusev, T I Yudintseva and D V Kuznetsov pg. 65

012013

Chromium-based catalyst for HFC-125 synthesis: promoters effect OPEN ACCESS S I Reshetnikov, L G Simonova, A A Zirka and R V Petrov pg. 73

012014

Separation of rare earth elements by zone recrystallization OPEN ACCESS D V Akimov, A N Dyachenko, N B Egorov and N A Zhuravlev pg. 78

012015

Surface layer composition of titania produced by various methods. The change of layer state under illumination OPEN ACCESS V Zakharenko, E Daibova, N Kosova and O Zmeeva pg. 85

012016

Nanostructured microtubes based on TiO_2 doped by Zr and Hf oxides with the anatase structure OPEN ACCESS VV Zheleznov, EI Voit, YV Sushkov, SA Sarin, VG Kuryavyi, DP Opra, SV Gnedenkov, SL Sinebryukhov and AA Sokolov pg. 92

012017

Investigation of supramolecular structure of the rare and rare-earth elements nanoparticles carrier when modified using microwave irradiation OPEN ACCESS I A Lysak, G V Lysak, T D Malinovskaya and L N Skvortsova pg. 100

012018

LiOH - H_2O_2 - H_2O trinary system study for the selection of optimal conditions of lithium peroxide synthesis OPEN ACCESS R A Nefedov, Yu A Ferapontov and N P Kozlova pg. 106

012019

Investigation of mechanical properties of masterbatches and composites with small additions of CNTs OPEN ACCESS I N Burmistrov, T I Yudintseva, I A Ilinykh, B B Khaydarov, I N Mazov, S M Anshin and D V Kuznetsov pg. 114

012020

Nanostructured $SmFeO_3$ electrophysical properties OPEN ACCESS K G Abdulvakhidov, S N Kallaev, M A Kazaryan, P S Plyaka, S A Sadikov, M A Sirota and S V Zubkov pg. 118

012021

The research of structure and mechanical properties of superhard electro-spark coatings for hardwearing mining tools OPEN ACCESS P A Bajin, A P Chijikov, D V Leybo, K O Chuprunov, A G Yudin, M A Alymov and D V Kuznetsov pg. 123

012022

Quantum chemical study of the structure and properties of isotopically pure lead chalcogenides OPEN ACCESS O Kh Poleshchuk, N B Egorov, D V Akimov, I I Zherin and N A Zhuravlev pg. 128

012023

Nucleation and crystallization behavior of RE - doped tellurite glasses OPEN ACCESS V Goncharuk, A Mamaev, V Silant'ev, P Starodubtsev and I Maslennikova pg. 137

012024

Hydrofluoride decomposition of natural materials including zirconium-containing minerals OPEN ACCESS N Laptash and I Maslennikova pg. 142

012025

Lead ferroniobat ceramic and films structures manufacture features and dielectric characteristic OPEN ACCESS S H Alikhadjiev, M A Kazaryan, A V Pavlenko and P S Plyaka pg. 149

012026

Experimental determination of energy transfer in Eu(III) complexes, based on pyrazole substituted 1, 3-diketones OPEN ACCESS E A Varaksina, S A Ambrozevich, N P Datskevich, I V Taydakov and A G Vitukhnovsky pg. 154

012027

Problem of the lithium peroxide thermal stability OPEN ACCESS R A Nefedov, Yu A Ferapontov and N P Kozlova pg. 160

012028

Estimation of the physico-chemical parameters of materials based on rare earth elements with the application of computational model OPEN ACCESS K Mamaev, A Obkhodsky and A Popov pg. 166

012029

Recent advances and future perspectives of nanosized zero- valent iron for extraction of heavy elements from metallurgical sludges OPEN ACCESS I Yu Mikhailov, V V Levina, E A Kolesnikov, K O Chuprunov, A A Gusev, A Yu Godymchuk and D V Kuznetsov pg. 175

012030

REM-containing silicate concentrates OPEN ACCESS V F Pavlov, O V Shabanova, I V Pavlov, M V Pavlov and A V Shabanov pg. 181

012031

Synthesis of micro-dispersed zirconium oxide for glass manufacturing OPEN ACCESS V Goncharuk, P Starodubtsev and I Maslennikova pg. 187

012032

Modification of Structural Phase State and Mechanical Properties of Poly-Grained Titanium Alloy Implanted by Aluminum Ions OPEN ACCESS A V Nikonenko, N A Popova, E L Nikonenko, M P Kalashnikov and A Kurzina pg. 191

012033

New technology of extracting the amount of rare earth metals from the red mud OPEN ACCESS G A Martoyan, G G Karamyan and G A Vardan pg. 197

012034

Titanium compounds as catalysts of higher alpha-olefin-based super-high-molecular polymers synthesis OPEN ACCESS K B Konovalov, M A Kazaryan, V N Manzhay and O V Vetrova pg. 202

012035

Prospects of lithium enrichment on ^{7}Li isotope by method of controlled ions electro-migration OPEN ACCESS G A Martoyan, M M Kalugin, A V Gabrielyan and A G Martoyan pg. 206

012036

Structure and phase composition of the superalloy on the basis of Ni-Al-Cr alloyed by Re and La OPEN ACCESS E L Nikonenko, N A Popova, N A Koneva and E V Kozlov pg. 211

012037

Effect of preparation conditions on physic-chemical properties of tin-doped nanocrystalline indium oxide OPEN ACCESS T D Malinovskaya, V I Sachkov, V V Zhek and R A Nefedov pg. 216

012038

Formation Structural Phase Gradients in Rail Steel During Long-Term Operation OPEN ACCESS Yu F Ivanov, K V Morozov, O Peregudov, V E Gromov, N A Popova and E N Nikonenko pg. 222

012039

The preparation of the Ti-Al alloys based on intermetallic phases OPEN ACCESS N Kosova, V Sachkov, I Kurzina, A Pichugina, A Vladimirov, L Kazantseva and A Sachkova pg. 226

012040

Study of rhenium production process by hydrogen reduction of rhenium hexafluoride OPEN ACCESS V I Vybyvanets, A V Kosukhin, A V Cherenkov and G S Shilkin pg. 231

012041

Investigation into Rhenium Fluorination Process OPEN ACCESS V I Vybyvanets, A V Kosukhin, A V Cherenkov and G S Shilkin pg. 236

012042

Production and investigation into properties of high-pure rhenium fluorides OPEN ACCESS V I Vybyvanets, A V Kosukhin, A V Cherenkov and G S Shilkin pg. 243

012043

Structural Transformations in heat resistant coatings containing rare earth elements OPEN ACCESS N I Afanasiev, O K Lepakova and N I Kosova pg. 249

012044

Deformation and Damage Accumulation in a Ceramic Composite under Dynamic Loading OPEN ACCESS M V Korobenkov, S N Kulkov, O B Naymark, U V Khorechko and A V Ruchina pg. 254

012045

Self-propagating high-temperature synthesis of the high- current emission lanthanum and niobium contained ceramics OPEN ACCESS O Yu Dolmatov, I V Shamanin and S S Chursin pg. 259

IOP Conf. Series: Materials Science and Engineering **112** (2016) 011001 doi:10.1088/1757-899X/112/1/011001

Preface

The **2nd International Symposium «Fundamental Aspects of Rare-earth Elements Mining and Separation and Modern Materials Engineering» (REES2015)** was jointly organized by Tomsk State University (Russia), National Academy of Science (Armenia), Shenyang Polytechnic University (China), Moscow Institute of Physics and Engineering (Russia), Siberian Physical-technical Institute (Russia), and Tomsk Polytechnic University (Russia) in September, 7-15, 2015, Belokuriha, Russia.

The Symposium provided a high quality of presentations and gathered engineers, scientists, academicians, and young researchers working in the field of rare and rare earth elements mining, modification, separation, elaboration and application, in order to facilitate aggregation and sharing interests and results for a better collaboration and activity visibility.

The goal of the REES2015 was to bring researchers and practitioners together to share the latest knowledge on rare and rare earth elements technologies.

The Symposium was aimed at presenting new trends in rare and rare earth elements mining, research and separation and recent achievements in advanced materials elaboration and developments for different purposes, as well as strengthening the already existing contacts between manufactures, highly-qualified specialists and young scientists.

The topics of the REES2015 were: (1) Problems of extraction and separation of rare and rare earth elements; (2) Methods and approaches to the separation and isolation of rare and rare earth elements with ultra-high purity; (3) Industrial technologies of production and separation of rare and rare earth elements; (4) Economic aspects in technology of rare and rare earth elements; and (5) Rare and rare earth based materials (application in metallurgy, catalysis, medicine, optoelectronics, etc.).

We want to thank the Organizing Committee, the Universities and Sponsors supporting the Symposium, and everyone who contributed to the organization of the event and to publication of this proceeding.

The Organizing Committee of the REES2015

Editors:

Prof. Levon Tavadyan, National Academy of Sciences of the Republic of Armenia
Prof. Viktor Sachkov, Tomsk State University, Russia
Dr. Anna Godymchuk, Tomsk Polytechnic University, Russia
Anna Bogdan, Tomsk Polytechnic University, Russia

Content from this work may be used under the terms of the Creative Commons Attribution 3.0 licence. Any further distribution of this work must maintain attribution to the author(s) and the title of the work, journal citation and DOI.
Published under licence by IOP Publishing Ltd

Peer review statement

All papers published in this volume of *IOP Conference Series: Materials Science and Engineering* have been peer reviewed through processes administered by the proceedings Editors. Reviews were conducted by expert referees to the professional and scientific standards expected of a proceedings journal published by IOP Publishing.

Content from this work may be used under the terms of the Creative Commons Attribution 3.0 licence. Any further distribution of this work must maintain attribution to the author(s) and the title of the work, journal citation and DOI.
Published under licence by IOP Publishing Ltd

IOP Conf. Series: Materials Science and Engineering **112** (2016) 012001 doi:10.1088/1757-899X/112/1/012001

Extraction of rare earth elements from hydrate-phosphate precipitates of apatite processing

M O Andropov[1,2], A V Anufrieva[1], A S Buynovskiy[1], Y N Makaseev[1], I N Mazov[1,2], R A Nefedov[1,2], V I Sachkov[1,2], O B Stepanova[2], and AV Valkov[1]

[1] Seversk Technological Institute of National Research Nuclear University "MEPhI", Tomsk 634050, Russian Federation

[2] Siberian Physical-Technical Institute of Tomsk State University, Tomsk 634050, Russian Federation

Email: oyuna12@yandex.ru

Abstract. The features of extraction of rare earth elements (REE) were considered from hydrate-phosphate precipitates of REE of apatite processing by nitric acid technology. The preliminary purification of nitrate solution of REE from impurities of titanium, aluminum, iron, uranium and thorium was suggested to obtain stable solutions not forming precipitates. Washing the extract was recommended with the evaporated reextract that allows to obtain directly on the cascade of REE extraction the concentrated solutions suitable for the separation into groups by the extraction method. Technical decisions were suggested for the separation of REE in groups without the use of salting-out agent.

1. Introduction

The apatite raw materials contain up to 1 % (wt.) of rare earth elements (REE) and present a very important potential source of cheap rare earth concentrates [1]. Depending on the accepted technology of processing the REE present in apatite are concentrated in a nitrogen-phosphate solution (NPS) or in phosphogypsum. The most promising variant is the extraction of REE at the opening of apatite by nitric acid. REE content in the resulting nitrogen-phosphate solution is 6 – 7 g / l. Standard extraction technology of REE from NPS provides (after freezing calcium nitrate and separation of fluorine) the selective extraction of hydrate-phosphate precipitates of REE at pH 1.1 - 2.0, filtration and washing the precipitate. Previously the technology has been tested in variant of the integrated tests on Dneprodzerzhisk "Azot" and Kirov-Chepetz factory [1-3].

The various methods of extraction of REE from the concentrate were suggested. The results of studies were published on the extraction of REE in the form of oxalate by precipitation followed by conversion of oxalates into hydroxides [4], by treatment of the precipitate by sulfuric acid with formation sulfates of REE [5] or by extraction with neutral organophosphorus compounds [6]. According to [6] the hydrate-phosphate precipitate containing 20 – 30 % of REE is dissolved in nitric acid to the residual content of nitric acid 4 - 6 mol/l, then the excess of nitric acid is neutralized with ammonium hydroxide to obtain a solution containing 60-70 g/l of REE (per oxides), 1.5 - 2 mol/l of nitric acid, 500 g/l of ammonium nitrate and a significant amount of phosphate and fluoride ions. In the process of countercurrent extraction the authors [6] noted the intense sedimentation, which

Content from this work may be used under the terms of the Creative Commons Attribution 3.0 licence. Any further distribution of this work must maintain attribution to the author(s) and the title of the work, journal citation and DOI.
Published under licence by IOP Publishing Ltd

increases with advancing aqueous solution by extraction steps. The resulting solutions are extremely unstable, affected to form precipitates which are formed during standing solutions or during the extraction process. Since phosphates of REE, iron, aluminum and titanium are not soluble, and then the increase of stability of the system is reached by introducing nitric acid or salting-out agent. However, this does not always lead to stop sedimentation.

In this paper the results of studies on the extraction purification of hydrate–phosphate precipitate of REE and the subsequent division into groups in the mode, decreasing significantly the possibility of sedimentation, are given.

2. Experimental procedure

2.1 Processing of the initial apatite by nitric acid technology with separation of the sum of REE as a precipitate

The nitrogen-phosphate solution was prepared by treatment of apatite with nitric acid at temperature of 50 - 70 °C for 3 hours, the precipitate of the unopened apatite on the filter was separated and calcium nitrate (strontium) was separated by freezing. After purification from the fluorine the solution contained 400 – 500 g/l of phosphoric acid, 80 - 120 g/l of nitric acid, 50 - 80 g/l of calcium (as the oxide) and 6 – 7 g/l of REE (as oxide). To isolate REE the solution was treated with gaseous ammonia to pH = 1.8 - 2.0. The resulting precipitate was badly filtered. The increase of the filtration rate was provided by delay or by the step neutralization. The total content of REE reached 24 – 26 % in good filter precipitate. The impurities of iron, aluminum, partially calcium, as well as uranium and thorium, contained in the apatite in minor amounts, transferred in the precipitate from the solution. The resulting precipitate contains (depending on the humidity) (wt.%): 20 - 25 % of REE oxides, titanium 0.16 %, thorium 0.05 %, uranium 0.01 %, impurities of iron (III) and aluminum.

2.2 Dissolution of the hydrate-phosphate precipitates of REE and precipitation affinage of the resulting solution

The hydrate-phosphate precipitate of REE was dissolved in nitric acid to neutralize the excess of nitric acid with the source precipitate and the salting-out agent was added – NH_4NO_3 (500 - 600 g/l) or $Ca(NO_3)_2$ (400 - 500 g/l). The pulp was filtered. To the filtrate the ammonium hydroxide was added to pH = 0.2 - 0.7, and the precipitate was separated by filtration. In the precipitate the impurities were concentrated: iron, aluminum, thorium, titanium, uranium and partially REE. The precipitate is removed from the process, and the filtrate was acidified up to 1 - 1.5 mol/l of nitric acid. When performing recommended parameters the thorium and the titanium are removed almost completely, and uranium by 70 – 80 %. Experimentally it is found that if you do not remove the thorium by chemical precipitation, it is extracted into the organic phase during the extraction with 100 % tributyl phosphate (TBP). In the absence of salting-out agent the distribution coefficient of uranium varied within values of 30 - 40, of thorium - 0.3 - 0.5, the distribution coefficient of REE - 0.2 - 0.3.

2.3 Investigation of extraction processes of separation of the sum of REE into groups

Experiments were carried out according to the method by shaking the organic and aqueous phases in separating funnels with subsequent slicing. The content of individual REE was determined by atomic emission spectrometry with the inductively coupled plasma on spectrometer "VISTA-PRO" of company "VARIAN".

REE were extracted in the presence of salting-out agents (the distribution coefficient of REE reached values of 1 - 3) by extraction with 100 % tributyl phosphate. As the wash solution the 15 - 20 % evaporated reextract of the purified sum of REM was used, concentration of 250 - 300 g/l. The remainder of the evaporated reextract was deduced as finished products to be separated into groups. Composition of total REE isolated from the apatite concentrate (wt. %): La_2O_3 - 25.5; CeO_2 – 46.4; Pr_6O_{11} – 4.0; Nd_2O_3 – 11.6; Sm_2O_3 – 1.7; Eu_2O_3 – 0.5; Gd_2O_3 – 1.4; Tb_4O_7 – 0.2; Dy_2O_3 – 0.7; Ho_2O_3 – 0.2; Er_2O_3 – 0.3; Yb_2O_3 – 0.4; Y_2O_3 – 7.0.

The division into groups was carried out by extraction in system 100 % TBP- nitrates of REE (300 - 350 g/l). As the wash solution the evaporated reextract was used, concentration of 250 - 300 g/l, the reextracted solution is 0.1 - 0.2 mol/l of nitric acid. From the rare earth nitrate solution the cerium was recovered by electrolytic oxidation. Extraction of cerium is 95 %. Composition of REE of solution without cerium (wt. %.): La_2O_3 – 45.0; CeO_2 – 5.3; Pr_6O_{11} – 7.7; Nd_2O_3 – 20.5; Sm_2O_3 – 3.0; Eu_2O_3 – 0.9; Gd_2O_3 – 2.5; Tb_4O_7 – 0.4; Dy_2O_3 – 1.2; Ho_2O_3 – 0.3; Er_2O_3 – 0.5; Yb_2O_3 – 0.7; Y_2O_3 – 12.0. To separate into groups the scheme of successive separation was adopted on the first cascade through line of neodymium - samarium, and on the second cascade - through line of praseodymium-cerium.

The calculated parameters of the first separation cascade: the capacity of the organic phase - 150 g / l, the separation coefficient samarium-neodymium $\beta = 2.0$, the number of stages in the extraction 35, in the washing 30, in the reextraction 10, the ratio phases Vorg : Vinit : Vprod: Vreextr = 4 - 6 : 1.0 : 1.8 : 3.0. The reextract was evaporated to concentration of 250 - 300 g/l and the calculated amount was used as the wash solution. The remaining amount was output from the process in the form of solution containing elements of the average and the yttrium groups. Composition of REE of the reextract (wt. %): La_2O_3 - < 0.05; CeO_2 - < 0.05; Pr_6O_{11} - < 0.05; Nd_2O_3 – 6.4; Sm2O3 – 12.7; Eu_2O_3 – 3.7; Gd_2O_3 – 10.5; Tb_4O_7 – 1.5; Dy_2O_3 – 5.2; Ho_2O_3 – 1.5; Er_2O_3 – 2.3; Yb_2O_3 – 3.0; Y_2O_3 – 52.

In the raffinate phase the less extractable elements are concentrated, composition (wt. %): La_2O_3 – 58.9; CeO_2 – 7.0; Pr_6O_{11} – 9.2; Nd_2O_3 – 24.5; Sm_2O_3 - <0.02; Eu_2O_3 - <0.02; Gd_2O_3 - <0.02; Tb_4O_7 - <0.02; Dy_2O_3 - <0.02; Ho_2O_3 - <0.02; Er_2O_3 - <0.02; Yb_2O_3 - <0.02; Y_2O_3 - <0.02. At the second separation cascade the division was carried out through line of cerium-praseodymium to yield pure didymium in the extract. The calculated parameters of the second separation cascade: the capacity of the organic phase - 150 g l, the separation coefficient praseodymium - cerium $\beta = 1.6$, the ratio of phases: Vorg : Vinit : Vprod: Vreextr = 5-6 : 1.0 : 2.3 : 3.0. The extractant was entered into the first stage, the starting solution in step 35, the wash solution in step 65, the reextracted solution was injected into step 75. Composition of REE in the reextract (wt. %): La_2O_3 - <0,01; CeO_2 - <0,01; Pr_6O_{11} - 25; Nd_2O_3 - 75; Sm_2O_3 - <0,02; Eu_2O_3 - <0,02; Gd_2O_3 - <0,02; Tb_4O_7 - <0,02; Dy_2O_3 - <0,02; Ho_2O_3 - <0,02; Er_2O_3 - <0,02; Yb_2O_3 - <0,02; Y_2O_3 - <0,02.

In the raffinate the lanthanum with residues of cerium and part of praseodymium was concentrated (wt. %): La_2O_3 - 88; CeO_2 – 10.3; Pr_6O_{11} – 1.5; Nd_2O_3 – 24.5; Sm_2O_3 - <0.02; Eu_2O_3 - <0.02; Gd_2O_3 - <0.02; Tb_4O_7 - <0.02; Dy_2O_3 - <0.02; Ho_2O_3 - <0.02; Er_2O_3 - <0.02; Yb_2O_3 - <0.02; Y_2O_3 - <0.02. Both in the first and second cascades of the extraction process was carried out in a transient mode, with accumulation distinguishable component in the first cameras of the extraction part of the cascade.

3. Research results and discussion

As a result of these experiments the nitrogen-phosphate solution has been obtained by nitric acid technology used in industry. To solve the problem of complex processing of apatite from the resulting solution the strontium nitrate and nitrate solution of the sum of REE were isolated for further processing.

Mode of extraction of REE is shown from this solution by precipitation with ammonia and subsequent dissolution of the resulting precipitate with nitric acid, to remove the impurities, preventing the extraction, wherein, titanium and thorium are removed almost completely, and uranium by 70 - 80 %.

The suggested scheme of the extraction separation of the sum of REE allows you to extract of cerium with the degree of extraction of 95 % and groups of REE on lines samarium-neodymium and cerium-praseodymium in the form of products. The obtaining didymium in the extract is suitable for production of alloys or the manufacture of magnets. The raffinate, containing 80 - 90 % of lanthanum, is suitable for producing lanthanum with content of 99.99 % of the main substance.

4. Conclusions

1. The features of REE extraction are found from the hydrate-phosphate precipitates of apatite processing by nitric acid technology.

IOP Conf. Series: Materials Science and Engineering **112** (2016) 012001 doi:10.1088/1757-899X/112/1/012001

2. It is proposed to output impurities by selective precipitation and to carry out the process of extraction purification with saturation of the organic phase by the evaporated reextract.
3. Possible technical decisions for the separation into groups of the evaporated reextract of the purification cascade were discussed.

Acknowledgment
This work was supported by the federal target program «Research and development on priority directions of scientific-technological complex of Russia for 2014-2020» (RFMEFI57514X0024).

References
[1] Goldinov A L, Moiseev S D, Novoselov F I 1980 *Rus. Chem. Ind.* **4** 33
[2] Kazak V G, Melnikova N S, Dmitriev V D, Konjuchenko N N 1983 *Chem. Technol.* **1** 13
[3] Lebedev V N, Popova L A, Rudenko A V 2002 *Chem. Technol.* **9** 18
[4] Lebedev V N, Lokshin E P, Masloboev V A 2007 *Method for Processing the Rare Earth Phosphate Concentrates"* RU Patent 2344079 dated from 25.04.2007
[5] Lokshin E P, Tareeva O A 2010 *The Purification Process of the Rare Earth Fluoridecontaining Concentrate* RU Patent 2429199 dated from 15.03.2010
[6] Moiseev S D, Semenov V A, Shelekhova G L 1981 *Rus. Chem. Ind.* **8** 24

IOP Conf. Series: Materials Science and Engineering **112** (2016) 012002 doi:10.1088/1757-899X/112/1/012002

Rare earth elements materials production from apatite ores

A V Anufrieva[1], O S Andrienko[2,3], A S Buynovskiy[1], Y N Makaseev[1], I N Mazov[1,2], R A Nefedov[1,2], V I Sachkov[1,2], O B Stepanova[2,3,4], and A V Valkov[1]

[1] Seversk Technological Institute of National Research Nuclear University "MEPhI", 65 Kommunisticheski avenue, Seversk, 634050, Russian Federation

[2] Siberian Physical-Technical Institute of Tomsk State University, 30 Lenina avenue, Tomsk 634050, Russian Federation

[3] Institute of Atmospheric Optics Named after Academician V.E. Zuev of the Russian Academy of Sciences, 1 Academician Zuev square, Tomsk 634050, Russian Federation

[4]E-mail: oyuna12@yandex.ru

Abstract. The paper deals with the study of processing apatite ores with nitric acid and extraction of the rare earth elements. The rare earth elements can be successfully separated and recovered by extraction from the nitrate- phosphate solution, being an tributyl phosphate as extraction agent. The developed scheme of the processing apatite concentrate provides obtaining rare earth concentrates with high qualitative characteristics.

1. Introduction

Rare earth elements (REEs) are a group of 15 chemical elements in the periodic table, specifically the lanthanides. Although relatively abundant in the earth's crust, they rarely occur in concentrated forms, making them economically challenging to obtain. These elements comprise critical components of many of our modern-day technological devices and everyday electronics. The increased use of REEs in magnets, modern electronics and in a variety of commercial products has led to a shortage of REEs for production purposes. Currently, REEs are being disposed in large quantities rather than being recovered and reused. The two major mineral sources of REEs are bastnasite and monazite and will be the focus of this section. However, it is important to note that production can come from a variety of minerals, such as xenotime, apatite, yttrofluorite, cerite, and gadolinite[1].

The general formula for apatite is $Ca_5F\ (PO_4)_3$, where the part of the calcium can be substituted mainly by Na, Sr, Mn, and the REEs. Apatite contains an average of 0.1–0.8 % of rare earth oxides [2, 3].

The two primary methods for extracting REEs from apatite are selective acid extraction and physical separation techniques. The use nitric acid for processing of apatite is the most perspective. The total dissolution of the apatite with nitric acid as per the following reaction:

$$Ca_5F(PO_4)_3 + 10\ HNO_3 \rightarrow 3H_3PO_4 + 5Ca(NO_3)_2 + HF.$$

All impurities, including REE, are dissolved:

$$LnPO_4 + 3H^+ + 3NO_3^- = Ln^{+3} + 3NO_3^- + H_3PO_4.$$

The formed calcium nitrate is separated by filtration. The rare earths are ended up in the nitrate-phosphate solution.

Content from this work may be used under the terms of the Creative Commons Attribution 3.0 licence. Any further distribution of this work must maintain attribution to the author(s) and the title of the work, journal citation and DOI.
Published under licence by IOP Publishing Ltd

IOP Conf. Series: Materials Science and Engineering **112** (2016) 012002 doi:10.1088/1757-899X/112/1/012002

For the extraction from nitrate solutions, organic extractants based on neutral organophosphorous compounds (preferably tributylphospate - TBP) are typically used. By choosing suitable process conditions (nature of the extraction system, the ratio of the organic extractant/aqueous solution, amount of extraction and re-extraction stages, REEs precipitation mode), up to 98% REEs contained in the solution can be extracted into concentrate [4, 5].

The purpose of this work is to study the complex processing apatite ores with nitric acid and to extract the rare earth metals and radionuclides.

2. Experimental procedure

2.1 The process of opening apatite concentrate

The decomposition of samples of apatite raw was carried out using excess of nitric acid with concentration of 56 wt % at 50 °C for 1.5 hours. Scheme of the processing apatite concentrate is shown in the Figure1.

The following measuring instruments, auxiliary equipment, materials and reagents were used:

- glass reactor with the main technical specifications: volume 5 L, rotational speed of the stirring device: 0-700 / min, working pressure: 0.2-1.2 kg / cm^2 , working temperature range (– 20) - 180 °C;
- cryostat to the reactor, providing a temperature of from – 30 to 100 ± 2 ° C;
- vacuum pump;
- Thermo Scientific iCAP 6200 inductively coupled optical emission spectrometer (ICP-OES);
- electronic scales GR 120 «A&D Company Ltd.»;
- laboratory shaker LOIP LS-110;
- nitric acid CAS № 7697-37-2;
- sodium bicarbonate CAS № 144-55-8;
- sulfuric acid CAS № 7664-93-9;
- ammonia CAS № 7664-41-7;
- tributyl phosphate CAS № 126-73-8.

Composition of apatite sample was studied with the help of Thermo Scientific iCAP 6200 inductively coupled optical emission spectrometer (ICP-OES). Results of this study are shown in Table 1.

Table 1. Composition of apatite sample

Oxide	Content, wt %	Oxide	Content, wt %	Oxide	Content, wt %
SiO_2	1.73	ΣREE_2O_3	0.90	TiO_2	0.45
CaO	51.64	La_2O_3	0.25	Eu_2O_3	0.01
P_2O_5	39.93	CeO_2	0.38	Gd_2O_3	0,022
MgO	0.10	Pr_2O_3	0.04	SrO	0.64
Fe_{all}	0.46	Nd_2O_3	0.14	Tb_2O_3	0.002
Na_2O	0.42	Sm_2O_3	0.02	Dy_2O_3	0.01
F	3.14	Y_2O_3	0.04	Yb_2O_3	0.00

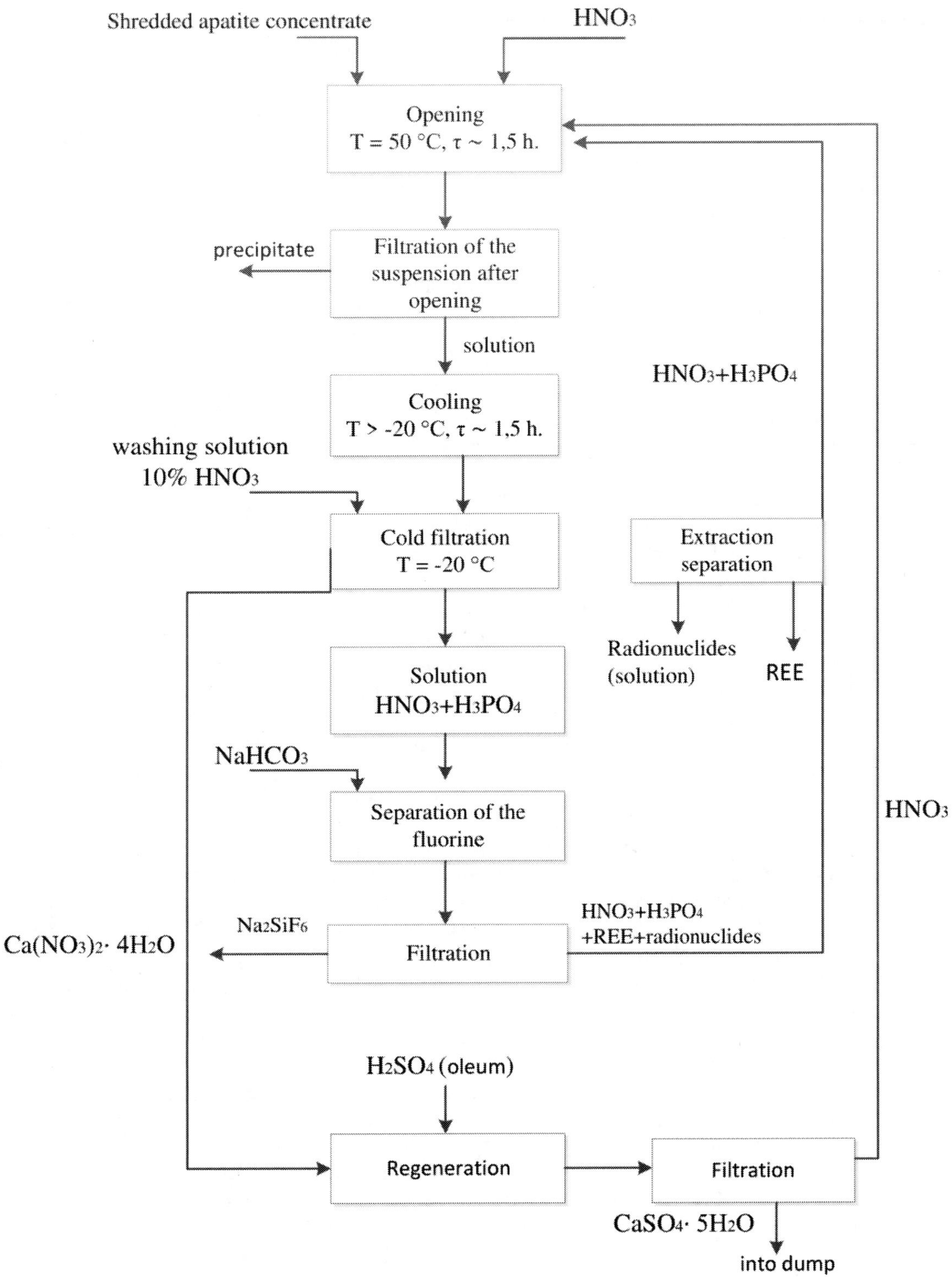

Figure 1. Scheme of the processing apatite concentrate.

The sample of apatite concentrate was placed in a reactor, the concentrated nitric acid with excess 10-15% by weight was added and at temperature of 50 °C it was stirred for 1.5 hours at frequency of rotation 120 rev / min.

The suspension was filtered after opening. The solid precipitate was separated and was dried in the oven at temperature of 60 °C to constant weight. The resulting solution after filtration of the suspension was placed in a glass reactor, it was cooled to – 20 °C and was stirred for 1.5 hours. Then it was washed by nitric acid with concentration of 10 % wt. Thus about 90 % of $Ca(NO_3)_2$ was allocated to a solid phase.

The formed crystals of calcium nitrate were separated from the mother liquor by filtration and were washed by nitric acid at temperature of – 10 °C. The concentrated sulfuric acid (oleum) was added to the filtered precipitate of calcium nitrate under stirring. The obtained precipitate of calcium sulfate was filtered at temperature of 20 °C, additionally was washed with distilled water at a ratio of S : L = 1 : 5 on the filter, the precipitate of calcium sulfate was dried. The washing solution (REE, HNO_3 + H_3PO_4, radionuclides) was directed to the stage of decomposition of apatite.

The mother liquor (HNO_3 + H_3PO_4) after the cold filtration was directed to obtain fluorine. The mother liquor at room temperature was mixed with twice excess of $NaHCO_3$, F – ion was precipitated as a poorly soluble silicic sodium fluoride, to increase degree of extraction of REE. After precipitation of Na_2SiF_6 the solution was neutralized by ammonia to pH 2.5 to obtain REE phosphates. After repeated water washing (decantation) the separated phosphates are the technical product containing 25-30 % of Ln_2O_3, 30 % of P_2O_5, 10 % of SiO_2, 5-8 % of $Ca(NO_3)_2$, 3 % of Fe_2O_3, 1.5 % or less of F, 5-10 % of H_2O.

The solution containing the composition of HNO_3 + H_3PO_4, rare earth metals and radionuclides was directed to the extraction of REM and radionuclides.

2.2 Extraction from the obtained solution of rare earth elements and radionuclides

The extraction was carried out in separating funnels of 100 ml. The undiluted tributyl phosphate (TBP) was used as an extractant. The extractant was previously saturated by nitric acid with a concentration corresponding to the working solution. After saturation the extraction process was carried out.

The shaking of funnels was carried out using a laboratory shaker LOIP LS 110. The composition of the initial solution was determined by the evaporation of the starting solution. For this purpose the single-stage extraction was carried out with a ratio of organic phase and water phase (O : W) = 1 : 1 (15 ml : 15 ml) by 100 % TBP in a volumetric flask of 50 ml. Re-extraction was carried out with 0.1 M solution of HNO_3, O : W = 1: 1.

Investigation of the process of re-extraction was carried out at temperature of 60 °C. After the purification steps of impurities from the solution REM were extracted with 100% TBP in ratio of O: W = 1: 1. Separation on a subgroup was performed in step of the washing of extract: for selective washout of light REE the concentrated solution of medium and heavy REE was used as wash solution. Next on stage re-extraction REE were washout by 0.1 M HNO_3. Solutions of the various steps were combined, thereby obtaining two solutions - solution after washing containing light REE and reextract, containing medium and heavy REE.

3. Research results and discussion

Nitric acid was selected to allow most complete dissolution of components contained in apatite. Use of HNO_3 at concentration of 56 wt % was especially preferred. The concentration of acid in the solution should prevent REM precipitation in form of phosphates, fluorides, double sulphates. The resulting nitrate - phosphate solution was suitable to isolate the rare earth elements.

As a result of researches it is established that the degree of precipitation and the content of the rare earth elements in the precipitate depend on the pH value. A series of experiments was carried out in the pH range 1 - 5 to study the effect of pH on precipitation degree of a number of elements (figure 1).

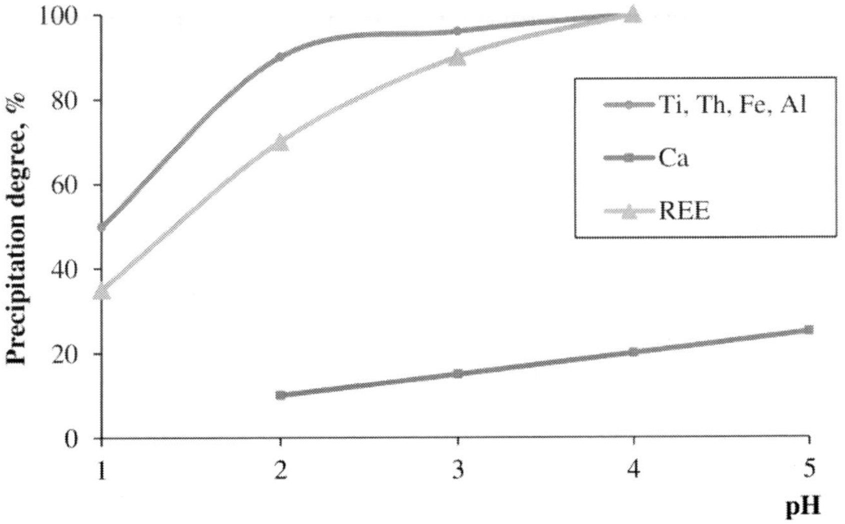

Figure 1. The precipitation degree of a number of elements depending on the pH

It was observed that about 90-95 % of the rare earth elements from nitrate - phosphate solution were precipitated at pH = 2.3-2.7, but the content of rare earth elements in this precipitate was decreased because the significant amount of calcium was co-precipitated. The degree of precipitation of the rare earth elements was about 50 % at pH 1.4-1.6, but the content of the rare earth elements in this precipitate was increased. In addition the filtration rate of the precipitate was decreased at higher pH. Thus, the precipitation of REE from solution was carried out under pH= 2.5.

In order to avoid any contamination problems due to the fluorinated acids (in particular, H_2SiF_6), it was necessary to eliminate such acids before extraction. On this purpose, the precipitation in the form of Na_2SiF_6 after addition of a sodium salt $NaHCO_3$ was chosen. The precipitation of Na_2SiF_6 was more effective at room temperature. Analysis of the obtained precipitate showed that more than 80 % of the fluorides initially present in the apatite can be recovered as Na_2SiF_6.

The use of the crystallization of calcium nitrate makes it possible to regenerate a significant part of nitric acid. The resulting calcium sulphate is a product suitable for additional industry.

The use of 100% of triburyl phosphate as an extractant significantly reduced the volumes of solutions and provided recovery of rare earth elements.

To determine the optimum ratio of the organic /water phase we conducted a single-stage extraction. We investigated the following ratios of O: W = 0.5:1; 1:1; 1.5:1; 2:1; 4:1; 6:1; 10:1. It was found that extraction of rare earth elements should be carried out under the following conditions: temperature - 22-25 °C, the ratio of organic/water phase = 1:1.

4. Conclusion

The method of opening of apatite ores involves the decomposition of the apatite concentrate by nitric acid, the separation of the insoluble residue, freezing and subsequent separation of crystals of calcium nitrate tetrahydrate, the precipitation of fluorine from the nitrate-phosphate solution, the neutralization of the purified solution to a final pH 2,5, the separation of the precipitate of phosphates of rare earth elements and radionuclides, the extraction of rare earth metals and radionuclides from the solution. The choice of the above conditions, the sequence of operations, modes and parameters of the process was based on the analysis of the results of studies and the data obtained from the tests. This scheme of the processing apatite concentrate provides obtaining rare earth concentrates with high qualitative characteristics.

Acknowledgment
This work was supported by the federal target program «Research and development on priority directions of scientific-technological complex of Russia for 2014-2020» (RFMEFI57514X0024).

References

[1] R J Weber, D J Reisman 2012 *Rare Earth Elements; A Review of Production, Processing, Recycling and Associated Environmental Issues* US EPA Region 7: Kansas, KS, USA.
[2] K A Bolshakov 1976 *Chemistry and Technology of Rare and Scattered Elements* (Moscow: Vysshaya Shkola)
[3] L Alders 1957 *Solvent Extraction* (Moscow: Foreign Literature Publishing House)
[4] D I Rjabchikov and V A Riabukhin 1966 *Analytical Chemistry of Rare Earth Elements and Yttrium* (Moscow: Nauka)
[5] S D Moses, V A Semenov and G L Shelekhova 1981 *Chem. Industry* **8** 24

IOP Conf. Series: Materials Science and Engineering **112** (2016) 012003 doi:10.1088/1757-899X/112/1/012003

Acoustoplasma Technique for Obtaining of Metal-Based Nanostructures

N A Bulychev[1], M A Kazaryan
Lebedev Physical Institute of RAS, 53, Leninsky ave., Moscow, 119991, Russia

[1]E-mail: nbulychev@mail.ru

Abstract. In this study, plasma discharge in a liquid at intensive ultrasonic field above the cavitation threshold has been proven to be of great interest for initiation of various physical and chemical processes. In such discharge, nanoparticles of tungsten and zinc oxides have been synthesized. Further exploration of synthesized nanoparticles has demonstrated that the factor of ultrasonic cavitation during the synthesis substantially affects physical and chemical characteristics of nanoparticles.

1. Introduction

In recent years, approaches to synthesis of nanomaterials and nanopowders in particular have been of increasing interest and have attracted a great deal of attention of many groups of researchers [1-3]. A range of techniques, including sol-gel method, high-energy dispersing, nanoscale biosynthesis have been being applied for several decades already that enabled to obtain nanoparticles of titania, silica, mica, carbon black and other materials. Along with traditional chemical methods of synthesis, some physical methods, such as electric discharges and ultrasound cavitation became increasingly important in view of creation of novel functional and composite materials.

One of the new promising ways to obtain nanoscale materials, including metal oxide powders is the combined effect of the elastic oscillations of high intensity ultrasound and pulsed or steady electric fields in a liquid medium. This type of plasma, being of great interest as a new object of physical study, has several advantages as a method for the synthesis of nanomaterials – a relatively narrow particle size distribution of the synthesized nanopowder, specific composition and the properties of nanomaterials [4-7].

More specifically, the traditional arc discharge in aqueous electrolytes, which is widely used in engineering, is the most well-known form of stationary plasma discharge in liquid media. For many years, such discharge is used in physicochemical studies and in the synthesis of various materials. The specific feature of arc discharge in liquid media is the localization of plasma region near the electrode ends and "falling" form of volt-ampere characteristics.

However, when ultrasonic cavitation is applied to a liquid, phase composition and physical properties of the liquid change abruptly, and this effect can lead to some specific features of the formation of electric discharges in the liquid. In the region of intensive cavitation, the fraction of gas-vapor component in the liquid has a significant value, therefore it can be assumed that the conditions of electric breakdown in the cavitation region should become easier, which can result in the initiation of different forms of discharges. Varying the parameters of ultrasonic field, it is apparently possible to exert an influence on the processes of plasma glow in a cavitating liquid. Furthermore, in this configuration, ultrasonic cavitation will interact with micro- and nanoparticles obtained in plasma and

Content from this work may be used under the terms of the Creative Commons Attribution 3.0 licence. Any further distribution of this work must maintain attribution to the author(s) and the title of the work, journal citation and DOI.
Published under licence by IOP Publishing Ltd

affect their physical properties and surface characteristics. The objective of the present work is to check this hypothesis and, moreover, to analyze the features of such acoustoplasma discharge and possible chemical conversions that can occur under its excitation in water and water-hydrocarbon mixtures.

2. Methods and materials

2.1. Plasma discharge in a cavitating liquid

Acoustoplasma technique involves the initiating of an electric discharge in a liquid that goes along with ultrasonic power assisted cavitation. In general, the specially designed acoustoplasma reactor includes a reservoir with two immersed electrodes and ultrasonic cavitator. Cavitation bubbles in the liquid phase provide the outstanding characteristics of a discharge and these characteristics have been proven to be governed by the regimes of ultrasonification [4-7]. Furthermore, ultrasonic power prevents the secondary agglomeration of nanoparticles being synthesized from metal electrodes in electric discharge. The scheme of experimental setup is shown in the figure 1.

The setup consists of a reaction chamber, where discharge electrodes and an ultrasonic irradiator are introduced, a generator of high-voltage pulses for discharge initiation, a power supply of discharge in liquid, an ultrasonic generator and blocks of control of electric and acoustic characteristics. The chamber is provided with quartz windows for the observation of dynamic processes and registration of optical spectra of visible discharge glow.

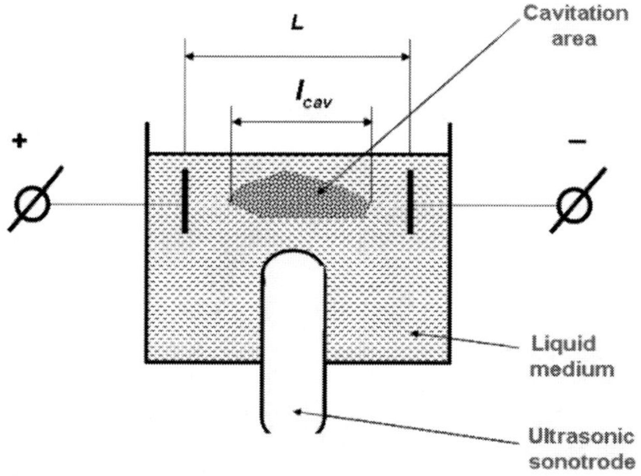

Figure 1. Scheme of experiment. L means the distance between the electrodes; Lcav means the size of cavitation zone.

The ultrasonic generator with a piesoceramic transducer provides the output acoustic power up to 2 kW in the frequency range 27-44 kHz. The parameters of acoustic equipment allow one to implement the intensity of an ultrasonic field in the volume of liquid up to 10 W/cm^2 and to vary cavitation regime in a wide range.

2.2. Synthesis of nanoparticles by acoustoplasma

In previous studies it was found that combined excitation of electric arc discharge and acoustic cavitation in water and organic liquids was the effective method to create free hydrogen and various kinds of solid nanoparticles. Chemical compositions, dimensions, morphologies, optical and other properties of these nanoparticles can be easily regulated by the plasma discharge parameters, materials of electrodes and liquids. In other works, it was demonstrated that ultrasonic cavitation itself is a promising way for modification of properties of solid nano- and microparticles [8-12].

IOP Conf. Series: Materials Science and Engineering **112** (2016) 012003 doi:10.1088/1757-899X/112/1/012003

In this work, the capability of plasma combined with ultrasonic cavitation for synthesis of novel nanoparticles was exemplarily investigated on tungsten and zinc oxide nanoparticles. For this purpose, electric discharge in liquid medium was initiated using tungsten and zinc electrodes.

3. Results

First, optical spectroscopy was applied for study of plasma at tungsten and zinc electrodes (figure 2 and 3).

At the initial stage of the plasma discharge, the light emission spectra consist from narrow lines of atomic emission of hydrogen and oxygen. However, in the process of the discharge, a broad band corresponding to the thermal equilibrium emission of tungsten oxide grows monotonously. During the growth of the tungsten oxide band, the hydrogen line superposed onto this band is broadened. This broadening of the hydrogen line can be attributed to the partial capture of the hydrogen atoms by nanoparticles followed by corresponding deformations of the emission line.

The light emission spectra of the plasma discharge in water with tungsten and zinc electrodes demonstrate distinct differences both in the broad bands and in positions of hydrogen lines. Differences in the spectral positions of narrow lines for tungsten and zinc electrodes reveal active interactions of hydrogen and other ions created in the plasma discharge with the nanoparticles created in the process.

Further on, scanning electron microscopy pictures of synthesized nanoparticles were taken (figure 4).

In the most of structures here, tungsten oxide has oxygen content varied from WO_2 to WO_3. These variations of stechiometry result in a wide multitude of crystalline structures (hexagonal, cubic, orthorhombic, tetragonal, monoclinic, triclinic). This variety of atomic structures and corresponding morphologies of nanoparticles determines tremendous multitude of agglomerations, well revealed by scanning electron microscopy (hexagons, fibers, plates, etc.).

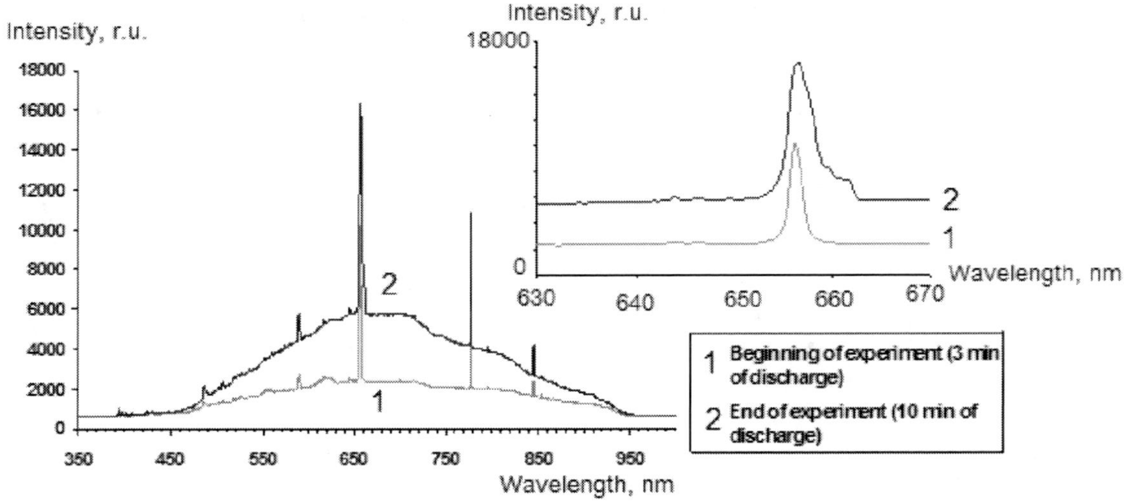

Figure 2. Optical spectra of plasma discharge at tungsten electrodes in water

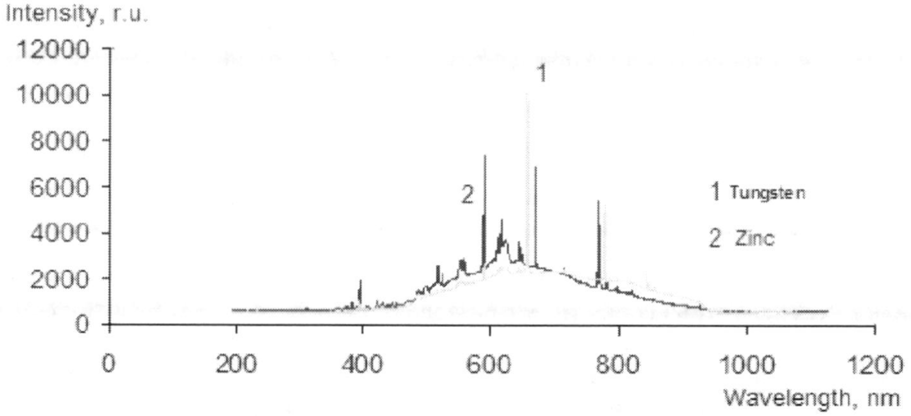

Figure 3. Optical spectra of plasma discharge at tungsten and zinc electrodes in water.

Figure 4. Various kinds of agglomerations of tungsten oxide and zinc oxide particles

Scanning electron microscopy of the particles of tungsten oxide synthesized by plasma discharge using tungsten electrodes in water revealed their dimensions from 10 nm to more than 100 nm. Morphologies of these particles differ in a wide range. Well faceted particles having either rectangular or triangular symmetry are observed simultaneously with the particles having smooth ellipsoid – like geometries.

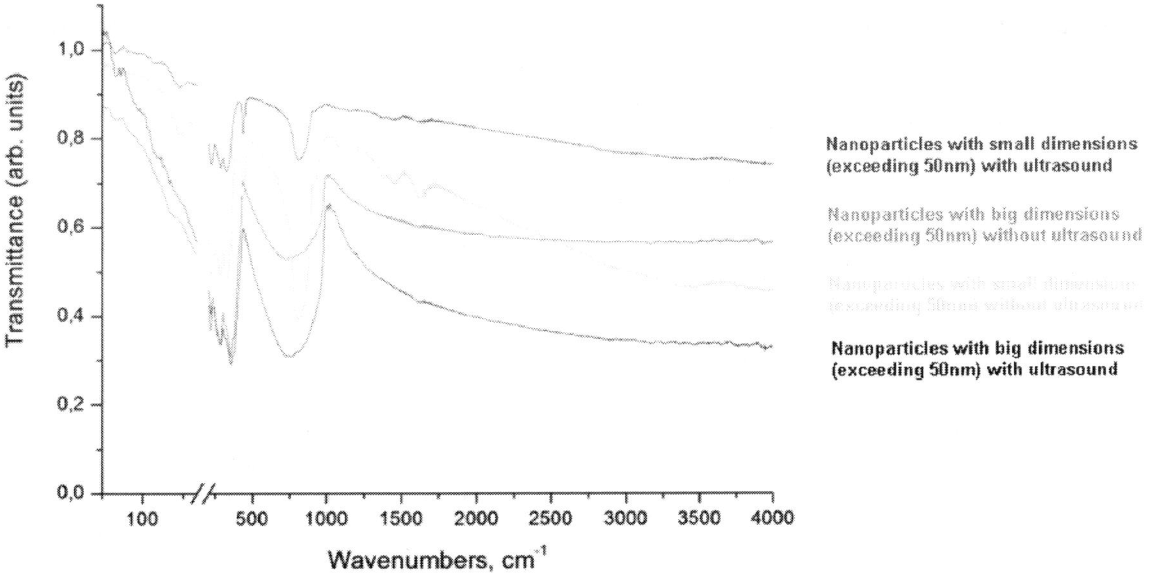

Figure 5. Infrared transmission spectra of nanoparticles of tungsten oxide created in the plasma discharge

The absorption bands between 500 and 1000 cm^{-1} are rather different for big and small nanoparticles (they correspond to vibrations in W–O bonds). For big particles these bands are wide and their center is shifted to lower frequencies. These features are explained by lower content of oxygen in comparison to WO_3.

Small particles (with dimensions in the range from 10 to 30 nm) are characterized by more strong and definite W–O bonds. In the spectral regions 1400 – 1750 cm^{-1} and 2800 – 3500 cm^{-1} several bands corresponding to vibrations of O–H and C–H are observed in the particles created under ultrasonic treatment. So, we can assume that ultrasonic cavitation treatment promotes entrainment of hydrogen into tungsten oxide particles.

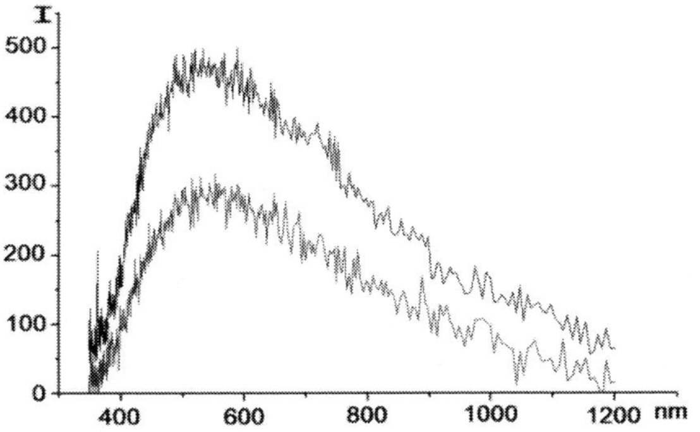

Figure 6. Photoluminescence spectra excited by 308 nm xenon – chlorine pulsed laser of tungsten oxide nanoparticles obtained by acoustoplasma technique with the intensity of ultrasonification below the cavitation threshold (lower curve) and above the cavitation threshold (upper curve).

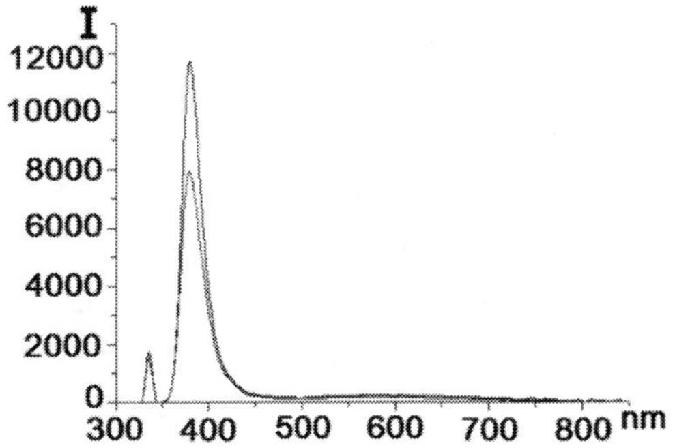

Figure 7. Photoluminescence spectra excited by 337 nm nitrogen pulsed laser on zinc oxide nanoparticles obtained by acoustoplasma technique with the intensity of ultrasonification below the cavitation threshold (lower curve) and above the cavitation threshold (upper curve).

Photoluminescence spectra of tungsten oxide nanoparticles excited by xenon – chlorine laser (308 nm emission) differ significantly from the emission spectra of the same particles in plasma (figure 1 and 2). The emission maxima are observed at about 550 nm (instead of 650 nm for plasma discharge). Narrow emission lines corresponding to hydrogen and other gases (clearly seen in the plasma discharge) are not observed at all. These spectra are observed for small particles which correspond to WO_3 composition.

The absence of hydrogen emission lines in the photoluminescence spectra does not reject the presence of hydrogen in tungsten oxide particles. Their absence is explained by non- absorption of the laser light by hydrogen atoms.

Important observation is that the intensity of ultrasonic power during synthesis substantially affects the optical properties of synthesized nanoparticles. It was also demonstrated on ZnO particles (Fig. 7).

Important feature is the pronounced difference between the luminescence intensity of particles obtained without ultrasonic treatment and particles obtained in plasma under ultrasonic treatment: higher intensity of particle in case of ultrasonic action during the synthesis can be attributed to the numerous defects in crystal structure and may be advantageous in view of possible applications for creation of optically active materials.

As a result of these experiments, acoustoplasma technique has been demonstrated to be new interesting approach to get nanomaterials in two ways in the same reaction – solid powders precipitating in the reaction vessel and gaseous aerosol containing oxide nanoparticles exhibiting the density comparable with that of air. Two perspective avenues of the utilization of these nanopowders can be performed: nanoparticles could first be deposited on a flat substrate and so create a surface modifying coating and the second option is an entrainment of the nanoparticles into the monomer matrix which can be polymerized afterwards yielding a polymer with immobilized nanoparticles.

4. Conclusions

Acoustoplasma technique based on combination of a discharge in liquid with acoustic cavitation treatment provides an effective route for synthesis of solid nanoparticles of metals, oxides and semiconductors. Nanoparticles, synthesized by acoustoplasma technique demonstrate interesting physical properties and can be effectively applied for different applications, such as chemical catalyst, radiation detectors, solar cells, lasers, medical therapy etc.

Acknowledgments

This work has been supported by RFBR, projects No. 15-02-03559, 14-02-92019, 14-02-31515 and 14-02-00602.

References

[1] Bang J Ho and Suslick K S 2010 *Adv. Mater.* **22** 1039
[2] Ghomi H, Yousefi M, Shahabi N and Khoramabadi M 2013 *Radiation Effects and Defects in Solids* **168/11** 881
[3] Askarinejad A, Alavi MA and Morsali A 2011 *Iran J. Chem. Chem. Eng.* **30/**3 75
[4] Bulychev N A, Kazaryan M A, Gridneva E S, Murav'ev E N, Solinov V F, Koshelev K K, Kosheleva OK, Sachkov V I and Chen S G 2012 *Bull. of the Lebedev Physics Inst.* **39/**7 214
[5] Bulychev N A, Kazaryan M A, Chaikov L L, Burkhanov I S, Krasovskii V I 2014 *Bull. of the Lebedev Physics Ins.* **41/**9 264
[6] Burkhanov IS, Chaikov LL, Bulychev N A, Kazaryan M A, Krasovskii V I 2014 *Bull. of the Lebedev Physics Inst.* **41/**10 297
[7] Klassen N, Krivko O, Kedrovб Shmurak V S, Kiselev A, Shmyt'ko I, Kudrenko E, Shekhtman A, Bazhenov A, Fursova T, Abramov V, Bulychev N and Kisterev E 2010 *IEEE Transactions on Nuclear Science* **57** (3) 1377
[8] Ganiev R F, Bulychev N A, Fomin V N, Arutyunov I A, Eisenbach C D, Zubov V P and Malyukova E B 2006 *Doklady of Chemistry* **407** 54

IOP Conf. Series: Materials Science and Engineering **112** (2016) 012003 doi:10.1088/1757-899X/112/1/012003

[9] Bulychev N A, Kisterev EV, Arutunov I A and Zubov V P 2008 *J. of the Balkan Tribological Association* **1**(14) 30
[10] Bulychev N A 2010 *Inorganic Materials* **46**/4 393
[11] Bulychev N, Dervaux B, Dirnberger K, Zubov V, Du Prez F E and Eisenbach C D 2010 *Macromol. Chem. Phys.* **9** (211) 971
[12] Bulychev N, Confortini O, Kopold P, Dirnberger K, Schauer T, Du Prez F E, Zubov V, Eisenbach C D 2007 *Polymer* **48**(9) 2636

The effects of liquid-phase oxidation of multiwall carbon nanotubes on their surface characteristics

I N Burmistrov[1], D S Muratov[1], I A Ilinykh[1], E A Kolesnikov[1], A Yu Godymchuk[1,2], D V Kuznetsov[1]

[1] National University of Science and Technology "MISiS", 4, Leninsky Prospekt, Moscow, 119049, Russia
[2] National Research Tomsk Polytechnic University, 30, Lenina Ave., Tomsk, 634050, Russia

E-mail: glas101@yandex.ru

Abstract. The development of new sorbents based on nanostructured carbon materials recently became a perspective field of research. Main topic of current study is to investigate the effect of different regimes of multiwall carbon nanotubes (MWCNT) surface modification process on their structural characteristics. MWCNT samples were treated with nitric acid at high temperature. Structural properties were studied using low temperature nitrogen adsorption and acid-base back titration methods. The study showed that diluted nitric acid does not affect MWCNT structure. Concentrated nitric acid treatment leads to formation of 2.8 carboxylic groups per 1 nm^2 of the sample surface.

1. Introduction

Intensive development of new science-driven technologies of rare-earth metal extraction is one of the strategic goals of today's industry. Main focus of research in this field is concentrated on the separation of so called allied substances. The problems of allied substances separation and purification are typically solved by ion-selective and ion-exchange processing. Ion-exchange technique shows good results when used for extraction and collection of low quantities of rare-earth metal contaminations. On the other hand the increasing amount of non-traditional mineral raw materials leads to development of new more efficient techniques and methods [1-3].

Promising perspectives in the field has been shown with carbon nanotubes [4-7]. Carbon nanotubes are known to have good sorption properties due to the high specific surface area which goes up to 2000 m^2/g for single walled carbon nanotubes. MWCNTs have lower values of specific surface area varying from 150 to 600 m^2/g but even this is quite high compared to other traditional sorbents. Graphene is also considered good sorption agent with specific surface area of 200-700 m^2/g and in some cases even up to 1000 m^2/g.

Ion-exchange activity of carbon nanomaterials could be achieved by surface functionalization which involves covalent and van der Waals bonding of different substances to MWCNT structure. Generally non-covalent bonding of different agents with the surface of MWCNT is not considered as functionalization process and is studied separately. MWCNT surface functionalization is usually done in stages. The first stage is oxidation which leads to formation of carboxyl surface groups. The second one is grafting of oxidized surface with various functional groups to obtain more specific surface reaction.

Content from this work may be used under the terms of the Creative Commons Attribution 3.0 licence. Any further distribution of this work must maintain attribution to the author(s) and the title of the work, journal citation and DOI.
Published under licence by IOP Publishing Ltd

MWCNT surface oxidation is the most typical functionalization treatment and is also used for surface cleaning [8-10]. As the result of such cleaning various oxygen-containing surface groups are formed. Functionalization effectiveness is often evaluated by functional group content relative to the sample weight.

There are two generally used oxidation techniques: gas-phase thermal oxidation in oxygen or ozone atmosphere; and liquid-phase oxidation in nitric acid or in mixture of HNO_3 and H_2SO_4. Present study is focused on effect of MWCNT functionalization on their surface characteristics.

2. Materials and methods

Present study involved MWCNT Taunit-M and polygraphene materials provided by Ltd. "NanoTechCenter" (Tambov, Russia). MWCNT oxidation treatment was done by boiling of MWCNTs in concentrated and diluted nitric acid (HNO_3) for 1–8 hours with MWCNT/acid ratio equal to 1/20. Oxidation process was carried out in two-necked flask with backflow condenser and a thermometer to control the reaction temperature between 100 and 105 °C Next the reaction products were rinsed with distilled water until neutral pH. Reaction products separation was done by decantation with subsequent vacuum filtering and drying at 60 °Cfor 48 hours.

Surface oxygen-containing group count was measured by acid-base back-titration. MWCNTs were then coverd with sodium hydroxide solution of the known concentration and left for 24 hours. The precipitate was filtered on the paper filter, and then the aliquot was mixed with hydrogen chloride solution with subsequent titration with sodium hydroxide. The pH value was measured by "Anion-4100" with absolute error of ±0.02. Specific surface measurements were done according to ASTM D3663 and BET method using Quantachrome Nova 1200 instrument.

3. Results and discussions

Specific surface measurement results are shown in the Table 1. SEM and TEM images are shown on the Figure 1.

Table 1. Specific surface for MWCNT and graphene samples

Sample	Material	Specific surface area, m^2/g
1	MWCNTs	170±7
2	Graphene	90±5

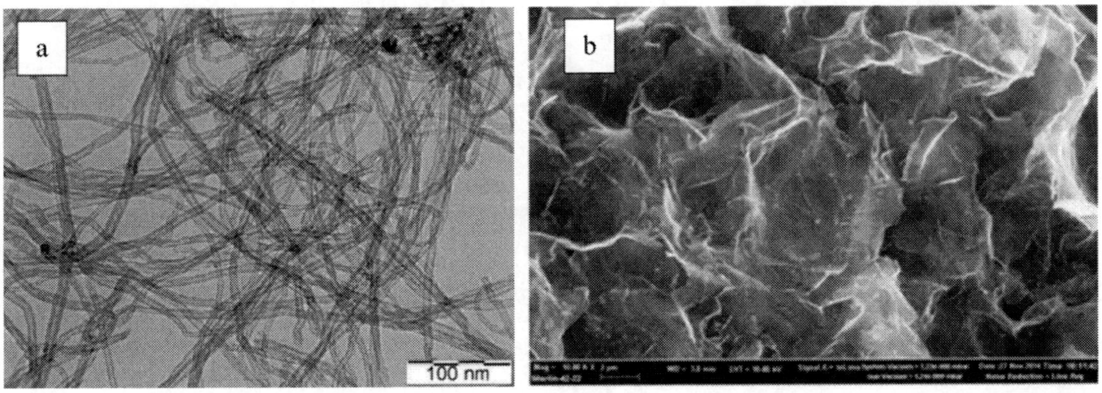

Figure 1. Microphotography of the materials: a) TEM image of MWCNT Taunit-M, b) SEM image of graphene sample

Specific surface area for MWCNT sample (table 1) correlates with nanotube diameter measured from TEM (fig. 1A). It should be noted that nanotubes in sample 1 are entangled and it is quite difficult to obtain good suspension for those samples without additional intensive processing.

The evaluation of surface oxidation processing is shown in the Table 2.

Table 2. Titration results for MWCNT samples

#	Treatment time, hours	Treatment type	Temperature, °C	Specific surface area, m^2/g	Amount of oxygen-containing groups, nm^{-2}
1	1	HNO_3 concentrated	100–105	160	1.5
			130–140	Sample degraded	
		Diluted HNO_3	100–105	170	0
2	2	HNO_3 concentrated	100–105	140	2.6
			130–140	Sample degraded	
		Diluted HNO_3	100–105	170	0
3	4	HNO_3 concentrated	100–105	130	2.8
			130–140	Sample degraded	
		Diluted HNO_3	100–105	170	0
4	8	HNO_3 concentrated	100–105	Sample degraded	
			130–140	Sample degraded	
		Diluted HNO_3	100–105	170	0

The specific surface area does not change at the treatment with a nitric acid and decreases in the case of treatment with concentrated nitric. According to table 2 data it is clear that diluted HNO_3 does not affect MWCNT surface characteristics even when the treatment time is 8 hours. In those cases the treatment just dissolves metal catalyst from the sample. In the treatment of a concentrated acid for 8 hours CNTs degraded and the specific surface area cannot be measured.

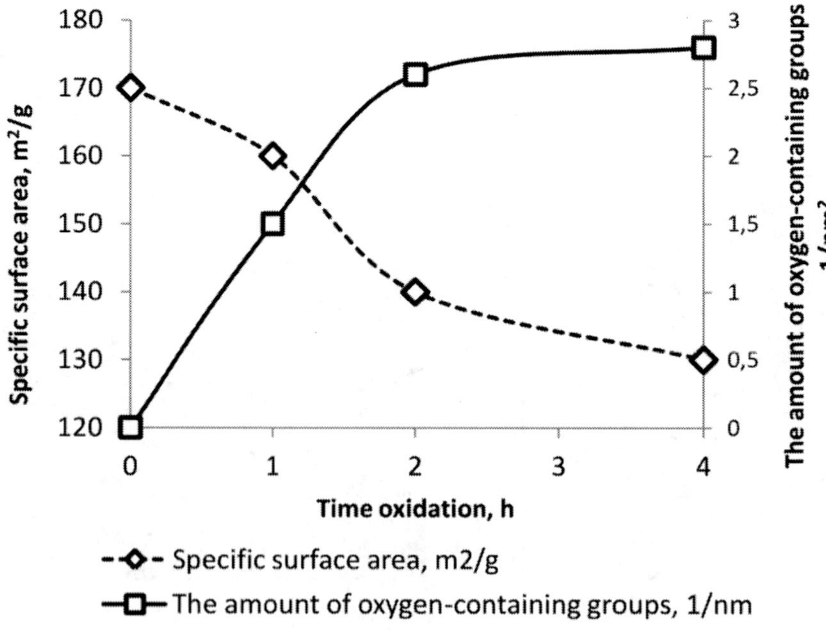

Figure 2. Specific surface area and amount of oxygen-containing surface groups for MWCNT sample against treatment time

Specific surface area and the amount of oxygen-containing groups are plotted against treatment time on the Figure 2. The specific surface area is significantly reduced during the first two hours of treatment. The content of carboxyl groups is also increased within this range treatment time. According to the Figure 2 the optimal treatment time is 2.5-3 hours.

On the other hand concentrated nitric acid treatment leads to formation of oxygen-containing groups with simultaneous decrease in specific surface area values. This is probably caused by decrease of highly reactive amorphous carbon amount due to treatment process. Amorphous carbon is present in the sample volume due to incomplete nanotube formation process at the synthesis stage.

With longer treatment times the oxygenated group amount per nm^2 steadily increases. When treatment time is longer than 8 hours surface of all the samples is considered to be oxygenated and yields a clear homogeneous solution.

4. Conclusions

The research of functionalization process effect on the structural properties of carbon nanomaterials showed that diluted nitric acid does not affect MWCNT structure even when treated for 8 hours. In those cases the treatment just dissolves mal catalyst from the sample.

Concentrated nitric acid treatment leads to formation of oxygen-containing groups with simultaneous decrease in specific surface area values. This is probably caused by decrease of highly reactive amorphous carbon amount due to the treatment process. Amorphous carbon is present in the sample volume due to incomplete nanotube formation process at the synthesis stage.

With longer treatment times the oxygenated group amount per nm^2 steadily increases. When treatment time is longer than 8 hours surface of all the samples is considered to be oxygenated and yields a clear homogeneous solution.

Acknowledgements

This research was supported by Russian Ministry of Education and Science in frames of grants ID RFMEFI57814X0084.

References

[1] Liao C, Wu S, Cheng F, Liu S, Wang S, Liu Y, Zhang B and Yan C 2013 *J. Rare Earths* **31**(4) 331
[2] Binnemans K, Jones P T, Blanpain B, Van Gerven T, Yang Y, Walton A and Buchert M 2013 *J. Cleaner Prod.* **51** 1
[3] Kato Y, Fujinaga K, Nakamura K, Takaya Y, Kitamura K, Ohta J, Toda R, Nakashima T and Iwamori H 2011 *Nature Geosci.* **4**(8) 535
[4] Baughman R H, Zakhidov A A and Heer W A 2002 *Science* **297**(5582) 787
[5] Cao J, Wang Q, Rolandi M and Dai H 2004 *Phy.s Rev. Lett.* **93**(21) 1
[6] Tibbetts G G and Beetz C P 1987 *J. Phys. D-Appl. Phys.* **20**(3) 292
[7] Callister W D 2003 *Materials Science and Engineering: Introduction* (New York: Wiley)
[8] Mazov I N, Ilinykh I A, Kuznetsov V L, Stepashkin A A, Ergin K S, Muratov D S, Tcherdyntsev V V, Kuznetsov D V, Issi J-P 2014 *J. Alloys and Compounds* **586**(1) S440
[9] Macutkevic J, Paddubskaya A, Kuzhir P, Banys J, Maksimenko S, Kuznetsov V L, Mazov I N, and Krasnikov D V 2014 *J. Nanosci. and Nanotechnol.* **14**(7), 5430
[10] Macutkevic J, Kuzhir P, Paddubskaya A, Shuba M, Banys J, Maksimenko S, Kuznetsov V L, Mazov I N, and Krasnikov D V 2013 *Phys. Status Solidi (A) Applications and Mater. Sci.* **210**(11) 2491

Study of sulfur isotopes by vibrational spectroscopy and quantum chemistry

O Kh Poleshchuk[1,2], N B Egorov[1], D V Akimov[1], I I Zherin[1] and N A Zhuravlev[1]

[1]National Research Tomsk Polytechnic University, 30 Lenina ave., Tomsk, 634050, Russia
[2] Tomsk State Pedagogical University, 60 Kievskaya street, Tomsk, 634061, Russia,

E-mail: egorov@tpu.ru

Abstract. The results show the dependence of the experimental and calculated frequencies in the IR- and Raman spectra from the molecular mass of the sulfur isotope. The effect of a sulfur isotope shifts on the change of thermodynamic parameters. The results obtained demonstrate that the quantum-chemical calculations are sensitive to the isotopic shifts in the vibrational spectra of sulfur isotopes and are able to assess changes in their thermodynamic properties.

1. Introduction

Natural sulfur consists of a mixture of stable isotopes with mass numbers 32, 33, 34 and 36, the relative contents of which are 95.04%, 0.75%, 4.20% and 0.015%, respectively [1]. Isotopes of sulfur are applied in biology [2], geology [3], geochemistry [4, 5], ecology [6-8], agrochemistry [9], archeology [10] and medicine [11]. It is obvious that information regarding the spectral and thermodynamic properties of stable sulfur isotopes is important for both applied and scientific problems. Other groups [12-15] have studied the vibration spectra of sulfur α-^{nat}S и α-^{32}S at low temperatures in detail, and have studied the effect of temperature on the infrared spectra of natural and isotopically-pure sulfur $^{32}S_8$. These experiments were performed using a single crystal of sulfur. However, in applied research, researchers often use the polycrystalline powders of sulfur rather than single crystals of sulfur. Therefore, in this paper, the infrared spectra and Raman spectra of and isotopically-pure polycrystalline α-^{32}S, α-^{33}S and α-^{34}S are presented. The experimental data are compared with theoretical calculations of the infrared and Raman spectra, obtained via the theory of functional density. In addition, this paper shows the relationship between the change in Gibbs free energy and the mass of an isotope of sulfur.

2. Experimental

The experimental work used isotopes of sulfur with an isotopic purity: ^{32}S – 99.9%; ^{33}S – 99.4%; ^{34}S – 99.9%. To obtain a chemically pure orthorhombic sulfur modification (α-S) the isotopes were subjected to triple distillation under vacuum with subsequent crystallization from carbon tetrachloride. Isotopes of sulfur obtained in this way have a chemical purity of more than 99.9%.
The IR spectra of polycrystalline samples of α-S in KBr tablets in the range 400-500 cm^{-1} were obtained using the FT-IR spectrometer Nikolet 5700 at room temperature, with a resolution of 2cm^{-1}. Raman spectra were recorded on a spectrometer Nikolet 5700 with a Raman module in the

Content from this work may be used under the terms of the Creative Commons Attribution 3.0 licence. Any further distribution of this work must maintain attribution to the author(s) and the title of the work, journal citation and DOI.
Published under licence by IOP Publishing Ltd

range 100-500 cm^{-1} at room temperature, with a resolution of 1cm^{-1}. In both cases a laser was use with Nd: YAG (λ=1064 nm, 514 mW).

3. Computational details

The calculations of the S_8 molecules containing the different sulfur isotopes were carried out using the GAUSSIAN'03 program package [16]. The geometries were first optimized at the gradient-corrected DFT level using the three-parameter fit of the exchange-correlation potential suggested by Becke [17] in conjunction with the LYP [18] exchange potential (B3LYP). This method is now generally accepted to best describe the thermodynamic properties and vibrational spectra, and gives results similar to the MP2 and GMP2 methods. All electron 6-31G(d), and extended cc-PVTZ basis sets were used. The absence of imaginary vibration frequencies confirmed the stationary character of the structures.

Energy calculations were corrected for zero-point vibrational energy and reduced to normal conditions (298.15K, 1 atm) using thermal corrections to enthalpy and free energy.

The optimized geometries of these molecules were also used for the calculations by the ADF'2004 (Amsterdam density functional) program. The OPTX exchange functional [19] was used in combination with the PBE correlation functional [20] using an all electron triple-ζ + polarization basis set of uncontracted Slater orbitals. Scalar relativistic effects were considered using the zero order regular approximation (ZORA) [21].

4. Results and discussion

Figure 1 shows the IR spectra, obtained for the polycrystalline samples of the α-^{32}S, α-^{33}S, α-^{34}S and α-^{nat}S isotopes. According to a previous study [22], the observed IR spectra absorption corresponds to the valence vibration (stretching vibration) - ν_5. Figure 1 show that the position of the lines of oscillation (ν_5) for the α-^{33}S and the α-^{34}S isotopes are shifted to the low-energy side, and a line corresponding to the α-^{32}S isotope has a high-energy shift with respect to the spectrum of α-^{nat}S.

This behavior can be explained by the fact that the provisions of the oscillation frequencies in the spectra are directly connected with the mass of the atoms oscillating around the equilibrium positions in the crystal lattice. This dependence has the form: $\nu = \propto m^{-½}$ (here and below m the mass of isotope).

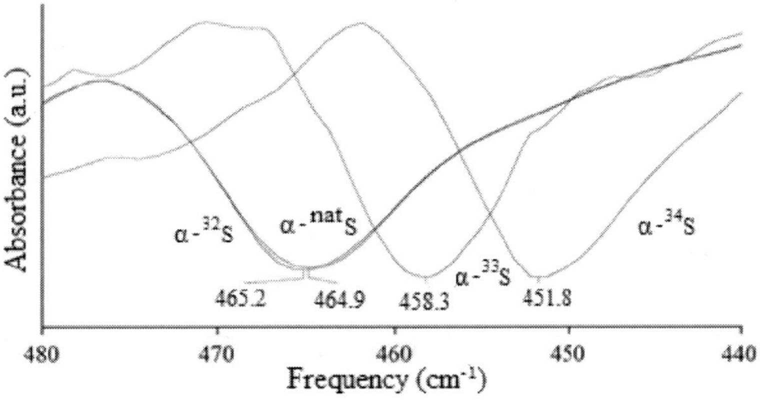

Figure 1. Raman spectra of the ^{32}S, ^{33}S and ^{34}S isotopes, and sulfur with the natural isotopic composition ^{nat}S

Isotopic dependence of the oscillation frequencies of α-S in the IR spectra are shown in figure 2. The dependence of the experimental vibration frequencies from the atomic weight of sulfur isotope in the IR spectra has the next form: ω = 680 - 6.7m (r=0.999). Here and below, r denotes the correlation coefficient.

Figure 3 shows the Raman spectra obtained for polycrystalline samples of the isotopes of α-S. The spectra of isotopes involves all absorption bands which are typical for α-S.

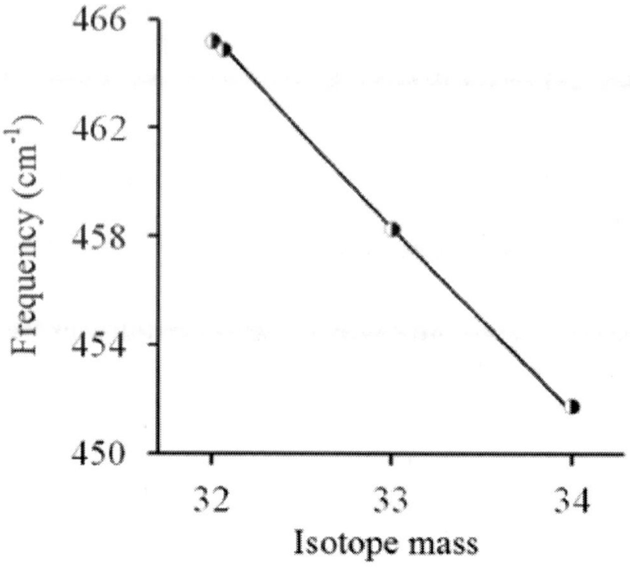

Figure 2. Isotopic dependence of the oscillation frequencies of α-S in the IR spectra

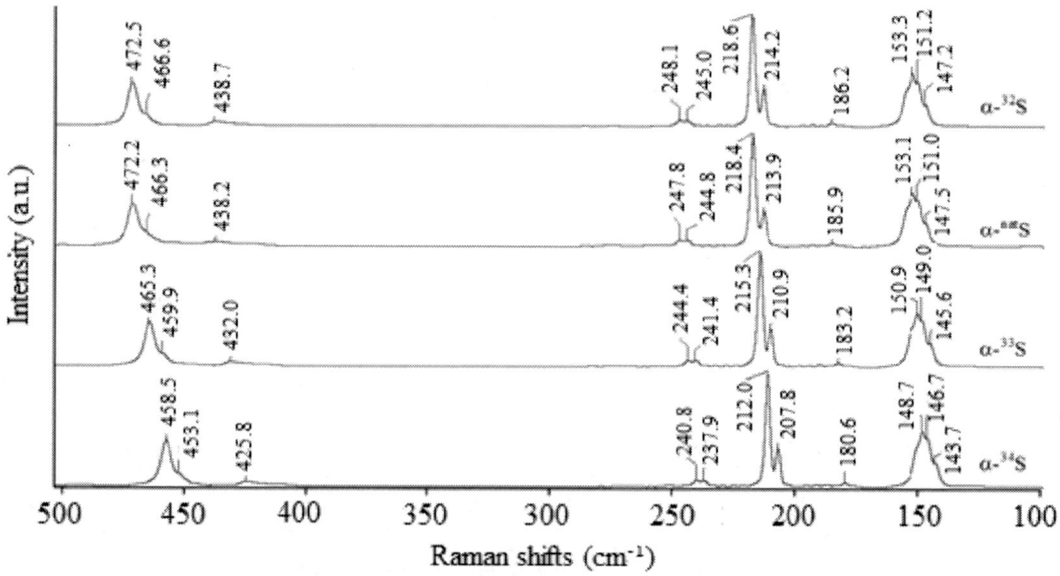

Figure 3. Raman spectra of the isotopes α-^{32}S, α-^{33}S, α-^{34}S and α-^{nat}S

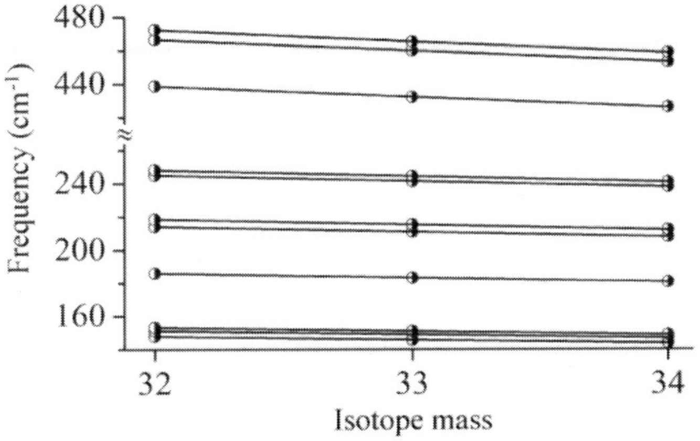

Figure 4. Isotopic dependence of the oscillation frequencies of α-S in the Raman spectra

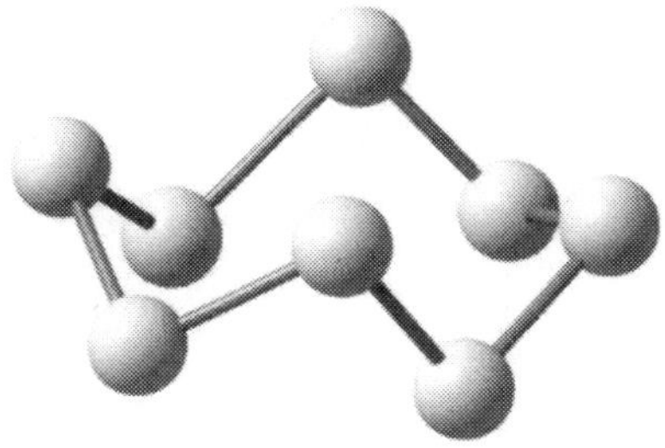

Figure 5. The model structure of α-S used for quantum-chemical calculations

Figure 4 shows the experimental vibrational frequencies in the Raman spectra of the atomic mass of the α-S isotope. As can be seen from the figure, depending on the maximal provisions of the bands in the Raman spectra of sulfur, isotopes are a linear function of the atomic masses of the isotopes.

The obtained linear dependence on the known frequencies of the α-^{32}S, α-^{33}S and α-^{34}S isotopes allows the determination of the oscillation frequency in the IR and Raman spectra of the sulfur isotope α-^{36}S, for which it is difficult to obtain the vibration spectra due to its high cost, which is associated with its low content in natural sulfur. The calculated vibration frequencies in the α-^{36}S isotope are shown in figure 1.

It is well known that cyclo-octa-sulfur can form three crystalline allotropic modifications, the first of which, orthorhombic α-S, is stable at room temperature [15]. The structure of this modified version was studied by X-ray analysis, and contains S8 as main structural elements, which are formed with each other as columns [16]. Figure 3 shows the model structure of α-S used in quantum chemical calculations.

As the figure 5 shows, the resulting structure is not truly orthorhombic, because it is a solid S_8 ring coordinated by adjacent rings. However, the resulting structure has allowed the calculations of the Raman spectra to be performed for the isotopes of sulfur for the unit cell. A full optimization of geometrical parameters was also done.

The calculated length of the S-S bonds (2.09Å) in the S8 cluster was close to the experimental value obtained (2.06Å [15]). Figure 1 shows the experimental (^{32}S, ^{33}S, ^{34}S) and calculated (^{32}S, ^{33}S, ^{34}S, ^{36}S) values of the basis states in the 6-31G* and cc-PVTZ frequency in the Raman spectra.

Data from figure 1 also show that, despite the small differences between the masses of the isotopes, the theoretical frequencies are in satisfactory agreement with the experimental data and almost completely describe the vibration spectra of sulfur isotopes. The data also indicates that the used calculation methods are sensitive to changes in the frequency spectra (a few cm^{-1}), depending on the mass of the isotope.

From the correlation ratios (in figures 4 and 5) it is shown that a significant expansion of the basis for using the cc-PVTZ does not improve the agreement between the calculated and experimental frequencies. The smaller basis 6-31G(d) gives even better results:

$$\omega_{experimental}=2.33+1.005\ \omega_{calculated}\ r=0.999;\ sd=5.6;\ n=32 \text{ for basis 6-31G(d)} \quad (1)$$

$$\omega_{experimental}=2.89+1.003\ \omega_{calculated}\ r=0.999;\ sd=5.8;\ n=24 \text{ for basis cc-PVTZ} \quad (2)$$

Here and below, *sd* signifies the standard deviation, and *n* indicates the number of connections.

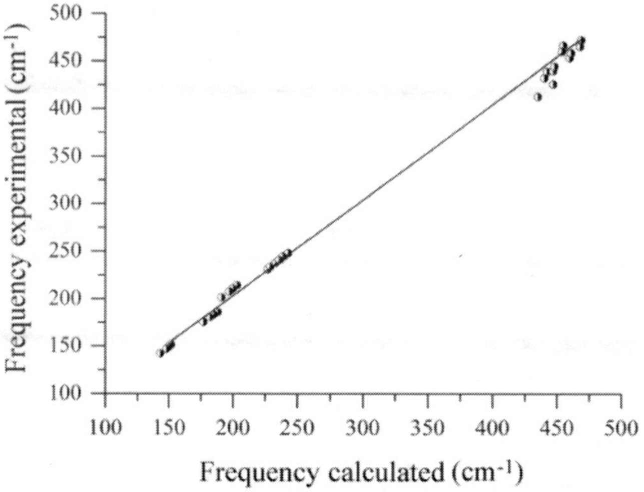

Figure 6. The relationship between the experimental and calculated by B3LYP/6-31G method (d) frequencies of oscillation Raman spectra of the sulfur isotopes α-^{32}S, α-^{33}S and α-^{34}S

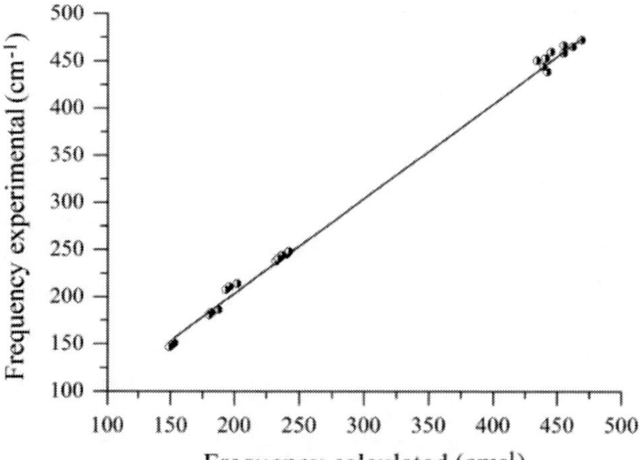

Figure 7. The relation between the experimental and calculated cc-PVTZ vibration frequencies of the Raman spectra of the sulfur isotopes α-^{32}S, α-^{33}S and α-^{34}S

α-^{32}S				α-^{33}S				α-^{34}S				α-^{36}S		α-[illegible]S	Mode and species [23]
exp.	6-31G*	cc-PVTZ	TZ2P+	exp.	6-31G*	cc-PVTZ	TZ2P+	exp.	6-31G*	cc-PVTZ	TZ2P+	exp.	6-31G*	exp.	
147.9				145.6				143.7				139.4		147.5	
151.2	152	153	143	149.0	150	151	140	146.7	148	149	138	142.2	143	151.0	$\nu_8(E_2)$
153.3				150.9				148.7				144.1		153.1	
186.1	188	187	186	183.2	185	182	183	180.6	182	180	181	175.1	177	185.9	$\nu_6(E_1)$
214.2	203	202	195	210.9	200	196	192	207.8	197	193	189	201.4	191	213.9	
218.6				215.3				212.0				205.4		218.4	$\nu_2(A_1)$
245.0	240	240	232	241.4	237	235	228	237.9	233	232	226	230.8	227	244.8	$\nu_4(B_2)$
248.1	243	242	234	244.4	239	237	230	240.8	236	234	227	233.5	229	247.8	$\nu_{11}(E_3)$
	426	433	420						433	428	408				
438.7	434	442	444	432.0	440	434	414	425.8	447	439	430	412.8	421	438.2	$\nu_{10}(E_3)$
466.6	461	455	450	459.9	454	445	444	453.1			437	439.6	435	466.3	$\nu_5(E_1)$
472.5	476	469	471	465.3	468	462	464	458.5	461	455	457	444.4	448	472.2	

Figure 8. The experimental and calculated frequencies in the Raman spectra of sulfur isotopes to molecules α-S, cm^{-1}

Similar correlations were obtained using the ratio of the program based on the ADF optimized by the B3LYP/6-31G(d) of the molecular structures (Figure 9):

$$\omega_{experimental}=14.7+0.979\ \omega_{calculated}\quad r=0.998;\ sd=7.9;\ n=24 \text{ for basis TZ2P+} \qquad (3)$$

Although the calculations using the software packages GAUSSIAN and ADF use different functionals and basis states, in principle the quantitative relationships are close enough to each other and the experimental frequencies (relative error of the estimate of frequencies in the Raman spectra is about 3-4%).

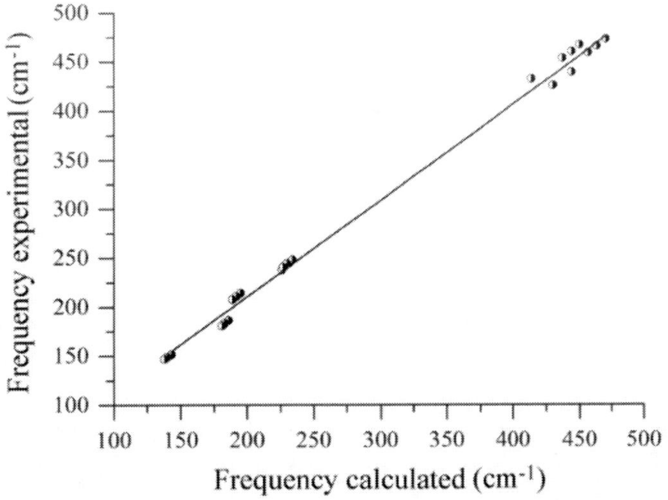

Figure 9. The relation between the experimental and calculated BP86/TZ2P+ oscillation frequencies of the Raman spectra for the sulfur isotopes α-^{32}S, α-^{33}S and α-^{34}S

The resistance of sulfur clusters is determined by their thermodynamic parameters. It should be noted that the experimental formation enthalpy of the transition from S_2 to S_8 was virtually unchanged, while the entropy of formation varies considerably [18]. Table 1 shows the experimental and calculated values of the entropy of sulfur cluster formation.

From table 1 the correlation ratio between the experimental and calculated values of the entropy of the formation of sulfur clusters was obtained:

$$S_{experimental}=31.3+0.86S_{calculated}\quad r=0.996;\ sd=17.6;\ n=8 \text{ (in GAUSSIAN)} \qquad (4)$$

$$S_{experimental}=15.0+0.905S_{calculated}\quad r=0.985;\ sd=21.2;\ n=8 \text{ (in ADF)} \qquad (5)$$

Table 1. The experimental and calculated entropy of formation of sulfur clusters (J/mol·K) using B3LYP/6-31G(d) and BP86/TZ2P + methods.

Cluster	Experiment [24]	GAUSSIAN B3LYP/6-31G (d)	ADF BP86/TZ2P+
S	168	152	175
S_2	228	219	234
S_3	270	291	276
S_4	311	313	304
S_5	309	349	356
S_6	354	374	380
S_7	408	417	418
S_8	430	456	460

The resulting correlation ratio indicates that the calculated thermodynamic parameters of the different methods of sulfur cluster formation are sufficiently close to the experimental values and that the calculation error is less than 6 %.

The resulting thermodynamic characteristics obtained using the B3LYP/6-31G(d) calculation method cannot be directly compared with the enthalpy of formation and the Gibbs free energy. However, the effect of isotope shift in the change of thermodynamic parameters can be estimated. From the data presented in Table 2 a relationship between the change in Gibbs free energy and the mass of a sulfur (S_8) isotope was obtained:

$$M_S = -1402 -2552\ \Delta G \quad r = 0.991;\ sd = 0.3;\ n = 4\ (1.0\ \%\ \text{error}) \tag{6}$$

A similar dependence is obtained using the change of formation enthalpy. In addition, the use of the values of zero-point energy (ZPE) (Table 2) for calculations in both programs, results in a very close dependence:

$$M_S = 102 -224\ ZPE \quad r=0{,}999;\ sd=0.06;\ n=4\ (0.2\ \%\ \text{error})\ (\text{in GAUSSIAN}) \tag{7}$$

$$M_S = 102 -233\ ZPE \quad r=0{,}999;\ sd=0.06;\ n=4\ (0.2\ \%\ \text{error})\ (\text{in ADF}) \tag{8}$$

Table 2. The thermodynamic characteristics of sulfur isotopes in the S_8.

Isotope	The free Gibbs energy change in relative to 3185 a.u.	The value change of ZPE (B3LYP/6-31G (d)), eV	The value change of ZPE (BP86/TZ2P+), eV
α-32S	-0.562094	0.3125824	0.301504
α-33S	-0.562314	0.3077952	0.296899
α-34S	-0.562694	0.3032528	0.292512
α-36S	-0.562598	0.2947120	0.284266

It can be said that, along with the frequencies of the Raman spectrum, the thermodynamic parameters, in spite of small changes (no more than 0.3%), are quite sensitive to sulfur isotopic shift.

4. Conclusion

IR spectra and Raman spectra of the polycrystalline isotopes α-^{32}S, α-^{33}S and α-^{34}S were obtained.

It was found that experimental frequencies of the vibrational spectra of the isotopes of sulfur decrease in a linear manner with an increasing mass of the isotope. From the experimental correlations, the oscillation frequencies for the isotope α-^{36}S were determined.

The results obtained demonstrate that the quantum-chemical calculations are sensitive to the isotopic shifts in the vibrational spectra of sulfur isotopes and are able to assess changes in their thermodynamic properties.

Acknowledgements

This work is financially supported by Tomsk Polytechnic University within the Program of Improving the Competitiveness.

References

[1] Ding T, Valkiers S, Kipphardt H, Bievre P, Taylor P., Gonfiantini R and Krouse R 2001 *Geochimica et Cosmochimica Acta* **65** 2433
[2] McCready R G L, Laishley E J and Krouse H R 1976 *Archives of Microbiol.* **109** 315
[3] Smith J W, Doolan S and McFarlane E F 1977 *Chem. Geol.* **19** 83
[4] Thode H G, Monster J and Durford H B 1961 *Geochimica et Cosmochimica Acta* **25** 159
[5] Seal R R 2006 *Reviews in Mineralogy & Geochem.* **61** 633
[6] Thode H G, Krouse H R and Grinenko V A 1991 *Sulphur Isotopes in Nature and the Environment: An Overview/Stable Isotopes: Natural and Anthropogenic Sulphur in the Environment* (New York: John Wiley & Sons) chapter 1 pp 1–26
[7] Krouse H R, Rundel P W, Ehleringer J R and Nagy K A 1989 *Stable Isotopes in Ecological Research* (New York: Springer-Verlag)

[8] Dawson T E, Mambelli S, Plamboeck A H, Templer P H and Tu K P 2002 *An. Rev. of Ecol., Evolution Systematics* **33** 507
[9] Blair G and Till R 2003 *Guidelines for the use of isotopes of sulfur in soil-plant studies* (Vienna: IAEA)
[10] Richards M P, Fuller B T, Sponheimer M, Robinson T and Ayliffe L 2003 *Int. J. of Osteoarchaeology* **13** 37
[11] Porras I 2008 *Phys. Medicine and Biol.* **53** 19
[12] Becucci M, Bini R, Castellucci E, Eckert B and Jodl H J 1997 J. *Phy. Chem.:* B **101** 2132
[13] Eckert B, Bini R, Jodl H J and Califano S 1994 *J. of Chem. Phy* **100** 912
[14] Becucci M, Castellucci E, Foggi P, Califano S and Dows D A 1992 *J. Chem.Phys.* **96** 98
[15] Bini R, Califano S, Eckert B, and Jodl H J 1997 *J. Chem. Phys.* **106** 511
[16] Frisch M J, Trucks G W, Schlegel H B, Scuseria G E, Robb M A, Cheeseman J R, Montgomery J J A, Vreven T, Kudin K N, Burant J C, Millam J M, Lyengar S S, Tomasi J, Barone V, Mennuci B, Cossi M, Scalmani G, Rega N, Petersson G A, Nakatsuji H, Hada M, Ehara M, Toyota K, Fukuda R, Hasegawa J, Ishida M, Nakajima T, Honda Y, Kitao O, Nakai H, Klene M, Li X, Knox J E, Hratchian H P, Cross J B, Adamo C, Jaramillo J, Gomperts R, Stratmann R E, Yazyev O, Austin A J, Cammi R, Pomelli C, Ochterski J W, Ayala P Y, Morokuma K, Voth G A, Salvador P, Dannenberg J J, Zakrzewski V G, Dapprich S, Daniels A Da, Strain M C, Farkas O, Malick D K, Rabuck A D, Raghavachari K, Foresman J B, Ortiz J V, Cui Q, Baboul A G, Clifford S, Ciolowski J, Stefanov B B, Liu G, Liashenko A, Piskorz P, Komatomi I, Martin R L, Fox D J, Keith T, Al-Laham M A, Peng C Y, Nanayakkara A, Challacombe M, Gill P M W, Johnson B, Chen W, Wong M W, Gonzalez C and Pople J A 2003 *Gaussian 03 Revision B03* (Pittsburg: Gaussian Inc.)
[17] Becke A D 1993 *J. Chem. Phys.* **98** 5648
[18] Lee C, Yang W and Parr R G 1988 *Phys. Rev. B* **37** 785
[19] Handy N C and Cohen A 2001 *J. Mol. Phys.* **99** 403
[20] Perdew J P, Burke K and Ernzerhof M 1996 *Phys. Rev. Let.* **77** 3963
[21] Velde G, Bickelhaupt F M, Baerends E J, Guerra C, Van Gisbergen S J A, Snijders J G and Ziegler T 2001 *J. Comput.Chem.* **22** 931
[22] Steudel R 2003 *Elemental Sulfur and Sulfur-Rich Compounds II* (Springer, Heidelberg)
[23] Müller U 2006 *Inorg. Struct. Chemistry* (New York: John Wiley & Sons)
[24] Linstrom P J and Mallard W G 2010 *NIST Chemistry WebBook NIST Standard Reference Database N 69* (Gaithersburg: National Institute of Standards and Technology)

IOP Conf. Series: Materials Science and Engineering **112** (2016) 012006 doi:10.1088/1757-899X/112/1/012006

Simulation of hydrogen adsorption on carbon nanotubes with different chirality parameters

S V Gromov*, I N Burmistrov, I A Ilinykh and D V Kuznetsov
National University of Science and Technology "MISiS", 4 Leninsky Prospekt, Moscow, 119049, Russia

*E-mail: gromov@asu.misis.ru

Abstract. The hydrogen adsorption properties and surface charge distribution of single-walled carbon nanotubes (SWCNTs) with chirality parameters of (5,5), (9,5), (9,9) were investigated using molecular dynamics methods. Relation between maximum hydrogen uptake and SWCNT structural properties was investigated. It was shown that surface charge distribution of chiral nanotubes could not be compensated for the same amount of steps that leads to fully neutral CNT with armchair structure.

1. Introduction

Rare-earth elements are actively used for formation of nanotubes at electrochemical synthesis from carbon fragments of graphene type in the environment of ions of rare-earth metal [1]. This way of production has considerable impact on structure of carbon nanotubes and is reflected in functional properties. It demands attentive studying, including methods of mathematical modeling.

The present level of computer science and software development provides for significantly wider application of computational simulation as compared to studies conducted just a decade ago. Until recently, one of the most popular basic simulation methods for various structures and compounds was the calculation of the system molecular dynamics under various external conditions. These conditions, in their turn, were simulated by limiting the computational region and variations of the system initial state through setting the force-field, simulated atoms and molecules.

A large number of software solutions and algorithms for the implementation of various simulation conditions were developed. For instance, the Monte Carlo method was proposed as a technique to improve the efficiency of calculations. This method allowed one to reduce significantly the calculation time without a considerable loss in the accuracy of the end result [2-4]. However, all molecular dynamics methods used to describe the force-field and interatomic interactions do not account to the full extent for the electronic structure of individual atoms. Therefore, in some cases, it is impossible to describe all the processes occurring in the system. For example, without considering the electronic properties of atoms it is rather difficult to describe accurately various chemical reactions and magnetic properties of various compounds. Furthermore, when considering nanostructured systems, various quantum effects can be very important, and their electronic structures may significantly differ from the electronic structures of solid bodies. There is also a limitation with regard to a range parameter in interaction. The wider range makes it difficult to select the conditions and parameters of calculation, and also reduces the universality of potential.

Content from this work may be used under the terms of the Creative Commons Attribution 3.0 licence. Any further distribution of this work must maintain attribution to the author(s) and the title of the work, journal citation and DOI.
Published under licence by IOP Publishing Ltd

Therefore, comparatively recently it has been proposed to use quantum-mechanical calculations for multinuclear and many-electron systems to describe more accurately the processes occurring in the system [5].

The analysis of the adsorption characteristics of various carbon adsorbents including those based on carbon nanotubes (CNT) raises the issue of how to account for their electronic properties, for example, in the case that a CNT structure contains other functional groups ($-NH_2$, -OH, etc.) in addition to carbon atoms, or when considering more complex interactions. Similar problems are currently solved using quantum chemistry methods such as the variations of the density functional theory (DFT) and the computations according to the Hartree–Fock method. At the moment, some results of such calculations for a limited number of structures based on CNT are available. For instance, in [6], it was shown that the use of CNT functionalized with OH and NH_2 groups can significantly improve the adsorption properties of these materials. There are also published data on the calculation of the adsorption properties of CNT using the quantum-mechanical methods that demonstrate how some parameters of the electronic and molecular structures can affect their end properties [7-8].

However, the present data on complex studies of the adsorption properties and electronic structure of CNTs with different chirality obtained using quantum-mechanical calculations and molecular dynamics method are insufficient to compare the calculated results with experimental data [9-12].

The approach proposed in this paper will help to provide a more complete picture of adsorption properties for the basic types of CNT structures.

2. Calculation Method

This paper presents molecular dynamics simulation for hydrogen adsorption process on the CNT structures calculated using LAMMPS [13] software with AIREBO [14] potential for carbon-hydrogen bond simulation. Surface charge distribution was obtained with Avogadro [15] molecular editor.

The adsorption characteristics and structural properties were simulated for three basic types of single-walled nanotubes 2 nm in length and the following chirality parameters: $n = m = 5$ (armchair type), $n = 9$, $m = 5$ (chiral type) and $n = 9$, $m = 9$ (armchair type). Figure 1 shows the structures.

The adsorption characteristics were simulated for three types of structures: a single nanotube (a), "square" configuration (b) and "triangle" configuration (c). Figs. 2 and 3 give the types of configuration.

The calculation was performed for a simulation cell of SWCNTs square array and SWCNTs triangle array as well as is shown in Figures 2-3. The modeling was performed for a temperature of 77 K, and the initial pressure of hydrogen varied from 0.1 to 50 MPa. The simulation step was 0.5 fs. The periodic boundary conditions were used in all three directions. The temperature of 77 K corresponds to the boiling point of liquid nitrogen. Therefore, it was used to compare the obtained results with the experimental data, which are often obtained at this temperature. An NPT assembly was used for calculation, because it was necessary to control the pressure in the system. Fifty randomly distributed hydrogen molecules were introduced into the cell so that the minimum distance between the molecules was not less than 0.5 nm. Then the structure was subjected to optimization for 5000· steps in order to determine the nearest energy minimum of the system, after which $1 \cdot 10^6$ steps of molecular dynamics were calculated.

Based on the simulated results, the adsorption isotherms were calculated within the pressure range of 0 to 5 MPa by the molecular dynamics methods and the corresponding energy minima were calculated by the quantum methods.

3. Results and discussions

Based on the calculated data, the adsorption isotherms were constructed for each type of the nanotubes (Figures 4-6).

The highest adsorption values are typical for the nanotubes with larger diameters, i.e. the structures with the parameters of (9,9) and (5,5). Moreover, the "chiral" nanotubes with the structure parameters

IOP Conf. Series: Materials Science and Engineering **112** (2016) 012006 doi:10.1088/1757-899X/112/1/012006

of (9,5) not only have larger diameter, but are characterized by a large number of uncompensated bonds located at both ends of the structure. This means that adsorption is much more probable due to an increased energy of carbon atoms having the smallest number of neighbors.

Isotherms for "square" and "triangle" pattern differ from single SWCNT mostly in highest hydrogen uptake values. The "triangle" configuration shows the lowest hydrogen uptake due to the smallest distance between neighbor CNTs.

Here, we consider the distributions of surface charge of carbon nanotubes before and after adsorption simulation.

Figure 7a shows the model of the (9,5) CNT structure before the adsorption with the generated map of surface charge. The uncompensated negative charge is shown red and the uncompensated positive charge is shown blue. The neutral atoms are shown gray. As it is seen from the Figure 7a, the structure is characterized by a large number of uncompensated negative charges. Therefore, favorable conditions for the adsorption of positively charged particles, including atomic hydrogen, are created.

Figure 7b demonstrates that, after the hydrogen adsorption, the compensation of negative charge is observed at the edges of the nanotube. This results in an overall reduction of the surface energy of the structure. Similar changes in the surface charge distribution are observed for other types of the simulated structures; however, the example of chiral nanotubes with the parameters of (9,5) most clearly demonstrates the partial compensation of the surface charge as a result of adsorption.

Figures 8a and 8b show the models of the surface charge distribution for the nanotubes with the parameters of (5,5) before and after the adsorption of hydrogen, respectively. It is seen that, in this case, after the adsorption, an excess negative charge of the structure is completely compensated.
This is also an evidence for the formation of stable carbon-hydrogen chemical bonds, which is the expected result due to the presence of uncompensated surface bonds in the nanotubes.

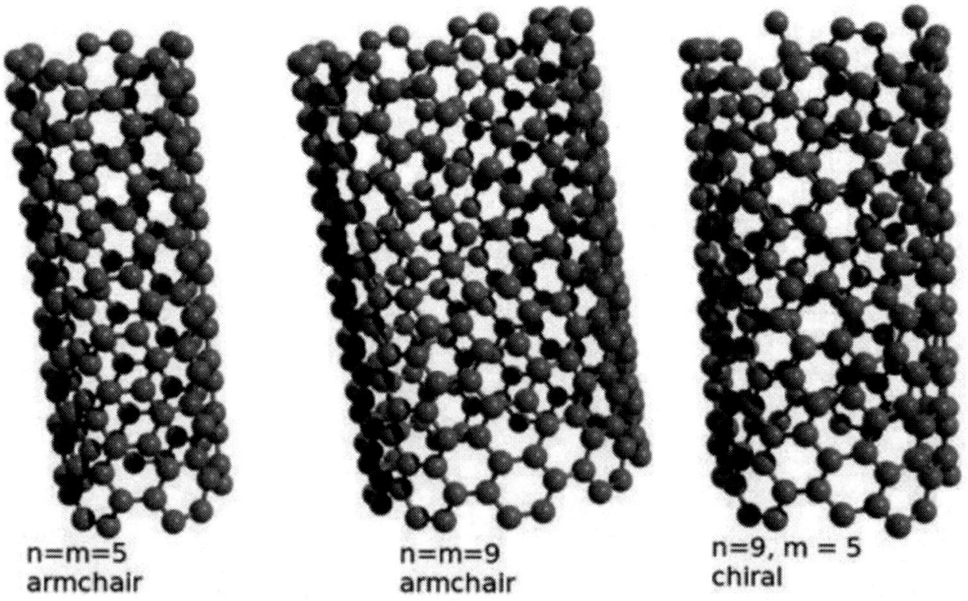

Figure 1. Models of nanotube structures (5,5), (9,9), (9,5)

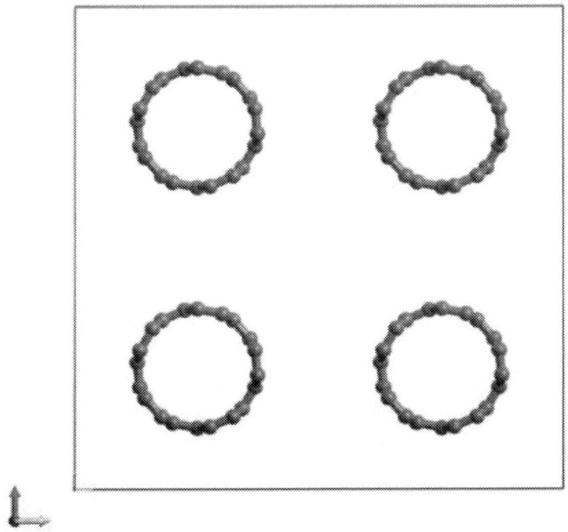

Figure 2. "Square" Configuration of Nanotubes

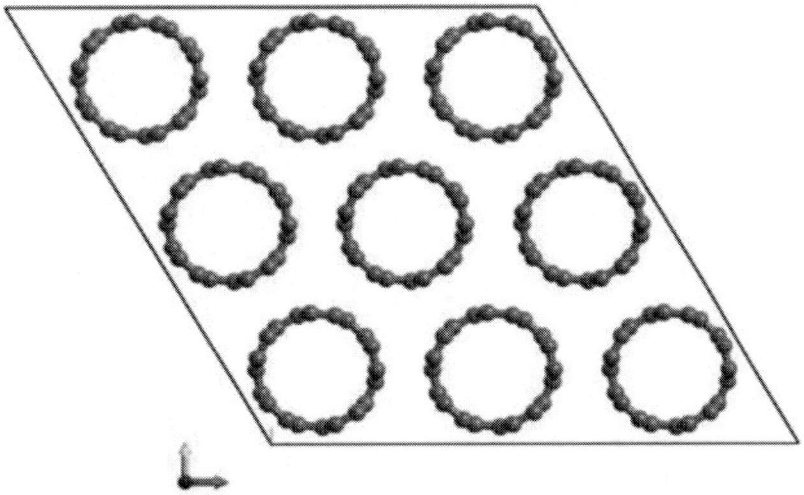

Figure 3. "Triangle" Configuration of Nanotubes

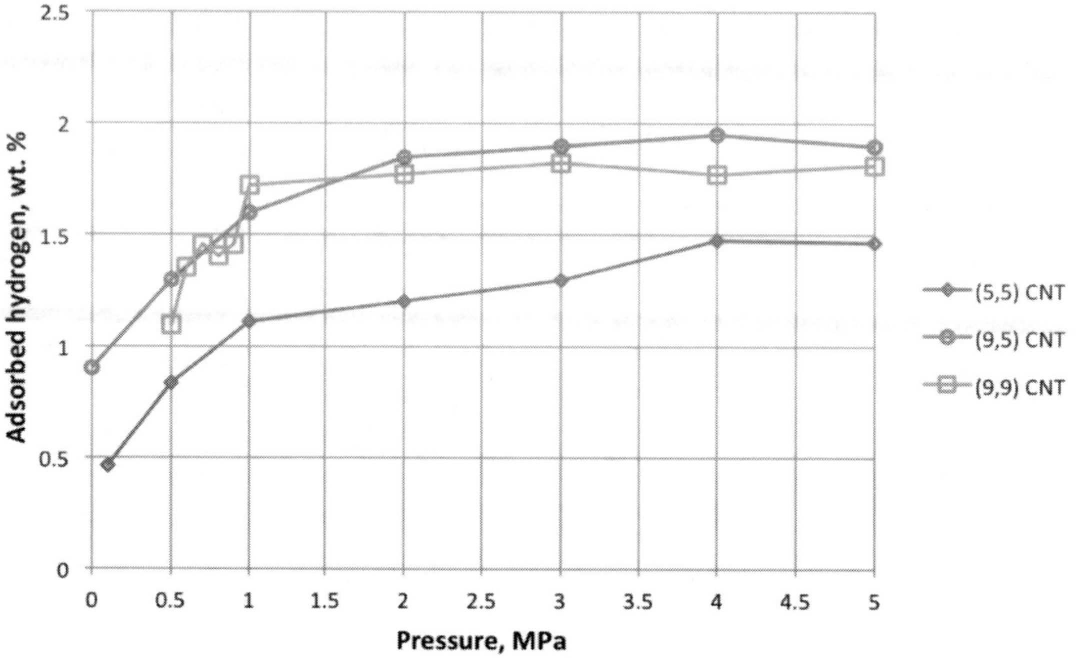

Figure 4. Adsorption isotherms for a single nanotube with the parameters of (5,5), (9,5), (9,9)

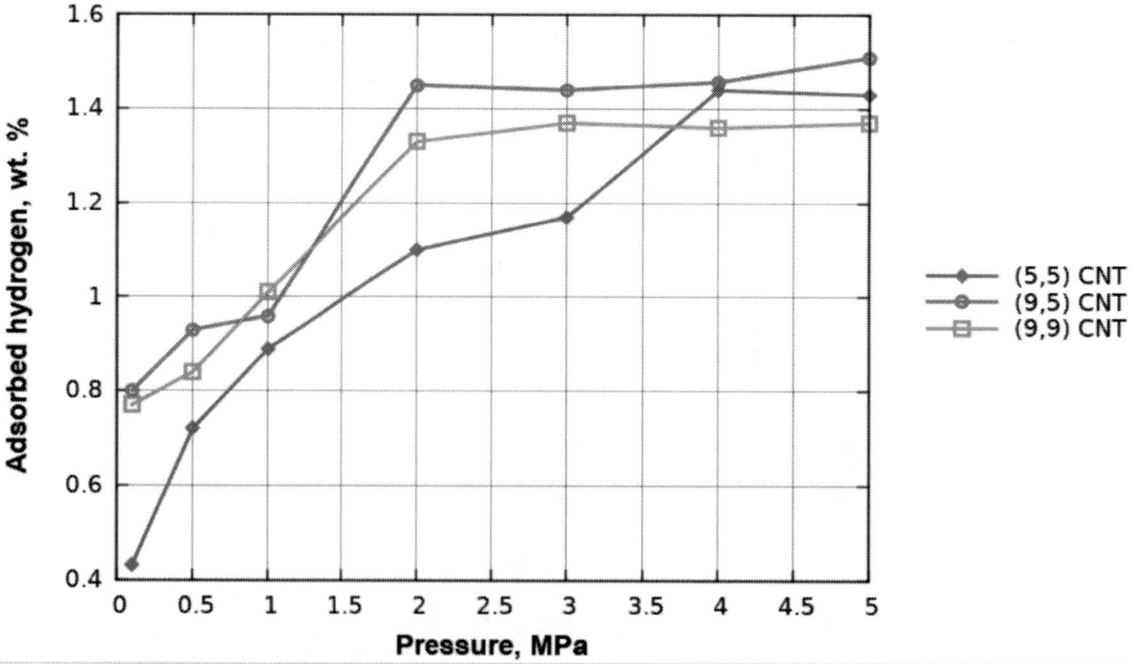

Figure 5. Adsorption isotherms for a triangle array of SWCNT (5,5), (9,5), (9,9)

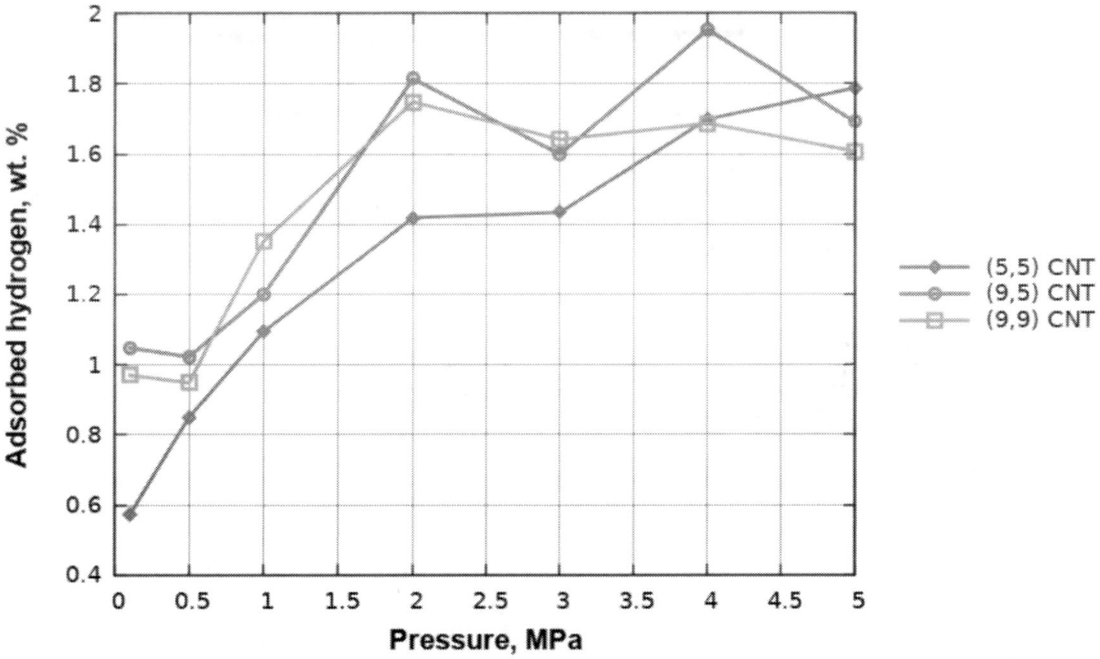

Figure 6. Adsorption isotherms for a square array of SWCNT (5,5), (9,5), (9,9)

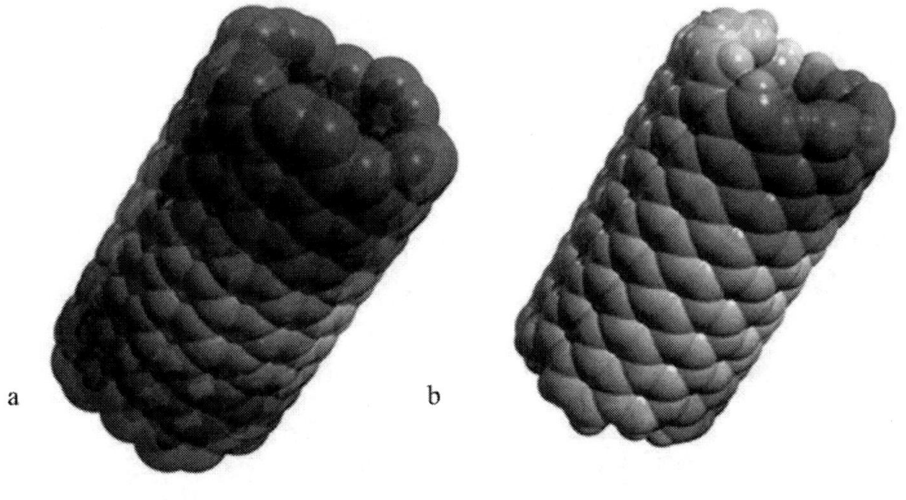

Figure 7. Model of surface charge (electrostatic potential) distribution for a carbon nanotube with the parameters of (9,5) a) before hydrogen adsorption and b) after the hydrogen adsorption

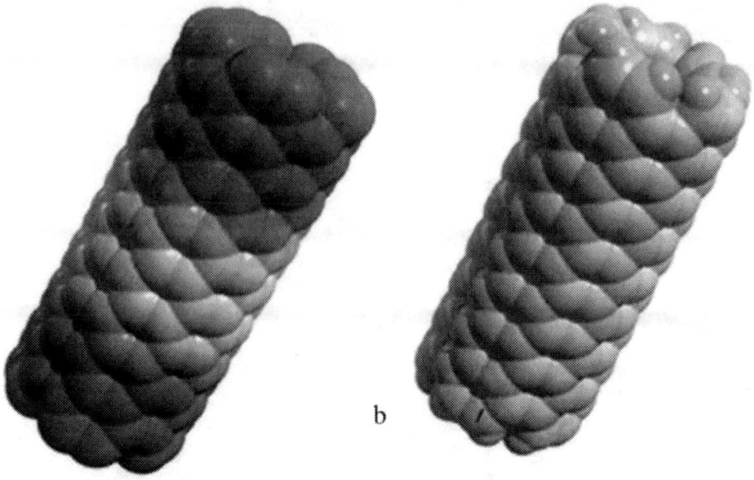

Figure 8. Model of surface charge distribution for a nanotube with parameters (5, 5) a) before hydrogen adsorption and b) after the hydrogen adsorption

4. Conclusions

Based on these results, we can conclude that the structures under consideration demonstrate rather similar properties in the context of the applied calculation methods. An increase in the adsorption with increasing diameter and number of defect of nanotubes (increasing number of uncompensated surface bonds) is also clearly seen. At the same time, when considering the simulated structures, it is seen that the hydrogen molecules are first located at the free ends of the nanotubes and then adsorption can be observed in the middle of the nanotube structure at sufficiently high pressure in the system. The inner surface of the nanotubes is practically unfilled.

As a result of the adsorption interaction of the simulated structures with hydrogen, the partial compensation of surface charge is observed in the case of chiral nanotube and full compensation in the case of nanotube of the armchair type.

5Acknowledgements

This research was supported by the Russian Federal Target Program "Research and Development in Science and Technology Priority Development Fields for 2014–2020" (grant agreement No. 14.575.21.0087 dated 21.10.2014, Project ID - RFMEFI57514X0087).

References

[1] Yao M, Liu B, Zou Y, Wang L, Cui T, Zou G, Li J, Sundqvist B 2006 Effect of rare-earth component of the RE/Ni catalyst on the formation and nanostructure of single-walled carbon nanotubes *J Phys Chem B* **110 31** 15284-90

[2] Mahdizadeh S J 2014 Hydrogen storage on silicon, carbon, and silicon carbide nanotubes: A combined quantum mechanics and grand canonical Monte Carlo simulation study *Int. J. Hydrogen Energy* **39** 1719-31

[3] Cheng H, Cooper A C, Pez G P, Kostov M K, Piotrowski P and Stuart S J 2005 Molecular dynamics simulations on the effects of diameter and chirality on hydrogen adsorption in single walled carbon nanotubes *J. Phys. Chem.* **109** 3780-86

[4] Ahadi Z, Shadman M, Yeganegi S and Asgari F 2012 Hydrogen adsorption capacities of multiwalled boron nitride nanotubes and nanotube arrays: a grand canonical Monte Carlo study *J. Molecular Model* **18** 2981-91

[5] Khan M Z and Khan M Z 2012 Computational study of hydrogen adsorption on potassium-decorated boron nitride nanotubes *In. Nano Letters* **2** 2-7

[6] Soleymanabadi H and Kakemam J 2013 A DFT study of H2 adsorption on functionalized carbon nanotubes *Physica E: Low-Dimensional Systems and Nanostr* **54** 115-117
[7] Ghasemi A S 2012 Highly Sensitive Single-Walled Carbon Nanotube (5, 0) for Adsorptions Gases CO_2 , CO and N_2 : A DFT Study *International Journal of Chemtech research CODEN* **4** 1302-08
[8] Ashrafi F 2010 Optimization of carbon nanotubes for nitrogen gas adsorption *J. Appl. Sci. Eng. Tech* **2(6)** 547-551
[9] Al-Ghamdi A A, Shalaan E, Al-Hazmi F S; Faidah A S, AlHeniti S; Husain M 2012 Adsorption Sites of Hydrogen Atom on Pure and Mg-Doped Multi-Walled Carbon Nanotubes *Journal of Nanomaterials* **2012** 1–5
[10] Barghi S H 2014 Chemisorption, Physisorption and Hysteresis during Hydrogen Storage in Carbon Nanotubes *Int. J. Hydrogen Energy* **39(3)** 1390-97
[11] Zhou L G and Shi S Q 2002 Molecular dynamic simulations on tensile mechanical properties of single-walled carbon nanotubes with and without hydrogen storage *Computational Materials Science* **23** 166-174
[12] Hong Z, Shaoqing W and Huiming C 2004 Effect of chemical potential on the computer simulation of hydrogen storage in single walled carbon nanotubes *Sci. in China Ser. B Chemistry* **47(3)** 222-227
[13] Plimpton S 1995 Fast Parallel Algorithms for Short-Range Molecular Dynamics *J. Comp. Phys* **117** 1-19
[14] Stuart S J, Tutein A B and Harisson J A 2000 A reactive potential for hydrocarbons with intermolecular interactions *J. Chem. Phys.* **112** 6472-86
[15] Hanwell M D, Curtis D E, Lonie D C, Vandermeersch T, Zurek E and Hutchison G R 2012 Avogadro: An advanced semantic chemical editor, visualization, and analysis platform *Journal Cheminformatics* **4** 1-17

IOP Conf. Series: Materials Science and Engineering **112** (2016) 012007 doi:10.1088/1757-899X/112/1/012007

Spectrophotometric Procedure for Fast Reactor Advanced Coolant Manufacture Control

O S Andrienko[1] , N B Egorov[2], I I Zherin[2] and D V Indyk[2,*]
[1] V.E. Zuev Institute of Atmospheric Optics 1, Academician Zuev square Tomsk 634055 Russia
[2] Tomsk Polytechnic University, 634050, Russia, Tomsk, Lenin Avenue, 30

*E-mail: stoya@mail.ru

Abstract. The paper describes a spectrophotometric procedure for fast reactor advanced coolant manufacture control. The molar absorption coefficient of dimethyllead dibromide with dithizone was defined as equal to 68864 ± 795 l·mole^{-1}·cm^{-1}, limit of detection as equal to 0.0583·10^{-6} g/ml. The spectrophotometric procedure application range was found to be equal to 37.88 – 196.3 g. of dimethyllead dibromide in the sample. The procedure was used within the framework of the development of the method of synthesis of the advanced coolant for fast reactors.

1. Introduction

Metal lead enriched by ^{208}Pb is considered to be an advanced coolant for fast reactors by several authors [1,2]. Therefore, there exists a demand in the method for obtaining metal lead enriched by ^{208}Pb isotope from gaseous lead tetramethyl ($Pb(CH_3)_4$), which is used as a working media in isotope separation process [2]. The authors [3] show that one of the compounds received after the tetramethyllead bromination is dimethyllead dibromide ($Pb(CH_3)_2Br_2$). While working with the method for obtaining the isotopically enriched lead as the advanced coolant for fast reactors from gaseous lead tetramethyl [4] there appeared a target to control the amount of $Pb(CH_3)_2Br_2$ during the $Pb(CH_3)_4$ bromination process.

There is a method for spectrophotometric determination of Pb^{2+}, R_3Pb^+ and R_2Pb^{2+} (R – alkyl radical) ions with 4-(2-Pyridylazo) resorcinol known [5]. The use of this method is complicated since no value of molar absorption coefficient of the ions under consideration is represented.

A procedure for spectrophotometric determination of Pb^{2+}, $Pb(C_2H_5)_2^{2+}$, $Pb(C_2H_5)_3^+$ ions is also known [6]. The direct application of this procedure is impossible since no value of molar absorption coefficient of the investigated ions is represented, and the ions under study are different from the ion yielded after the reaction between dithizone and $Pb(CH_3)_2Br_2$.

So, there is no data related to the procedure of the determination of the amount of $Pb(CH_3)_2Br_2$ in literature.

The aim of the work is to develop a spectrophotometric procedure for determination of $Pb(CH_3)_2Br_2$.

Content from this work may be used under the terms of the Creative Commons Attribution 3.0 licence. Any further distribution of this work must maintain attribution to the author(s) and the title of the work, journal citation and DOI.
Published under licence by IOP Publishing Ltd

2. Experimental

2.1. Reagents and equipment

The quality grade of $Pb(CH_3)_2Br_2$ (Siberian Chemical Plant, Russia), dithizone (Sigma-Aldrich, Switzerland), CCl_4 (Reachem, Russia), ammonia (Reachem, Russia), sodium sulfite (Reachem, Russia), was for analytical purpose. The quality grade of ammonia citrate (Reachem, Russia), bromine (Reachem, Russia) was synthesis reagents. The buffer solution #1 was prepared according to procedure [7], buffer solution #2 was prepared according to procedure [6]. Spectrophotometer "Thermal visio Evolution 600" (Thermo Scientific, USA) was used in the work.

2.2. Procedure

To prepare a standard solution, 5 ml of 33% of water solution of ammonia citrate, 50 ml. of buffer solution #1 and a weighted before sample containing $Pb(CH_3)_2Br_2$ was put into 500 ml. measuring flask and was mixed until all the components were dissolved. Then, the solution in the flask was diluted with water up to 500 ml. To prepare the dithizone solution, 60 mg. of dithizone were dissolved in 1 liter of CCl_4. 15 ml. of buffer solution #2 were put into 50 ml. glass reactor, then, an aliquot of standard solution was added. The extraction was carried out for 130 minutes with agitation. After the extraction the organic phase was separated with the help of a funnel and gathered in a 25 ml. flask. The solution in the flask was diluted with CCl_4 up to 25 ml. and was shaken. Spectrophotometric determination of $Pb(CH_3)_2Br_2$ ran in the 10.02 mm. quartz cell in comparison with CCl_4 between $350 \cdot 10^{-9}$ and $650 \cdot 10^{-9}$ m.

3. Discussion

Figure 1 represents the absorption curve of dithizone and $Pb(CH_3)_2Br_2$.

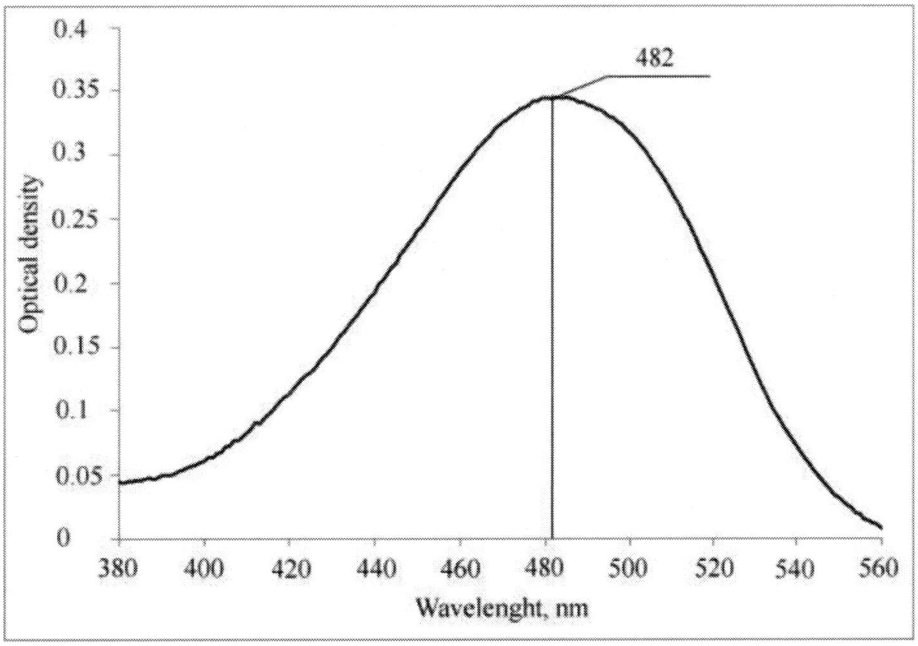

Figure1. The adsorption curve of the complex of $Pb(CH_3)_2Br_2$ with dithizone.

The curve shows that complex of dithizone and $Pb(CH_3)_2Br_2$ represents absorption maximum at 482 nm.

Figure 2 reflects the dependence between the optical density and concentration of $Pb(CH_3)_2Br_2$ in the sample.

IOP Conf. Series: Materials Science and Engineering **112** (2016) 012007 doi:10.1088/1757-899X/112/1/012007

To decrease the influence of random error on the result, each measurement was repeated 5 times. Based on the data represented at figure 2 it may be concluded that Beer's law is kept within the range of concentration of $Pb(CH_3)_2Br_2$ equal to 1.5 – 7.9 mg/l of $Pb(CH_3)_2Br_2$ that is equal to 37.88 – 196.3 g. of dimethyllead dibromide in the sample. The linearization line equation is $A = 0.1704 \times C(Pb(CH_3)_2Br_2) + 0.0119$ ($R^2 = 0.9992$).

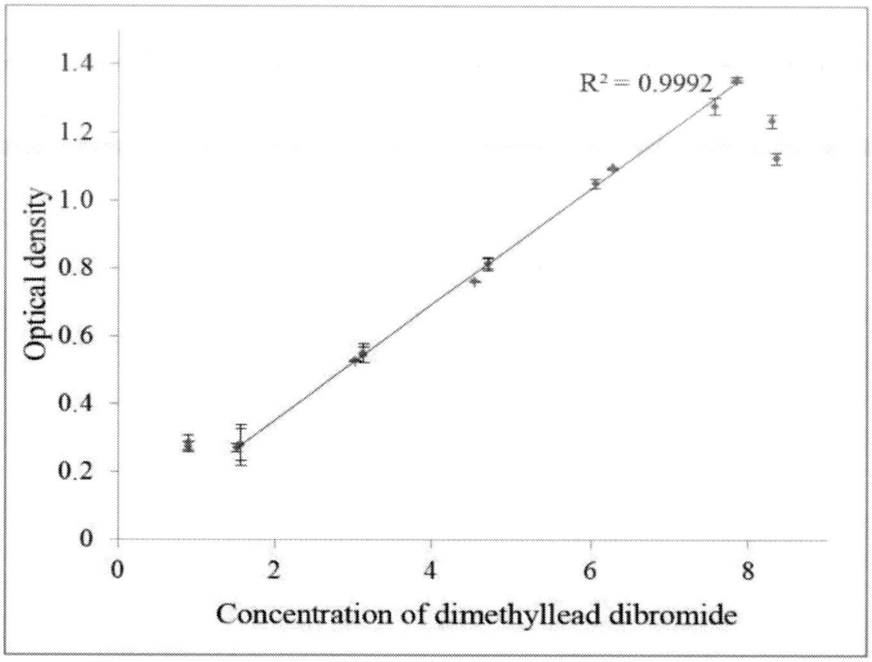

Figure 2. Dependence between the optical density and concentration of $Pb(CH_3)_2Br_2$.

Three standard solutions with measured concentration of $Pb(CH_3)_2Br_2$ were prepared to calculate the average molar absorption coefficient of dimethyllead dibromide with dithizone. Then, the concentration of $Pb(CH_3)_2Br_2$ was measured by means of the developed procedure. The results are represented in table 1.

Table 1. ε $Pb(CH_3)_2Br_2$ calculation data

Mass of $Pb(CH_3)_2Br_2$, $g \cdot 10^{-6}$	D	l, cm	ε $l \cdot mole^{-1} \cdot cm^{-1}$	$\bar{\varepsilon}$	ω, %
35.6	0.267	1.002	68589	68344	0.4
71.2	0.536		68889		0.8
106.8	0.801		68675		0.5
142.4	1.063		68375		0.0
178	1.332		68538		0.3
35.72	0.266		68188		0.2
71.44	0.534		68444		0.1
107.16	0.807		68986		0.9
142.88	1.065		68231		0.2
178.6	1.334		68410		0.1
37.32	0.283		69518		1.7

74.64	0.552	67759	0.9
111.96	0.825	67446	1.3
149.28	1.094	67084	1.8
186.6	1.386	68029	0.5

The average molar absorption coefficient of dimethyllead dibromide with dithizone was defined with the help of statistical methods and was found to be equal to 68864 ± 795 l·mole^{-1}·cm^{-1}, relative measurement error did not exceed 1.8% for each of the measurements. The detection limit of $Pb(CH_3)_2Br_2$, calculated with the use of the 3s - criterion (P = 0.95, n = 10) was found to be equal to $0.0583\cdot10^{-6}$ g/ml. To calculate the within-laboratory divergence, the Kohren criterion was calculated and was found to be equal to 0.3551 which is less than the tabulated point ($S_{tabulated} = 0.6838$; $P = 0.95$; $f = 3$, n=5). This means that the measured results are reproducible.

Five control specimens with known concentration of $Pb(CH_3)_2Br_2$ were prepared to verify the results. The concentration of $Pb(CH_3)_2Br_2$ in each sample was measured by means of the developed procedure. The results are shown in table 2.

Table 2. Results of verification of the procedure (n = 5, P = 0.95)

	$Pb(CH_3)_2Br_2$	Inserted mg/l	Detected mg/l	%	t^*
Sample number	1	39.2	38.58	1.58	1.78
	2	78.4	77.56	1.07	1.54
	3	117.6	115.84	1.5	2.53
	4	156.8	158.16	0.87	2.29
	5	196	195.26	0.38	0.67

It may be noted that the relative error is small (not more than 1.58 %) within the range of concentrations to which the procedure can be applied. Some deviations of the results between the known and measured concentration may be connected with a random error, since the value of calculated Student's coefficient (t*) is less than the tabulated point ($t^* = 2.78$; $n = 5$; $P = 0.95$).

4. Conclusion

A spectrophotometric procedure for the determination of dimethyllead dibromide during the manufacture of fast reactor advanced coolant has been developed. The molar absorption coefficient of dimethyllead dibromide with dithizone was found to be equal to 68864 ± 795 l·mole^{-1}·cm^{-1},the limit of detection was equal to $0.0583\cdot10^{-6}$ g/ml. The range of use of the spectrophotometric procedure was found to be equal to 37.88 – 196.3 g. of dimethyllead dibromide in the sample. Relative error within the range of application of the procedure didn't exceed 1.58 %.

Reference

[1] Khorasanov G L, Blohin A I and Valter A.A 2012 *Nuclear Reactors* Amir Zacarias (InTech) 57
[2] Khorasanov G L et al 2013 *Application of stable lead isotope Pb-208 in nuclear power engineering and it's acquisition techniques* (NY: Nova publishers)
[3] Akimov D V, Andrienko O S, Egorov N B, Zherin I I, Indyk D V, Kazaryan M A and Obmuch K V 2013 *J. For Alternative En. and Ecol.* **4/1** 77

[4] Andrienko O S, Afanasyev V G, Egorov N B, Zherin I I, Indyk D V, Kazaryan M A, Poleshuk O H and Sokovikov V G 2011 Method for obtaining metal lead *Patent RUS #2415185*
[5] Schmidt U, Huber F 1978 *Analytica Chimica Acta* **98** 147
[6] Henderson S R and Snyder L J 1961 *J Anal. Chem.* **33/9** 1172
[7] Shvarcenbah G and Flashka G 1970 *Chelatometric Titration* (Moscow: Chemistry)

The efficiency of rare elements used for hydrogen production from polymer waste

S A **Garelina**[1], **A L Gysev**[2], **R A Zakharyan**[3], **M A Kazaryan**[4,*], **G Martoyan**[5], **V I Sachkov**[6], **I N Feofanov**[4]

[1] Academy of Civil Protection, Ministry of Emergency Situations, 141435, Khimki, Novogorsk microdistrict, Russia
[2] Scientific Technical Centre «TATA», 29 Moscow Str., Sarov, Nizhnij Novgorod Reg., 607181, Russia
[3] Tarusskii branch of A.M. Prokhorov General Physics Institute of RAS, 6 Engels Str., Tarusa, Kaluga Reg., 249100 Russia
[4] P.N. Lebedev Physical Institute of RAS, 53 Lenin Av., Moscow, 119991, Russia
[5] NP OOO «ECOATM», Republic of Armenia
[6] Siberian Physical-Technical Institute of Tomsk State University, 1 Novosobornaya sq., Tomsk, 634050, Russia

[*]E-mail: kazarmishik@yahoo.com

Abstract. The authors have proposed the plasma-chemical method of the polymer waste processing and developed a schematic diagram of the reactor. The paper contains the results of the theoretical calculation of the polymer waste processing products. The reagents that allow obtaining hydrogen and other liquid products from polymer waste are selected. It has been proposed to use rare elements for increasing the efficiency of hydrogen production from polymer waste. The results of the calculation of the efficiency of hydrogen production from polymer waste using molybdenum have been revealed in the paper.

1. Introduction

It is well known that the environmental threat to the whole world has become the places of polymer waste (PW) storage. The degradation products of PW are substances that have harmful effects on the environment. Every year about 4.8 million tons of PW is formed in the Russian Federation [1]. World production of plastics is increasing annually by 5 – 6%, which leads to the increasing of the PW [2].

The universal solution of the PW destruction problem is not found yet, because the destruction of the PW is as complex and expensive as their production [2]. The Russian system of waste utilization is based on waste disposal (more than 80%) in landfills and unorganized dumps [3 – 5].

Using PW for energy systems and complexes will ensure the prevention of large-scale environmental crisis caused by the lack of PW processing technology that meet the economy and resources saving modern requirements, as well as the expanding the raw material base of energy by using the energy potential of these polymer waste.

Nowadays, an alternative method [6 – 8] of using the PW for energy systems and complexes is hydrogen production from waste. Hydrogen is a highly environmentally friendly fuel for transport and the development of hydrogen energy. Currently production, storage and using of hydrogen as an

Content from this work may be used under the terms of the Creative Commons Attribution 3.0 licence. Any further distribution of this work must maintain attribution to the author(s) and the title of the work, journal citation and DOI.
Published under licence by IOP Publishing Ltd

energy fuel receive considerable attention both from the government and large commercial structures [7 – 13].

The evaluation showed that buried PW can annually produce up to 1 million tons of hydrogen. According to various data, summarized in [14], there are two groups of world hydrogen production estimation. According to the first, "the annual world production of hydrogen by the end of 1990 was 40 – 45 million tons or 450 – 500 billion m3" According to the second group, world production of hydrogen was 60 million tons in 1990 and 80 million tons in 1998. As emphasized in [14], "in any case, these amounts are significant, corresponding to 20 – 25% of annual world production of natural gas".

It is recognized that the advanced PW processing technology is plasma technology (environmentally friendly and cost-effective).

Plasma technology for processing the PW to hydrogen is under active research and development for large-capacity recycling [8, 15, 16]. However, nowadays there is nowell-functioning waste processing with the using of plasma chemical methods [16]. According to [16], such plant is in the "setting up" stage in Canada (does not work reliably for a long time), South Korea has a pilot production plant for plasma processing of household and industrial waste up to 1,000 kg per hour. Experimental studies of this area were also conducted in France.

2. Schematic diagram of the plasma chemical reactor for polymer waste processing

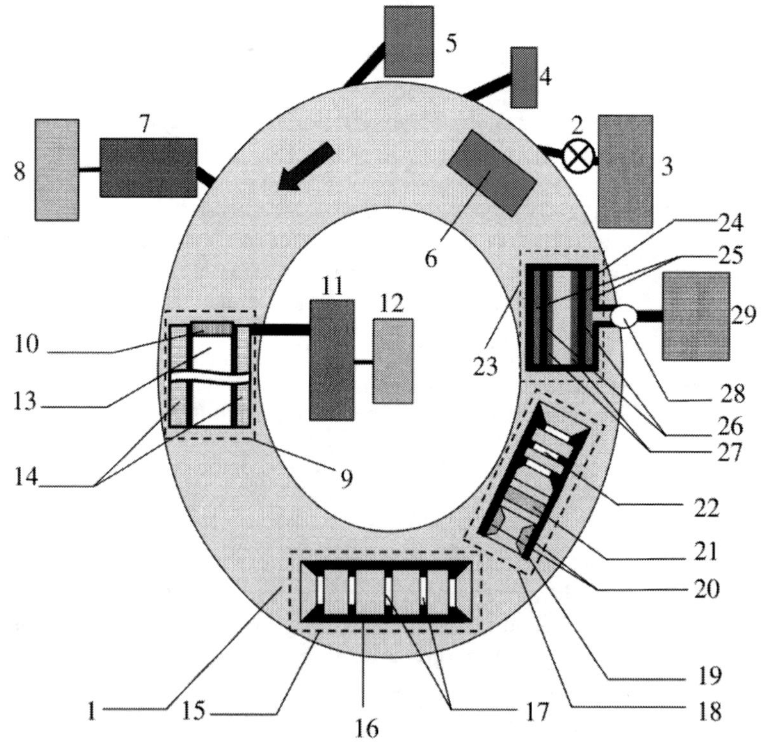

Figure 1. A schematic diagram of the reactor for plasma chemical PW processing

The figure 1 shows a schematic diagram of the reactor for plasma chemical PW processing. In the reaction chamber 1 the inert gas enters the valve 2 from the cylinder 3, the operating pressure of which is measured with a vacuum gauge 4. The pump 5 for pumping inert gas from the reaction chamber 1 sets a predetermined pressure of inert gas. Circulation of the inert gas in the reaction chamber 1 in the direction shown in Figure 1 by arrow occurs via pump 6. PW, converted into the gaseous state by the heater-cooler 8, adds continuously or portionwise to the inert gas from the reservoir 7 into the reaction chamber 1. The flow of inert gas entrains and displaces PW vapours at the reaction chamber 1 in the direction indicated by the arrow. The mixture of inert gas and vapours enters the pyrolysis unit 9. The initiation of pyrolysis occurs in the pyrolysis initiation device 10 by injection of metal vapour from the tank 11. The predetermined pressure metal vapour is provided by the heater 12. Within the movement of a gas mixture along the heat-resistant channel 13, (14 – thermal insulation) its temperature will increase and more PW vapours will be pyrolysed. As a result, gas mixture, most of which is PW processing products, is formed at the end of heat-resistant channel 13 of the pyrolysis unit 9. To reduce the time of pyrolysis requires preheating of the gas mixture at the inlet or discharge, which will create products involved in Association with the release of energy.

Processing products with an inert gas flow come into the removable condensation device 15 (16 – 15 device node). Condensed PW processing products from vapour phase accumulate between the diaphragms 17. Remaining in the reaction chamber 1, gaseous products such as carbon monoxide and hydrogen move with the inert gas stream in the removable carbon monoxide output device 18 (19 – 18 device node). At node 19 in the interaction of carbon monoxide molecules with a metal insert 20 metal carbonyl is formed (21 – a filter for removing microparticles that can break away from the walls of nodes). It is collected and stored in the system 22. The gaseous hydrogen with a stream of an inert gas comes into hydrogen output device 23 (24 –23 device node). Hydrogen, that was separated from membrane 25 (26 – mesh that protects it against aggressive chemicals) and heated by the heater 27 from the gas mixture circulated in the reaction chamber 1, is moved out from the reaction chamber 1by pump 28 and is collected in reservoir 29.

3. The results of theoretical studies of the composition of PW procession products

A detailed analysis of PW procession end products requires physical and numerical experiments. However, the main end-products can be defined for a given character of PW molecules dissociation with data of the binding energies of the atoms in the molecules of the primary products of processing and of their ability to react chemically with each other.

Within the analysis (on the base of polyethylene (PE), polyvinyl chloride (PVC) and polyethylene terephthalate (PET)), it has been suggested that the breaking of chemical bonds, the energy of which is less than 5 eV is carried out in repetitively pulsed discharge in the polymer waste molecules. Chemical reagent is calcium vapours.

The table 1 shows the results of evaluation of basic products of PE, PVC and PET processing and reagent costs for processing of 1 kg PW.

Table 1. Of evaluation of basic products of PE, PVC and PET processing and reagent costs for processing of 1 kg PW

PW	Processing products	Reagent cost, kg
PE	CaC_2 H_2	1.43
PVC	CaC_2 $CaCl_2$ H_2	0.96
PET	CaC_2 CO H_2	0.63

4. The efficiency of hydrogen production from PW

To reduce the cost of energy for processing PW proposed to use the energy released during association of the products of dissociation PW with each other and with the reagent.

First, it is necessary to estimate the amount of energy G_d that must be expended on the dissociation of one molecule of PW with T_{max}:

$$G_d = \sum_i (E_i n_i \alpha_i),$$

where i – the type of coupling in the molecule, n_i – number of similar bonds in the molecule, E_i – i-type chemical bond breaking energy, α_i – degree of dissociation of bonds of type i at T_{max}.

Then estimate the amount of energy that is released during the formation of refined products from one molecule of PW with T_{max}:

$$G_a = \sum_l n_l \sum_j (E_j n_j (1-\alpha_j)),$$

where j – bond type of association products, n_j– number of similar bonds in the association products, E_j – bond breaking energy of j-type, l – type of molecules resulting from association, n_l – number of similar molecules or radicals in the products of the association, j – degree of dissociation of type i bonds at T_{max}.

The amount of energy stored in the chemical bonds that were not dissociated by pyrolysis, is:

$$G_d' = G_{d\Sigma} - G_d,$$

$$G_{d\Sigma} = \sum_i (E_i n_i),$$

where $G_{d\Sigma}$ – energy of complete dissociation of a PW single molecule.

In the paper the maximum temperature (T_{max}), corresponding to the maximum value of the relative amount of products formed as a result of the dissociation products association with each other and reagent, was defined in the pyrolysis for PE, PVC and PET (concentrations $n = 10^{15}, 10^{16}, 10^{17}$ cm^{-3}) (Table 2). Estimates are made under the assumption that during the pyrolysis such bonds as C = O and C = C are remained and the reagent is calcium.

The assessment of minimum efficiency of hydrogen production (Table 2) is made on the assumption that electrical energy is wasted on gas heating to T_{max} and unexploded bond dissociation by pyrolysis:

$$\delta'_{H1} = \frac{W_H}{G_\partial' + G_{T\max} - \Delta G},$$

where $\Delta G = G_a - G_d$ (G_d– energy that must be expended on dissociation under T_{max}, G_a – energy that is released during the dissociation of association products with each other and the atoms of the reagent), G_d' – energy stored in the undissociated bonds during the pyrolysis, W_H – energy released by burning the hydrogen produced during the PW processing, $G_{T\max} = \frac{3}{2}k(T_p - T_0)(n_M N + YN)$ – energy that must be expended in heating the mixture of inert gas and molecules to the PW to T_{max} (T_0 = 300 K – initial temperature, n_M– number of atoms in the molecule, N – number of PW molecules in one kg, Y = 5; 10 – ratio of the concentration of inert gas atoms to the concentration of PW software.

Maximum efficiency of hydrogen production (Table 2) is estimated on the assumption that electrical energy is wasted on gas heating to T_{max}:

$$\delta'_{H2} = \frac{W_H}{G_{T\max} - \Delta G}.$$

It is clear that the real value of the efficiency of hydrogen production will be between δ'_{H1} and δ'_{H2}.

Table 2. The results of calculation: the maximum temperature T_{max}, reached in the pyrolysis zone (figure 1), the cost of energy G_d for the dissociation under T_{max}, energy G_a that is released during the formation of processed products under T_{max}, the energy G_d', stored in the bonds that are not braked at the result of pyrolysis under T_{max}, power consumption G_T' on heating the mixture of inert gas and molecules of processing substances under T_{max}, the minimal δ'_{H1} and maximal δ'_{H2} efficiency of hydrogen production, based on 1 kg of PW

PW	n, cm^{-3}	T_{Max}, K	$G_{d'}$ kW·h	$G_{a'}$ kW·h	G_d', kW·h	G_T (Y=5), kW·h	G_T (Y=10), kW·h	δ'_{H1} (Y=5)	δ'_{H1} (Y=10)	δ'_{H2} (Y=5)	δ'_{H2} (Y=10)
PE	10^{15}	2200	17.99	17.56	2.14	3.76	5.94	0.9	0.65	1.5	0.87
	10^{16}	2450	17.15	16.12	2.98	4.16	6.91	0.68	0.51	1.34	0.71
	10^{17}	2800	16.54	14.82	3.59	4.94	8.03	0.54	0.42	1.13	0.57
PVC	10^{15}	2200	7.62	7.33	1.70	1.16	1.7	0.63	0.54	1.71	1.00
	10^{16}	2500	7.62	7.05	1.69	1.34	1.69	0.55	0.47	1.48	0.79
	10^{17}	2850	7.36	6.55	1.95	1.56	1.95	0.46	0.39	1.3	0.65
PET	10^{15}	2200	7.66	7.9	1.18	0.93	1.18	1.07	0.39	2.94	1.48
	10^{16}	2450	7.2	7.19	1.65	1.05	1.65	0.73	0.79	1.87	1.59
	10^{17}	2750	6.53	6.24	2.32	1.19	2.23	0.52	0.68	1.35	1.16

Thus, the proposed method enables the production of hydrogen from PW with efficiency greater than one. The proposed method for hydrogen production can be used for hydrogen production from fossil hydrocarbons.

5. Improving the efficiency of hydrogen production from PW - rare elements

Improving the efficiency of hydrogen production can be achieved by selection of reagents, providing even greater energy release during the formation of the corresponding processed products, such as molybdenum.

Table 3 shows the calculation results of the same parameters as in Table 2, but with the molybdenum reagent.

The data in Table 3 indicate that at the optimum temperatures the costs to the dissociation by electric discharge of one kilogram of processed materials are significantly reduced as compared with the case discussed above.

Moreover, Table 3 shows that the amount of energy released during the formation of PW processing products in the case of using the molybdenum as a reagent exceeds the total cost of electric power for heating the mixture of inert gas and polymer waste vapour to T_{max} and the energy stored in the undissociated links as a result of the energy released in the formation of processed products.

Table 3. The results of calculation: G_d – energy expended on the dissociation of molecules contained in one kg, G_a – energy released during the formation of processing products from one kg, G_d' – energy stored in unbroken bonds for one kg, the reagent – molybdenum

PW	n, cm^{-3}	T_{max}, K	G_d, kW·h	G_a, kW·h	ΔG, kW·h	G_d', kW·h	G_T (Y=5), kW·h	G_T (Y=10), kW·h
PW	10^{15}	2200	17.99	33.4	15.41	2.14	3.76	6.10
	10^{16}	2500	17.79	31.38	13.59	2.33	4.35	7.07
	10^{17}	2850	17.12	28.25	11.13	3.01	5.04	8.19
PVC	10^{15}	2250	7.99	14.02	6.03	1.32	1.19	1.73
	10^{16}	2500	7.62	12.74	5.12	1.69	1.34	1.95

10^{17}	2900	7.63	11.95	4.31	1.67	1.59	2.31

6. Conclusion

The authors have proposed technical solution and found conditions (temperature and concentration are defined) at which efficiency of energy for hydrogen production that required in addition to the heat generated in the process of association is greater than one. Environmentally friendly and high energy efficiency in hydrogen production certainly provides the high competitiveness of the proposed plasma chemical PW processing with hydrogen production with the rare elements using. Moreover, on the base of this method well-functioning, environmental and technical safety mobile unit can be created. Calculations have shown that the performance of the mobile unit capacity of 250 kW/h for 1 hour ensure the production of 8 kg of hydrogen from 55 kg of polyethylene processing, 5.5 kg of hydrogen – from 125 kg of PVC, 8 kg of hydrogen – from 163 kg of PET. It is expediently to use the hydrogen produced in these mobile units to obtain the energy required for dissociation of recyclable materials.

Thus, there are all conditions for a qualitative leap in environmental and energy problems solution with the using of rare elements.

References

[1] Ostaeva G U and Potapov I I 2002 *Ecology Systems and Instruments* **12** 51 (in Russian)
[2] Ponomareva V T, Lihacheva N N and Tkachik Z A 2002 *Plastic Masses* **5** 44 (in Russian)
[3] *The concept of solid domestic waste treatment in Russian Federation Methodical document in construction 13-8.2000* 1999 (Moscow: Federal Centre of the accomplishment and city and rural settlement ecology safety) #17 (in Russian)
[4] Lobacheva K G, Zeltobruhov V F and Prockopov I I 2005 *The state of the issue of the waste and the modern ways of their processing* (Volgograd: Volgograd State University) (in Russian)
[5] Tugov A, Escin I, Litun D and Fedorov O 1998 *Science and Life* **5** 7 (in Rusian)
[6] Homkin K A 2005 *Experimental research in basing of the technology of the organic waste and natural gas complex processing into the hydrogen and carbonic materials.* PhD Thesis in Tech. Sci. 2005 Moscow (in Russian)
[7] Shamardin I M 2009 *Analitical Note* **1** 39 (in Russian)
[8] Mesyats G A and Prochorov G A 2004 *The bulletin of the Russian Academy of Science* **7** 575 (in Russian)
[9] Lepesov K K, Lepesov S K and Mukaev S B 2012 *Industry of Kazakhstan* **3(72)** 26 (in Russian)
[10] Bishimbaev V K and Lepesov K K 2012 *Energy and Fuel Resources of Kazakhstan* **4** (in Russian)
[11] Kazakov A V, Zavorin A S, Novosel'cev P Ju and Tabakaev R B 2013 *J. of Sci. of Siberia* **4(10)** 13 (in Russian)
[12] Konoplev V N and Shupljakov V S 2009 *Theoretical and Applied Problems of Service* **3(32)** 3 (in Russian)
[13] Kuzyk B N and Jakovec Ju V 2007 *Russia: Strategy of Transition to Hydrogen Energy* (Moscow: The Institute for economic strategies) (in Russian)
[14] Tarasov V P 2006 *Alternative Energy and Ecology* **8 (40)** 72 (in Russian)
[15] Tsvetkov U V 2006 *Power-metallurgical Complex on the Base of Plasma Technics. Russian Energy: Problems and Perspective* (Moscow: Nauka) (in Russian)
[16] Anshakov A 2010 *About the Energy from Waste*. Available at http://www.forum.ruboard.ru

Purification of rare-earth metals as the approach to improving properties of hard magnetic $Nd_2Fe_{14}B$-based materials

N B Kolchugina[1,*], G S. Burkhanov[1], A G Dormidontov[2], A A Lukin[2], Yu S Koshkid'ko[3,4], K Skotnicová[3], H Drulis[5] and B Smetana[3]
[1] Baikov Institute of Metallurgy and Materials Science, Russian Academy of Science, Leninskii pr. 49, 119991 Moscow, Russia
[2] JSC "Spetsmagnit", Dmitrovskoe sh. 58, 127238 Moscow, Russia
[3] Vysoka Skola banska - Technical University of Ostrava, 70833, 17 Listopadu, 15/2172, Ostrava-Poruba, Czech Republic
[4] International Laboratory for High Magnetic Fields and Low Temperatures, Polish Academy of Sciences, Gajowicka 95, 53-421 Wroclaw, Poland
[5] Institute of Low Temperature and Structure Research, Polish Academy of Sciences, Okolna 2, 50-422 Wroclaw, Poland

[*]E-mail: natalik014@yandex.ru

Abstract. Purification of rare-earth metals, namely, Nd, Pr, Tb, Dy used in manufacturing $Nd_2Fe_{14}B$-based magnets was realized. The metals were purified by vacuum distillation/sublimation. Conditions of the process were optimized and the structure of distilled metals was studied. Distilled terbium and dysprosium were used to prepare hydrides TbH_2 and DyH_2. Peculiarities of the decomposition of terbium and dysprosium hydrides were studied with the view of the use of the compounds as efficient additions, which allow the high-coercivity state of sintered magnets to be formed. Terbium hydride additions (to 4 wt %) favor the marked increase in the magnetization coercive force without excessive attenuation of the remanence (${}_jH_c$ = 1940 kA/m, (BH_{max}) = 292 kJ/m^3). Dysprosium hydride additions increase the stability of high-coercivity state (${}_jH_c$ = 1310 kA/m, (BH_{max}) = 322 kJ/ m^3 at 2 wt% DyH_2).

1. Introduction

Requirements for the purity of metals and substances, as a rule, are determined experimentally and depend on their end use. Rare-earth metals 95-99% purity (with respect to other rare-earth metals) are known to be suitable for manufacturing hard-magnetic, magnitostrictive, and hydrogen-storage materials. For these applications, mixtures of rare-earth metals also can be used. However, in recent years, investigators trend to discuss the properties of permanent magnets in relation to the purity of starting materials and substantial dependence of magnetic characteristics on manufacturing conditions. In accordance with [1], three grades of rare-earth metals (REMs) can be defined; these are (1) ultrapure, (2) routine metals prepared under laboratory conditions, and (3) commercial. The purity of high pure REMs is > 99.0 at % with respect to all elements in the periodic table. The purity of metals prepared under laboratory conditions is 99% with allowance for all elements. The content of base metal in commercial REMs is <98 at %. The transition from the commercial-purity metal to the metal

Content from this work may be used under the terms of the Creative Commons Attribution 3.0 licence. Any further distribution of this work must maintain attribution to the author(s) and the title of the work, journal citation and DOI.
Published under licence by IOP Publishing Ltd

corresponding to the next purity grade leads to about tenfold increases in the metal cost. The most important application of neodymium is Nd-Fe-B magnets. Sintered Nd-Fe-B magnets find wide applications for technology owing to the high maximum energy product ($\boldsymbol{BH}_{max}$), remanence ($\boldsymbol{B}_r$), and magnetization coercive force ($_j\boldsymbol{H}_c$) [2]. To increase $_j\boldsymbol{H}_c$, which is responsible for the time-temperature stability of sintered magnets, the basic $Nd_{14\text{-}15}Fe_{ocr.}B_{6\text{-}8}$ composition is alloyed with Pr, Dy, Tb (to increase the anisotropy field of the 2-14-1 phase); moreover, the following technological processes are used: "strip-casting", hydrogen decrepitation, mechanical alloying, diffusion saturation of rare-earth magnet surface with Dy and Tb, and multi-stage heat treatments at 450-1000°C [2-7]. The coercivity of Nd–Fe–B magnets is structure sensitive and dependent on the continuity and composition of REM-rich phases, grain-boundary defects, average grain size, exchange interaction between $Nd_2Fe_{14}B$ hard magnetic grains, etc. The coercivity of sintered magnets is reached by post-sintering annealing at the optimum temperature (around 500°C). The role of the low-temperature heat treatment consists in the formation of continuous non-magnetic REM-rich layers between $(Nd,REM)_2Fe_{14}B$ grains.

The experiments with Dy and Tb additions are carried out to achieve the optimum microstructure, which involves confining the Dy distribution to the grain boundary regions [4,7]. If Dy can be confined in the grain boundary regions, then: (1) it should be possible to increase locally the coercivity and hence reduce the probability of reverse domains forming at the grain boundaries; (2) limit the substitution of Nd by Dy in the matrix phase and thus limit the reduction of the magnetization and hence the remanence; and (3) reduce the amount of Dy required to achieve a particular increase in the coercivity and hence reduce the overall cost of the alloy.

The oxygen content and character of distribution of oxide inclusions are known to determine to a great extent the quality of Nd-Fe-B magnets. Thus, the probability of formation of oxide compounds in the course of production of Nd-Fe-B magnets is sufficiently high.

The alloying of the basic composition with rare-earth metals added in the form of hydrides was performed in [8-12]. According to data of [10], a relatively Dy-rich phase formed in the Nd-rich grain boundary phase and this slowed down the substitution of Dy for Nd in the $Nd_2Fe_{14}B$-based matrix phase, possibly enhancing local coercivity without excessive attenuation of the remanence. The magnetic and microstructural modification of the Nd–Fe–B sintered magnet by mixed DyF_3/DyH_x powder doping was studied in [11]. The possibility of application of dysprosium hydride nanoparticles for wastes processing is discussed in [12]; in this case, the properties of prepared magnets are: B_r = 1.296 T, $_jH_c$ = 1360 kA/m, $(BH)_{max}$ = 292.3 kJ/m^3.

There are a few studies devoted to the investigation of the kinetics of metal hydrides, in particular, rare-earth metal hydrides in the literature [13, 14]. Temperatures of the onset of the hydride decomposition demonstrate variations.

The aim of the present study is to demonstrate the improvement of Nd-Fe-B permanent magnet characteristics by using hydrides of heavy REMs, the purity of which is retained and controlled during hydride synthesis and application as the component of powder mixture. The role of the hydride additions in the formation of magnet characteristics is demonstrated. In the course of the study, we state the efficiency of vacuum distillation for the purification of REMs, which is accompanied by the formation of nontraditional structural morphology of the metals.

2. Experimental Procedure, Results and Discussion

2.1. Purification of REMs

Technology for purification of rare-earth metals by vacuum distillation (Y, Pr, Nd, Gd, Tb, Lu) and sublimation (Sc, Dy, Ho, Er, Tm) and optimum regimes of the purification have been elaborated at the Baikov Institute of Metallurgy and Materials Science [15]. The purification is performed at a residual pressure of 10^{-4}-10^{-5} Pa in a resistor furnace equipped with a graphite heater. A REM is evaporated from a tantalum crucible and deposited on a water-cooled copper condenser in the form of a druse (150-250 g) of small crystal growing together. Figure 1a shows the appearance of a distilled REM. The structure of distilled REMs was studied in air by atomic force microscopy using a Solver Pro EC (NT MDT) atomic-force microscope and semicontact regime. The studies allow us to conclude the existence of the unique morphology of distilled metals. Their structures are characterized by structural components differing in the dispersivity: elongated crystals (single crystals) are from 0.5 to 2.5 μm in lateral size and nano-sized grains, which are observed within spaces between elongated crystals and fasten them together. The developed surface of distilled metal is characterized by steps and terraces, which are formed in the course of deposition from vapor. According to atomic-force microscopy (AFM) data, the height of steps is from 15 to 80 nm, which can correspond to a grain size of ~30 nm. Figures 1b,c show AFM images of distillate structure (transverse and longitudinal sections), which confirm the existence of nano-sized structure [16].

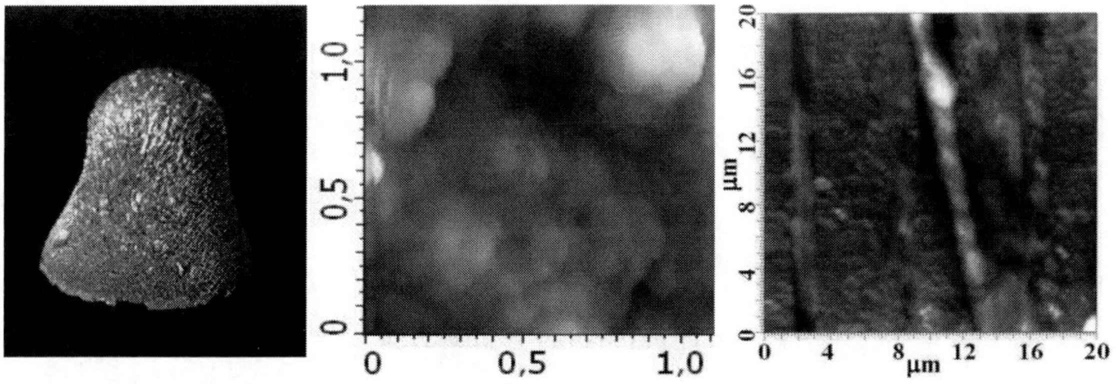

Figure 1. (a) Appearance of distilled REMs and AFM images of the surface of (a) transverse and (b) longitudinal sections of distilled gadolinium.

REMs purified by vacuum distillation/sublimation in accordance with the process are characterized by low contents of (1) gas-forming elements, (2) volatile and refractory elements, (3) other rare-earth metals, and (4) Fe, Cu, Al contents of $\sim 10^{-3}$ wt %. Figures 2a,b shows impurity-distribution histograms for distilled praseodymium, neodymium, and terbium. We can state that distillation process is more efficient for neodymium. The metals can be classified among high-purity REMs.

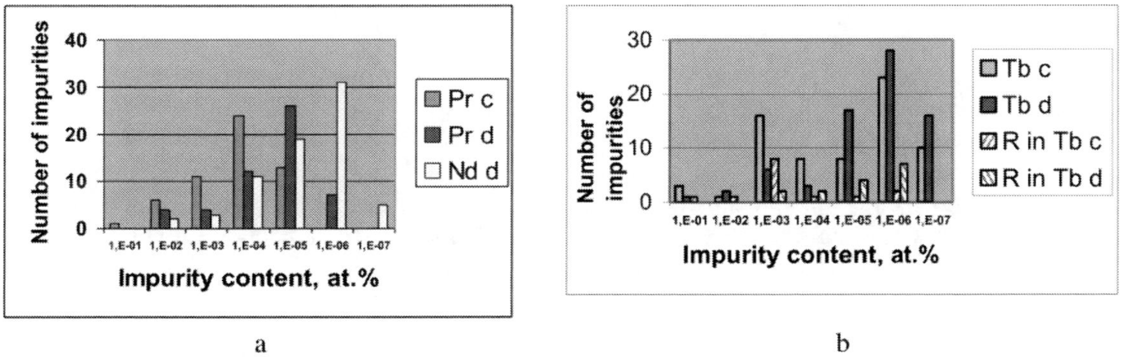

Figure 2. Impurity distribution histograms for (a) Pr commercial (Pr c) and distilled (Pr d) and distilled Nd (Nd d) and (b) Tb commercial (Tb c) and distilled (Tb d); R = REMs.

We have demonstrated the opportunities of distillation technology for the preparation of considerable quantities of high-purity REMs. Thus, the process can be recommended for application in small-capacity works since it realizes the optimum relationship metal purity, its cost, and process efficiency. The purity of REMs can be increased when distillation/sublimation is performed in decreasing the evaporation rate via varying the evaporation temperature.

2.2. Study of REMs hydrides

REM dihydrides samples ($TbH_{\sim 2}$ and $DyH_{\sim 2}$) were prepared by direct reaction of gaseous hydrogen with distilled metals using a glass Sieverts-type apparatus. The hydrogenation was carried out using pure hydrogen gas under pressures up to 0.1 MPa obtained from TiH_2 hydrogen storage. The amount of absorbed hydrogen was determined by measurements of hydrogen pressure change in the reactor chamber after finishing the hydrogenation reaction. This allows one to calculate the hydrogen concentration in the sample with an accuracy of ±0.02 H/f.u. After synthesis, the hydride samples were homogenized (annealed) for 72 h at 350°C. The obtained hydrides were breakable.

Kinetics of molecular hydrogen evolution from the terbium hydride was studied by Dr. V.S. Petrov using a MS-200 standard gas mass-spectrometer equipped with an evaporating attachment to operated in a temperature range of from 20 to 1000°C (the residual pressure in the cold system is $5 \cdot 10^{-5}$ Pa). Weight losses were controlled using a balance. The decomposition of dysprosium was studied by gravimetric analysis. Figure 3a shows data on the kinetics of decomposition of terbium dihydride.

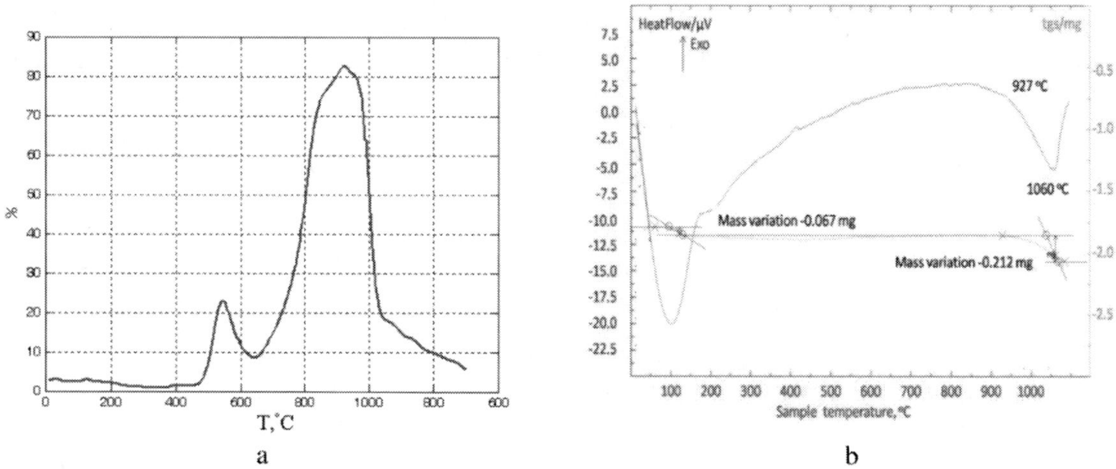

Figure 3. (a) Density of hydrogen desorption flow during TbH_2 decomposition; (b) DTA and TGA curves for $DyH_{\sim 2}$ dysprosium hydride decomposition.

The studies allowed us to determine the temperatures of the onset of terbium hydride decomposition and temperature range corresponding to intense hydrogen evolution. The hydrogen evolution is characterized by two peaks. The temperatures of the peaks do not correlate with those determined earlier in [14] (the onset, peak and end of decomposition correspond to 620, 750, and 800°C, respectively). It is likely that the discrepancies are related to different both purity of initial metals, conditions of sample preparation, composition of hydrides, and hydrogen-evolution recording techniques. In our opinion, the low-temperature peak in the hydrogen desorption flow is related to the surface and near-surface hydrogen evolution as a result of decomposition of terbium trihydride that is present in small amounts. The maximum hydrogen-evolution is reached at ~900°C. The decrease in the hydrogen flow at ~650°C is related to the hydrogen depletion of the surface and near-surface layers. We cannot relate specifically each of peaks or each component of high-temperature peaks with decomposition processes (hydride decomposition or hydrogen solid solution decomposition), but we can recommend to hold samples at temperatures above 1000°C for 1 h to ensure the more complete hydrogen evolution. Earlier, we have found that, after the decomposition of terbium dihydride, reduced terbium actively fixes oxygen to form the terbium oxide. Thus, the use of terbium particles or terbium hydride on the surface of magnets will favor the increase in the oxygen content in magnets. According to differential thermal analysis (DTA+TG) data (heating in an argon atmosphere (Ar 6N) at a rate of 10 C/min, the dysprosium hydride (DyH_2) decomposition takes place at 930°C (figure 3b).

IOP Conf. Series: Materials Science and Engineering **112** (2016) 012009 doi:10.1088/1757-899X/112/1/012009

2.3. Study of magnetic characteristics of $Nd_2Fe_{14}B$ magnets prepared with REM hydrides

The initial alloy in the form of flakes 300 μm in thickness was prepared by "strip-casting" technique [5]. The chemical composition (wt %) of the alloy is Nd-24.0, Pr-6.5, Dy-0.5, B-1.0, Al-0.2, Fe-balance. Flakes were subjected to hydrogen decrepitation in dry-hydrogen flowing atmosphere at 100°C for 1 h and subsequently passivated in gaseous nitrogen. After cooling the powder to room temperature, terbium hydride (2 and 4 wt %) or dysprosium hydride (2 wt %) were added. A mixture was subjected to fine milling for 40 min to an average particle size of 3 μm using a vibratory mill and isopropyl alcohol medium. After compaction in a magnetic field and sintering at 1070°C (for 2 h), the sintered blanks prepared with the terbium hydride were subjected to heat treatment (HT) 900°C (2 h) + 530°C (1 h) + 500°C (1h). The optimum heat treatment for magnets prepared with the dysprosium hydride is 1-h annealing at 500°C (HT1). The stability of hysteretic properties of the magnets prepared with $DyH_{\sim 2}$ was tested using heat treatments (HTs): HT-2 (500°C (1 h) + quenching + 500°C (2h) + cooling for 2 h to 400°C + 400°C (6 h) + furnace cooling); HT-3 (500°C (1 h) + quenching + 500°C (2h) + cooling for 2 h to 400°C + 400°C (6 h) + furnace cooling + 550°C (1h) + quenching); HT-4 (HT-1 + HT-2 + HT-3 + 1050°C + 500°C (2 h) + quenching in gaseous nitrogen).

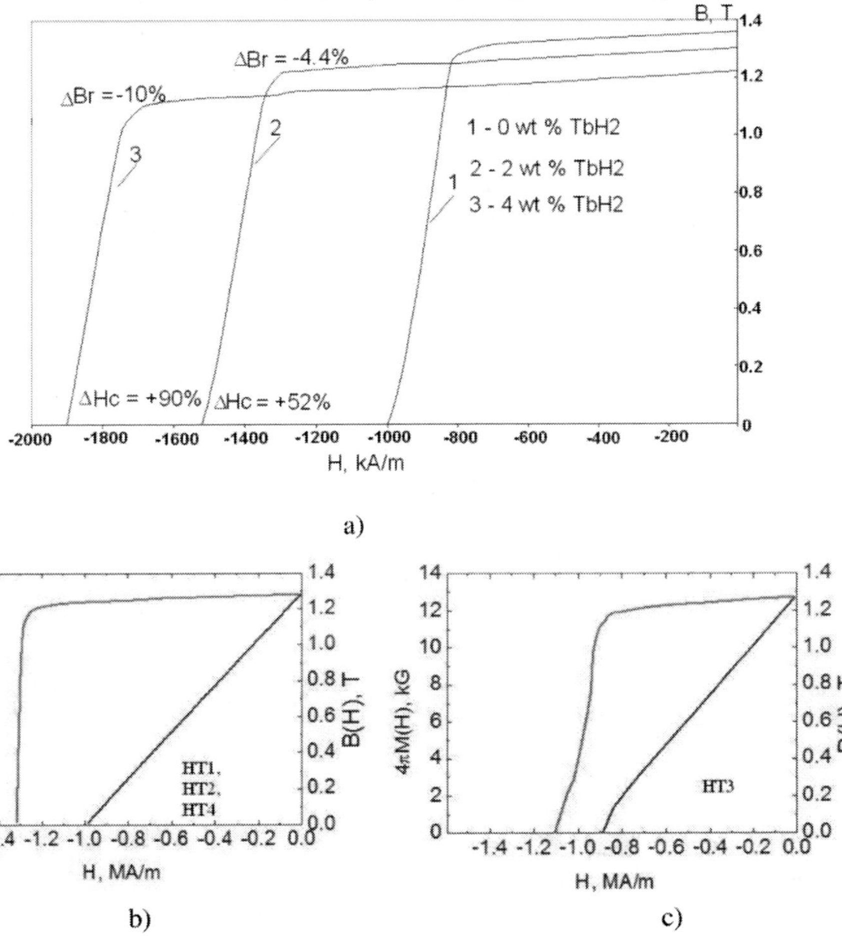

Figure 4. Magnetization reversal portions of hysteresis loops for the Nd-Fe-B sintered magnets prepared with (a) 0, 2, 4 wt % TbH_2 and (b,c) 2 wt % DyH_2 after (b) optimum (HT1), stepped (HT2), restoring (HT4) and (c) degrading (HT4) heat treatments.

Magnetic characteristics of samples were measured after grinding and saturation magnetization.

The measurements were performed at room temperature in magnetic fields up to 3 T using a close circuit of hysteresimeter. Figures 4a, b, c demonstrates the variations in the hysteresis loop, which are caused by the application of hydride additions and heat treatments.
For all samples, curves of magnetization from the thermally demagnetized state and dependences of ${}_{j}\boldsymbol{H}_{c}$ и $\boldsymbol{B}_{r}$ on the magnetized field are typical of magnets whose magnetization reversal mechanism is due to the delayed nucleation of reversed domains. The magnetic parameters correspond to permanent magnets prepared in the oriented magnetic field (1200 kA/m) perpendicular to pressing direction.

When applying the terbium hydride for the Nd-depleted Nd-Fe-B-based compositions, we found some peculiarities of REM distribution within the magnet structure [17]. The formation of the high-coercivity state of the Nd-Pr-Dy-Fe-B sintered magnets prepared by blending technology in applying TbH_2 in the powder mixture is discussed from the point of view of nonuniform distribution of REMs within 2-14-1 phase grains. This effect manifests itself more substantially with increasing contents of alloying elements (in total case, heavy rare-earth metals) and in using some manufacturing processes, ("strip casting", mechanical alloying, hydrogen decrepitation, etc.). The nonuniformity of principal hard-magnetic phase leads to the increases in structurally sensitive magnetic parameters, such as $\boldsymbol{H}_{k}$, $\boldsymbol{B}_{r}$, ${}_{j}\boldsymbol{H}_{c}$. Moreover, the data obtained for magnets with DyH_2 addition demonstrate the higher thermal stability of hysteretic properties. The observed effect also is attributed to the formation of nonuniform Dy distribution in $(Nd, Pr, Dy)_2Fe_{14}B$ principal magnetic phase grains, which is formed during the preparation of material from hydride-containing powder mixtures and determines the specificity of diffusion processes occurred during low-temperature annealing; the effect also is discussed in terms of continuity and discontinuity of grain boundaries [18]. These data are of importance for deciding the optimum heat treatment of Nd-Fe-B magnets and prediction of their service stability.

3. Conclusions

Possibilities of distillation purification of REMs and peculiarities of the structure of distilled metals are discussed. Hydrides TbH_2 and DyH_2 were synthesized and used as efficient additions to powder mixtures for manufacturing sintered Nd-Fe-B permanent magnets. The hysteretic characteristics of the magnets were reached in using the initial Nd-depleted strip-casting alloy and economical alloying with heavy REMs (Dy,Tb). The results obtained allow us to conclude that the suggested modification of the manufacturing process of sintered magnets (the use of REMs' hydrides) corresponds to the modern world tendency in their production, which consists in the saving of deficient expensive REMs. Moreover, the modified variant of the manufacturing process allows us to prepare magnets with the wide range of hysteretic characteristics in using a unified initial Nd-Fe-B alloy.

Acknowledgments

This study was supported by the ERA_NET_RUS_PLUS, project no. INNO_146, Russian Academy of Sciences (program no. IV.5.6) and project No. LO1203 (Czech Republic).

References

[1] Gschneidner K A Jr. 1989 *Proceeding of the I[st] Rare Metals Forum* (Tokyo, Japan: Society of Non-Traditional Technology) 13
[2] Glebov V A and Lukin A A 2007 *Nanocrystalline Rare-Earth Hard Magnetic Materials* (Moscow: Publishing House "FGUP VNIINM"), in Russian
[3] Li W F, Sepehri-Amin H, Ohkubo T and Hono K 2011 *Acta Materialia* **59** 3061
[4] Sepehri-Amin H, Ohkubo T and Hono K 2010 *J. Appl. Phys.* **107** 745
[5] Glebov V A, et al 2011 *Fiz. Chim. Obrab. Mater.* 3 16
[6] Watanabe N, Umemoto H, Ishima M, Itakura M, Nishida M, and Machida K 2009 *J. Microscopy* **236** (2) 104
[7] Seperhri-Amin H, Ohkubo T and Hono K 2013 *Acta Materialia* **61** 1982
[8] Kianyashi A, Mottram R S and Harris I R 1999 *J. Alloys Comp.* **287** 206
[9] Wang H, Li A and Li W 2007 *Rare Met. Mater. Eng.* **36**(6) 1088

[10] Gaolin Y, McGuiness P J, Farr J P G and Harris I R 2010 *J. Alloys Comp.* **491** L20
[11] Kim T-H, Lee S-R L, Kim H-J, Lee M-W and Jang T-S 2014 *J. Appl. Phys.* **115** 17A763
[12] Liu W Q, Li C, Zakotnik M, Yue M, Zhang D T and Zuo T Y 2014 *Proceeding REPM2014* (Annapolis, Maryland, USA) 116
[13] Yartys V A, Gutfleish O, Panasyk V V and Harris I R, 1996 *J. Alloys Comp.* **253** 128
[14] Gabis I, Evard E, Voyt A, Chernov I and Zaika Yu 2003 *J. Alloys Comp.* **356/357** 353
[15] Devyatykh G G and Burkhanov G S 1997 *High-Purity Refractory and Rare-Earth Metals* (Cambridge: Int. Sci. Publ.)
[16] Burkhanov G S, Kolchugina N B, Tereshina E A, Tereshina I S, Politova G A, Chzhan V B, Badurski D, Chistyakov O D, Paukov M, Drulis H and Havela L, 2014 *Appl. Phys. Lett.* **104** 242402
[17] Lukin A A, Kolchugina N B, Burkhanov G S, Klueva N E and Skotnicová K 2013 *Inorg. Mater.: Appl. Res.* **4** 256
[18] Burkhanov G S, Lukin A A, Kolchugina N B, Koshkid'ko Yu S, Dormidontov A G, Skotnicová K, Zivotsky O, Čegan T and Sitnov V V 2014 *Proceedings REPM2014* (Annapolis, Maryland, USA)

Corundum ceramic materials modified with silica nanopowders: structure and mechanical properties

M. A. Kostytsyn[1], D. S. Muratov, D. V. Lysov, K. O. Chuprunov, A. G. Yudin, and D. V. Leybo

National University of Science and Technology "MISiS," 4 Leninsky Prospekt, Moscow, 119049, Russia

[1]E-mail: slimak@mail.ru

Abstract. Filtering elements are often used in the metallurgy of rare earth metals. Corundum ceramic is one of the most suitable materials for this purpose. The process of formation and the properties of nanomodified ceramic materials, which are proposed as filtering materials with tunable effective porosity, are described. A silica nanopowder is used as a porosity-increasing agent. Vortex layer apparatus is used for mixing of precursor materials. The obtained results show that nanomodification with the vortex layer apparatus using 0.04 wt. % silica nanopowder as a modifying agent leads to an increase in the compression strength of corundum ceramic by the factor of 1.5.

1. Introduction.

The production of rare earth metals becomes more important due to rapid industrial development: this development is impossible without new materials, in particular, ceramic filter elements. Various nanosized additives are known to be used to increase the effectiveness of filtering materials. However, one of the main practical difficulties related to the use of nanopowders on the industrial scale is to achieve a homogenous mixture of micro- and nanosized powders [1-5]. This problem leads to the development of new techniques for forming homogenous powder mixtures.

A vortex electromagnetic field is one of the effective solutions to achieve a homogenous nanopowder distribution in the volume of a coarse-grained material [6]. A homogenous particle distribution in the final product can be reached by high-speed rotation of the material to be treated in combination with impact and attrite processes in vortex layer apparatus.

2. Experimental

Mullite with particle sizes of 0 – 500 μm, 0.5 – 1.0 mm, and 1 – 3 mm; corundum with a particle size of 100 μm; and calcium monoaluminate as high-aluminous binding agent were used as precursors. Silica nanoparticles with an average particle size of 15 – 20 nm were used as a modifying agent.

A standard ceramic sample preparation scheme was used; it included weighting, mixing, water addition, drying, and annealing. Standard samples consisted of a mixture having the following proportions: mullite size 0-500 μm, 29 wt. %; 0.5 – 1.0 mm, 12%; 1 – 3 mm, 44%; corundum 10%; and calcium monoaluminate 5%. The standard samples were mixed in a laboratory paddle binder. Modified samples were prepared by the addition of 0.05 wt. % silica nanopowder. This optimal

Content from this work may be used under the terms of the Creative Commons Attribution 3.0 licence. Any further distribution of this work must maintain attribution to the author(s) and the title of the work, journal citation and DOI.
Published under licence by IOP Publishing Ltd

concentration was investigated. The modified samples were mixed in vortex layer apparatus for 2 min, and 5.5 wt. % water was added after mixing. The samples were first dried at a temperature of 25°C for 24 h and, then, at 110°C. Some samples were annealed at a temperature of 1100°C. Cubic samples with 50-mm edges were then prepared for analysis. The samples were analyzed by compression strength measurements, scanning microprobe spectroscopy, X-ray diffraction, and electron microscopy.
Compression strength tests were carried out on a Zwick Z250 (Germany, Zwick) device. With this device, we performed tensile, compression, and three-point bending tests in the load range from 2 kg to 25 t. The measurement error was at most 1% and the testing rate was 0.0005--600 mm/min. We tested flat, cylindrical, and cubic samples. The frequency of read speed measurement and control electronics was 500 Hz. Compression strength was analyzed at constant loading rate of 2 mm/min.
The uniformity of the distribution of nanoscale additives was analyzed by X-ray fluorescence on an RAM-30 (Russia, St. Petersburg, Scientific Instruments) analyzer. We performed elemental mapping of a sample in an area of 50 by 50 mm at a micron range voltage of 15 - 45 kV using Mo-kα radiation. The samples for analysis were prepared in form of tablets compressed in a hydraulic hand press at a pressing force of 5 t. Zirconia nanopowders were used for calibration.
XRD analysis of samples was performed using an Ultima IV (Japan, Rigaku) X-ray diffractometer and copper radiation. The X-ray tube current was 40 mA and the voltage was 40 kV. X-ray diffraction patterns were recorded in the angular range $2\theta = 0\text{-}100°$ at a step of 0.1° and an exposure time of 2 s. The Bragg-Brentano focusing scheme with two Soller slits was used.
Micrographs were taken with a JSM-6700F (Japan, JEOL) field emission scanning electron microscope.

3. Results and discussion

We used scanning X-ray fluorescence to compare the samples treated in the vortex layer apparatus and traditionally prepared samples (figure 1). As is seen in figure 1, the vortex layer apparatus produces a more homogenous zirconia nanopowder mixture as compared to a standard laboratory binder.

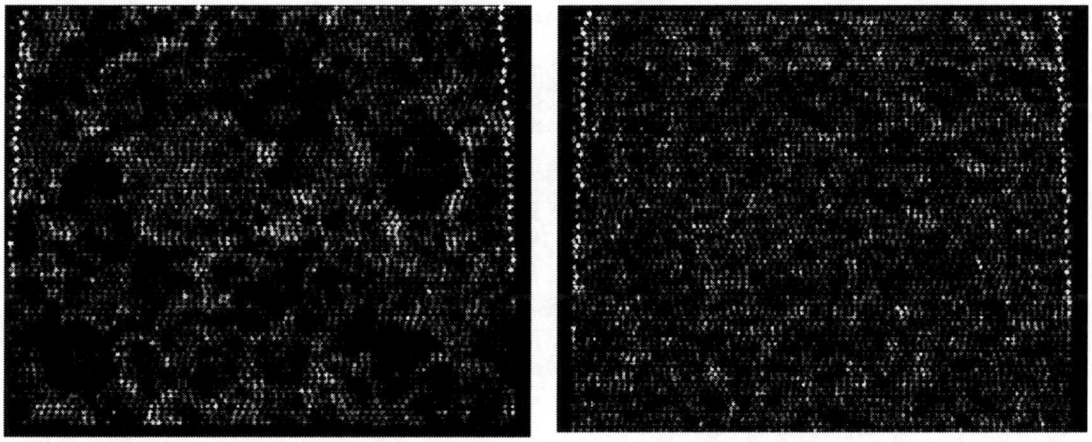

a b

Figure 1. Distribution of zirconia nanoparticles: a – standard laboratory binder mixing, b – vortex layer apparatus.

According to the experimental data, the optimal silica concentration is 0.05 wt. % (figure 2). The compression strength of the sample with 0.05 wt. % SiO_2 increased by 52 % as compared to the sample with no SiO_2 modifier. It is possible that the material strength increases due to both mechanical activation process and chemical modification at the stage of hydratation.
The XRD results (figures 3 – 4) of studying the modified samples show the formation of metastable calcium hydroaluminate $CaAl_2O_4\ 8H_2O$ at room temperature. The XRD data obtained for a reference

sample demonstrate the formation of cubic calcium hydroaluminate $3CaO{\cdot}Al_2O_3{\cdot}6H_2O$. This difference can be explained by good ability of hydrophilic silica nanoparticles to retain moisture in ceramic; as a result, the hydratation of aluminate is longer than that of the reference sample. The longer hydratation time leads to a decrease in the heat release per unit time, decreasing the probability of self-heating to the temperature of transformation of metastable $CaAl_2O_4$ $8H_2O$ into stable $3CaO{\cdot}Al_2O_3{\cdot}6H_2O$.

Based on the XRD data, we can conclude that the formation of the metastable $CaAl_2O_4$ $8H_2O$ phase with a less steep hydratation process improves the structure of the material. The XRD data of the sintered samples show that their phase composition is the same, namely, Al_2O_3, Al_2SiO_5, and $CaSiO_3$, and the measured patterns are quite similar.

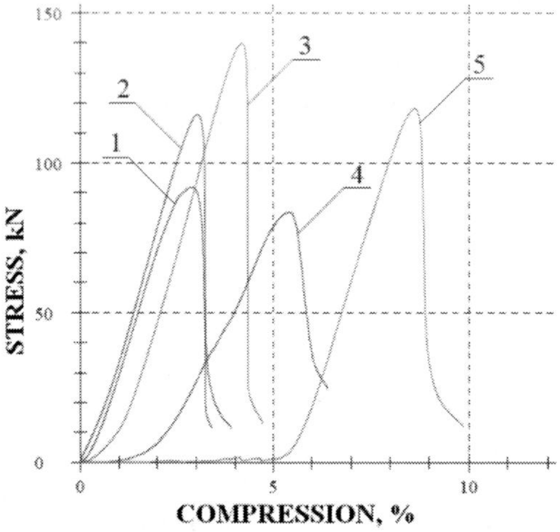

Figure 2. Loading curves of mullite-corrundum samples: 1 – 0% SiO_2; 2 – 0.01% SiO_2; 3 – 0.05% SiO_2; 4 – 0.1% SiO_2; 5 – 0.2% SiO_2

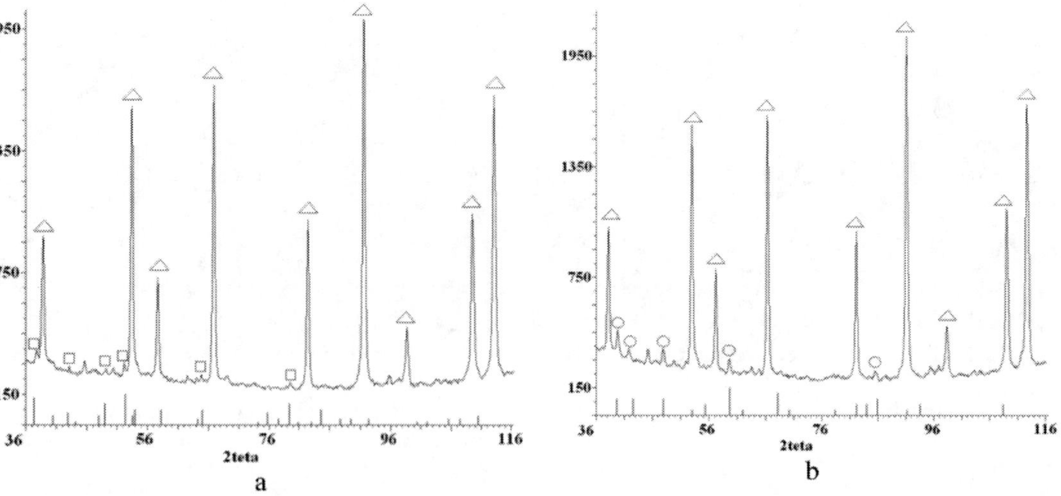

Figure 3. XRD data of mullite-corundum mixture at the first drying stage: a – reference; b –SiO_2 0.04 wt. %; Δ - Al_2O_3; □ - $3CaO{\cdot}Al_2O_3{\cdot}6H_2O$; o - $CaAl_2O_4{\cdot}8,5H_2O$

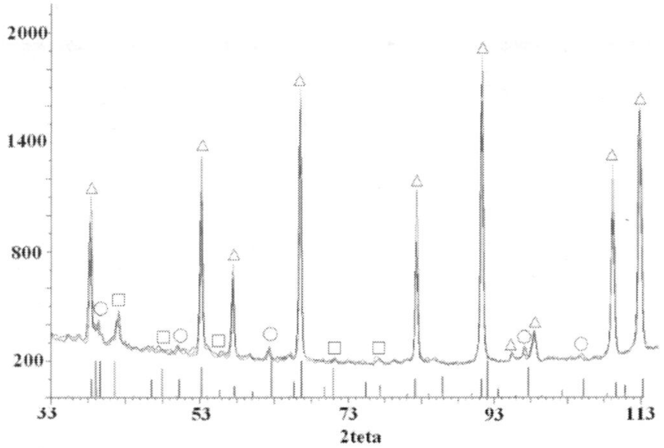

Figure 4. XRD data of sintered samples: Δ - Al_2O_3; o - Al_2SiO_5 and □ - $CaSiO_3$

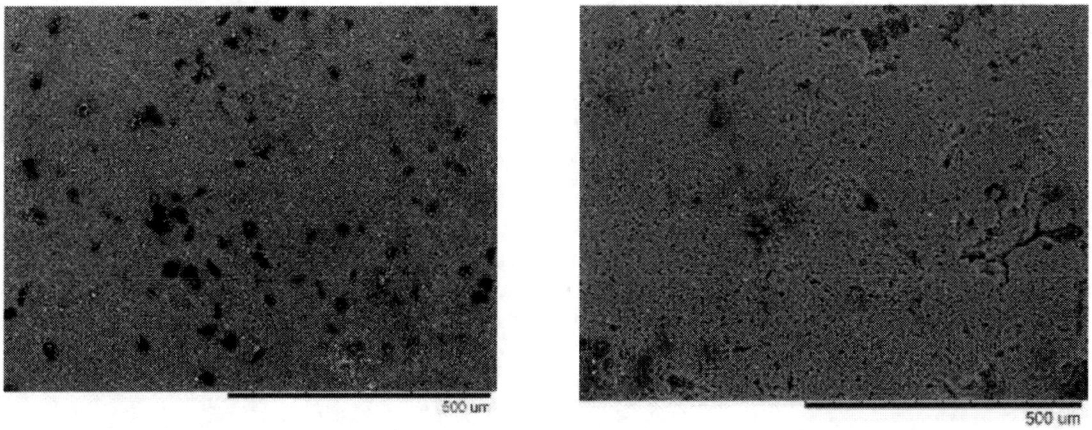

a b

Figure 5. SEM images of mullite-corundum ceramics: a – reference; b –SiO_2 0.04%

a b

c

d

Figure 6. SEM images of mullite-corundum ceramic samples: a, b – reference; c, d –SiO_2 0.04 wt. %

Scanning electron microscopy showed that the samples containing nanostructured silica are less porous as compared to the reference sample, which is due to an inhibited hydratation process and high water-binding properties at the first stages of drying. The reference sample has a much more porous structure.

Before annealing, the structure of the reference samples mostly consists of calcium hydroxide nanoparticles filling the pores between corundum particles. The binding phase has an average particle size of 50 – 150 nm. The silica-modified samples have a coarser structure due to a longer hydratation time, when the particle growth rate is higher than the nucleation rate.

4. Conclusions

The obtained results show that the addition of 0.05 wt % nanostructured silica increases the compression strength of ceramic by 52%. Treatment in vortex layer apparatus results in the formation of a more homogenous mixture as compared to standard mechanical mixing methods. The addition of 0.1 wt. % silica nanopowder increases the hydratation period, which results in the formation of metastable high-strength calcium hydroaluminate.

Acknowledgements

This research was supported by the Russian Ministry of Education and Science (project # 14.575.21.0004).

References

[1] Kascheev I D 2007 *Chemical technology of refractory*, ed. Strelkov K K and Mamykin P S (Moscow: Intermet Engineering)

[2] Allenshtein Y 2010 *Refractory materials. Structure, properties, tests: handbook* (Moscow: Intermet Engineering)

[3] Rubakhin A I and Savelev V G 2009 *Physical chemistry of refractory non-metal and silicate compounds: Textbook* (Moscow: INFRA)

[4] Gorokhovsky A V 2010 *Technical ceramics based on mixtures of bentonite and potassium polytitanate*, ed. Scherbakova N N and Burmistrov I N (Refractory and technical ceramic vol 11-12) pp 9-12

[5] Antonovic, V. A 2010 *Review of the possible applications of nanotechnology in refractory concrete* ed. I. Pundiene, R. Stonys, J. Cesniene and J. Keriene (Journal of civil engineering and management vol 16(4)) pp 595–602.

[6] Logvinenko D D and Shelyakov O P 1976 Intensification of technological processes in vortex layer (Kiev: Technics)

Energy effective approach for activation of metallurgical slag

I N Mazov[1], B B Khaydarov, S L Mamulat, D S Suvorov, Y S Saltikova, A G Yudin, D V Kuznetsov
National University of Science and Technology "MISiS", 4 Leninsky prospect, Moscow, 119049, Russia

[1]E-mail: ilya.mazov@gmail.com

Abstract. The paper presents results of investigation of the process of mechanical activation of metallurgical slag using different approaches – ball milling and electromagnetic vortex apparatus. Particle size distribution and structure of mechanically activated slag samples were investigated, as well as energetic parameters of the activation process. It was shown that electromagnetic vortex activation is more energy effective and allows to produce microscale milled slag-based concrete using very short treatment time. Activated slag materials can be used as clinker-free cement in civilian and road construction, providing ecology-friendly technology and recycling of high-tonnage industrial waste.

1. Introduction

Technological wastes of metallurgical industry are produced in significant amount every year. Slag output during production of ferrous metals results in increasing yearly per ca. 5-7 M tons, providing high environmental load and waste of nature.

One of the promising ways of utilization of metallurgical slag is its use as a component for construction materials, such as clinker-free concrete. Note that nowadays there are few technologies allowing energy effective use of metallurgical slag in the building industry. Most of approaches use slag for production of gravel, as filler for concrete and road pavements etc. [1,2]. Mining and separation of rare-earth metal produces various slags, which can be recycled for additional extraction of rare metals and used for production of construction materials.

Energy effective utilization of metallurgical slag with subsequent production of construction materials for civil and road applications is of great importance and needs usage of special mechanical activation technologies with low energy consumption and high efficiency. This paper presents preliminary results in the development of energy effective and cheap approach to production nanoscale-milled materials basing of blast-furnace slag as a model industrial raw material. The same approach can be used for activation of other slags, produced during rare earth extraction and processing.

2. Experimental

The process of formation of clinker-free cement materials was investigated using blast furnace slag produced by Joint Stock Company Severstal (Russia). Mechanical activation of the slag and mixtures was performed in the Vortex Electromagnetic Activation setup, consisting of non-magnetic chamber, loaded with treated material and magnetic bodies (cylinders with 2 mm diameter and 15-20 mm length). Traveling magnetic field is circular to the chamber, forcing milling bodies to move and

Content from this work may be used under the terms of the Creative Commons Attribution 3.0 licence. Any further distribution of this work must maintain attribution to the author(s) and the title of the work, journal citation and DOI.
Published under licence by IOP Publishing Ltd

activate loaded sample. Typical load of slag and/or additives was 100 g and 200 g of milling bodies were loaded. Typical activation time was 2-7 minutes.

Reference samples were produced by milling of slag particles in ball mill loaded with steel balls (12 mm dia.). PQ-N2 planetary ball mill was used for experiments with typical activation time 120-180 minutes.

Size distribution and morphology of mechanically activated samples were investigated using Scanning Electron Microscopy (Tescan Vega 3, Czech Republic) and Dynamic Light Scattering (Fritsch Analysette 22, Germany). For SEM investigation powder samples were glued on conductive carbon tape and placed in microscope chamber without additional treatment. DLS experiments were performed with dry powders dispersed in the DLS setup using integrated dispersion system.

3. Results and discussion

Scheme of the vortex layer activator (VLA) is shown in figure 1. VLA setup is designed for production of homogeneous mixtures of metallic, oxide and ceramic powders using travelling magnetic field and ferromagnetic agitators. Mechanical activation and mixing is provided due to high-speed rotation of ferromagnetic rods in rotating magnetic field, resulting in formation of pseudo-fluidized bed regime.

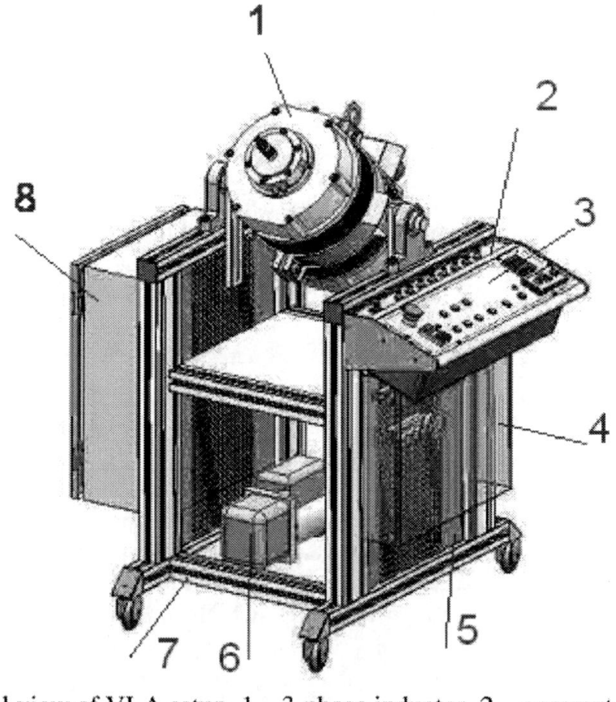

Figure. 1. General view of VLA setup. 1 – 3-phase inductor, 2 – commutation panel, 3 – control board, 4 – screen, 5 – pneumatic panel, 6 – vacuum pump or compressor, 7 – body of the setup, 8 – electrical wiring.

Laboratory setup used for experiments consisted of reactor with internal diameter 67 mm, and length of the work zone 180 mm. Typical size of the agitating rods was Ø2×20 mm. Technological parameters of the setup were calculated using common theory of the constrained motion.

The volume of the reaction chamber V_{react} is 634 cm^3, volume of the agitating rod V_{rod} is 0.0628 cm^3. Thus knowing typical critical parameter for the VLA setup K = 0.04…0.05 amount of rods can be easily calculated as: $x = K\times(V_{react}/V_{rod}) = 400…600$.

Various loading of rods as compared to slag was investigated. The main goal was to obtain the lowest size of activated slag particles and at the same time to keep uniform distribution of particles.

Details of the experiment are shown in Table 1 for 100 g of slag with different amount of steel agitating rods and various treatment times.

Table 1. Parameters of VLA mechanical activation of metallurgical slag.

Treatment time	3 minutes	5 minutes	7 minutes
Slag : rod ratio	1 : 2	1 : 2	1 : 2
	1 : 2.5	1 : 2.5	1 : 2.5
	1 : 3	1 : 3	1 : 3
	1 : 3.5	1 : 3.5	1 : 3.5

a

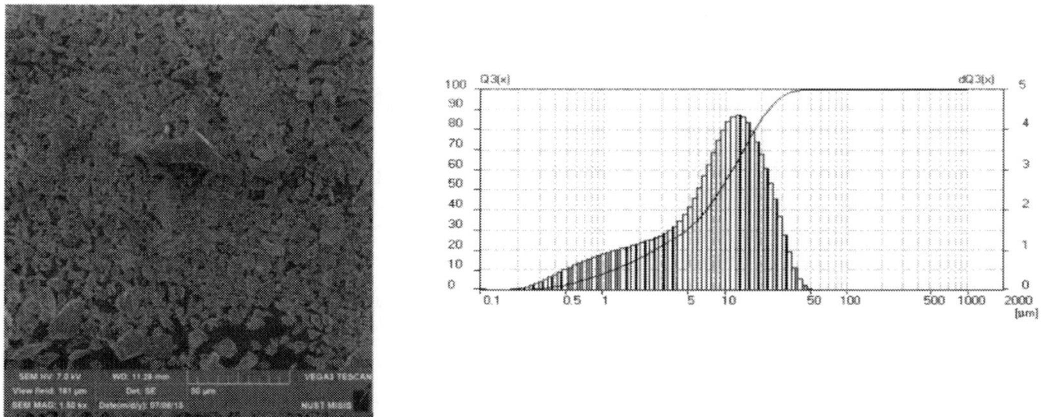

b

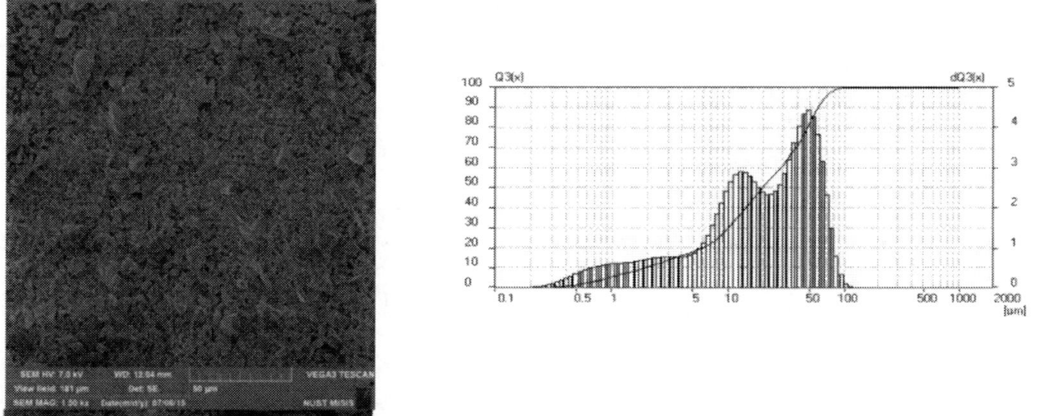

c

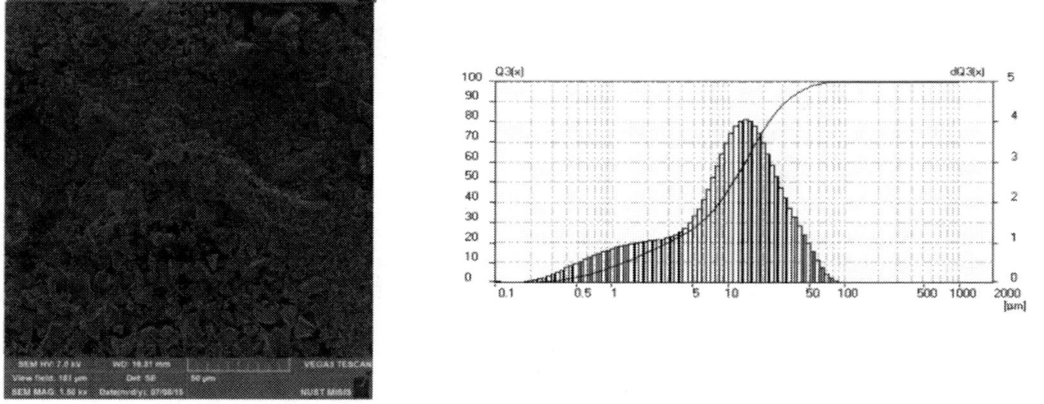

Figure. 2. SEM images (left) and particle size distribution (right) for VLA-activated samples with "slag:rod" = 1:3: a) – 3 minutes, b) – 5 minutes, c) – 7 minutes.

SEM images and particle size distribution of typical activated samples with "slag : rod" ratio 3 and various milling time are shown in figure 2. As it can be seen from the data shown in figure 2, mechanical activation during 3 minutes allows one to obtain uniform particle size distribution with maximum in range 7-10 μm. Increase of the time of the treatment results in polymodal dispersion of particles with increase of amount of particles with size higher than 50 μm. This may occur due to aggregation of particles and high-energy impacts during activation process.

Power parameters of VLA mechanical activation process and planetary ball milling were compared for samples with similar particle size distribution. Specific surface area was measured for typical samples using BET technique.

Using the VLA approach allows to reach SSA values in range 6-7 m^2/g. Note that the same process realized by planetary ball mill results in SSA values much lower, namely 1-1.5 m^2/g.

Energy consumption data for both cases is shown in table 2. As it can be seen, it is necessary to use longer period of time with planetary ball mill in order to produce samples with similar parameters as compared with VLA-produced materials.

As one can see from the data presented in Table 2, VLA process is 3 times more energetically effective as compared with planetary ball mill, which may be due to higher specific energy transferred by agitating rods as compared with relatively low-speed movement of milling ball in planetary mill.

Table 2. Energy consumption data for VLA and planetary ball milling mechanical activation.

Type of setup	Nominal power, W	Treatment time, min	SSA value, m^2/g	Energy consumption, KWh/kg
VLA	2100	3	6.8±0.4	1.050
Planetary ball mill PQ-N2	750	120	1.2±0.06	3.750

4. Conclusion

Vortex-layer activation was successfully adopted for production of microscale mechanically activated metallurgical slag. Obtained materials show uniform particle size distribution and can be used as a source for clinker-free cements for construction applications.

Technological parameters of VLA-process were optimized, showing 3 minutes as an optimal time to produce good distribution and microscale particles.

VLA process was shown to be 3 times more energetically effective as compared with traditional planetary ball mill. Thus, this approach can be of future importance for industrial-scale applications.

Acknowledgments

The authors gratefully acknowledge the financial support of the Ministry of Education and Science of the Russian Federation (project no. RFMEFI57814X0099).

References

[1] Zelic J 2005 *Cement Concrete Res.* **35** 2340
[2] Pasetto M and Baldo N 2010 *J. Hazard. Mater.* **181** 938
[3] Xuequan W, Hong Zh, Xinkai H and Husen L 1999 *Cement Concrete Res.* **29** 1103
[4] Ondova M, Stevulova N and Meciarova L 2013 *Procedia Eng.* **65**45

Ultrasound-assisted synthesis of nanosized zero-valent iron for metal cations extraction and wastewater treatment applications

I Yu Mikhailov, D V Lysov, V V Levina, I N Mazov, A A Gusev, T I Yudintseva, D V Kuznetsov

National University of Science and Technology "MISiS", 4 Leninsky prospect, Moscow, 119049, Russia

E-mail: ivan.mikhailov@misis.ru

Abstract. Nanosized zero-valent iron has shown good results in wastewater treatment and activation of physicochemical processes. Its applications in modern industry are complicated by high production costs of nanomaterials produced via existing synthesis routes. Therefore there is a need of cheap and high-productive methods of nanosized zero-valent iron with advanced functional properties. Improvement of oxidative conditions with additions may find its place in extraction of rare-earth metals, where high cost of nanomaterials could be viable. In this paper we studied an effect of ultrasonic irradiation on specific surface area and particle size of nanosized zero-valent iron synthesized by methods of chemical precipitation with high-temperature reduction in hydrogen flow and sodium borohydride reduction. Obtained results showed significant decrease of particle size and differences in particles morphology depending on presence of ultrasonication during synthesis and on chosen method. For ultrasonic-assisted synthesis with 100 % amplitude, particle size calculated from specific surface area was 70 nm for sample synthesized by chemical precipitation with high-temperature reduction and 35 nm for borohydide reduction method compared to 63 nm for reference sample without ultrasonication.

1. Introduction

Rare-earth metals have critical role in production of almost all electronic devices, magnets, batteries etc. On the other hand, raw sources of rare-earth metals are geographically dispersed and concentrated irregularly depending on countries. Recycling of waste electrical and electronic equipment (WEEE), fluorescent lamp waste, industrial wastes, municipal solid waste (MSW) and other rare-earth containing wastes became one of the most important issues in modern industry [1-4]. According various researchers, recovery of rare-earth methods is often performed by acid leaching and takes a few hours. Thus there is a need for activation of leaching kinetics and thereof improvement in performance and yield capacity of existing methods.

On the other hand Fenton-like processes and nanosized zero-valent iron (nZVI) as Fenton-like catalyst have presented promising potential for applications requiring strong oxidative conditions. Formation of extremely reactive hydroxyl radical (OH^*) and high reactivity of nZVI during Fenton-like reactions established nZVI applications as rapidly growing [5-10].

Content from this work may be used under the terms of the Creative Commons Attribution 3.0 licence. Any further distribution of this work must maintain attribution to the author(s) and the title of the work, journal citation and DOI.
Published under licence by IOP Publishing Ltd

IOP Conf. Series: Materials Science and Engineering **112** (2016) 012012 doi:10.1088/1757-899X/112/1/012012

For wastewater treatment, at the moment there is a common route of treatment consisting of four main steps: pretreatment, primary treatment, secondary treatment and fining. Most of existing technologies require construction of wastewater treatment purification plants as well as Great scientific interest for research in this area is represented by recovery agents and Fenton catalysts, such as nanoparticles based on iron, which has high values of reactive surface area. First of all, formation of hydroxyl radicals allows oxidizing of organic pollutants. On the other hand, strong sorption and reducing properties allows sequestration of cations from the solution. Unlike conventional methods, nZVI treatment allows to treat water and soil in situ via direct injection into the hot spots of the pollution [6]. At the same time, rapidly developing technologies based on nZVI usually associated with the expansion of persistent organic pollutants (POPs) in the presence of iron cations (Fenton-like processes) [11]. However, possible applications of nZVI are not limited to wastewater treatment. There could also be included: ultrasonic disinfection of groundwater («sono-Fenton process»), treatment of leachates from municipal landfills, pharmaceutical wastewater treatment and others. Therefore, recent advances and trends in the development of water treatment technologies allow us to conclude that nZVI has undoubted relevance in today's industry.

In this study we investigated the effect of ultrasonic cavitation during synthesis via chemical precipitation with high-temperature reduction and sodium borohydride reduction methods on specific surface area, particle size and morphology of formed nZVI.

2. Materials and methods

In present research we focused on synthesis of nZVI by two different methods: chemical precipitation with high-temperature reduction in continuous hydrogen flow and subsequent passivation in technical grade nitrogen and sodium borohydride reduction. Comparison of experimental procedures for each of investigated methods is presented in figure 1.

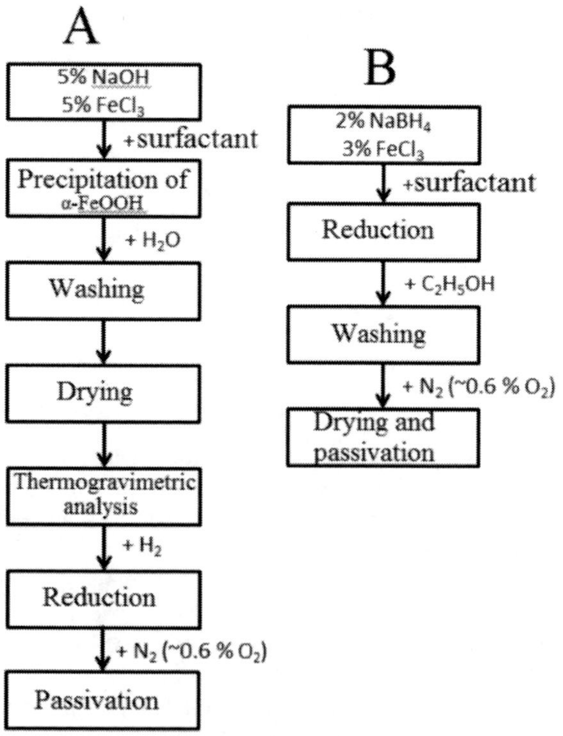

Figure 1. Experimental procedures: A: Chemical precipitation with high-temperature reduction; B: sodium borohydride reduction

As it could be seen from experimental procedures (figure 1), method of chemical precipitation with high-temperature reduction (figure 1A) includes 7 main steps. The first one is chemical precipitation (α-FeOOH is preferable due to higher values of dispersity compared to other iron hydroxide phases). Next step is washing with bi-distilled water from ions of Na^+ and Cl^- (from 5 to 10 times, controlled by pH value). After that washed precipitate should be dried at temperature of 60 °C for 10-12 h. In order to determine an optimal reduction temperature, dried powder is analyzed by thermogravimetric analysis (TA Instruments, USA, model SDT Q600, heating rate 10-15°C/min, in $Ar+H_2$ flow). Reduction is made in tube furnace made by Carbolite, UK (model HZS 12/900 with quartz reactor, linear heating and holding at reduction temperature, continuous H_2 flow with rate approximately 80 dm^3/h) with subsequent passivation in technical grade N_2 (room temperature, approximately 0.6 % of O_2). Such a long procedure allows us to produce air-stable nZVI with different particle size (from 20 to 100 nm, depending on pH, surfactant additions, reduction temperature) and morphology (i.e., chain or dumbbell shaped structure, depending on reduction temperature) with high reproducibility.

Compared to chemical precipitation with high-temperature reduction, sodium borohydride reduction (figure 1B) is rather unsophisticated. It includes only direct reduction of metallic iron from the solution, washing with 98 % ethyl alcohol (C_2H_5OH, from 3 to 5 times), drying and passivation in technical grade N_2 (2-3 h at 40-60 °C and 10-12 h at room temperature, approximately 0.6 % of O_2). Besides of apparatus and procedure simplicity, sodium borohydride reduction allows us to control the particle size (from 3-5 to 100 nm, depending on feed rates, concentrations of reactants and surfactant additions) and morphology (depends on external pressure, feed rates and concentrations of reactants). Main disadvantage of sodium borohydride reduction method is low yield of nZVI production.

All chemicals used in the experiments were purchased from local supplier (Labteh, Moscow, Russia), were analytical grade and used without any purification.

Experimental parameters used during synthesis of nZVI samples are shown in table 1. During synthesis as surfactant 0.1 % additions of sodium dodecylsulfate (SDS) was used.

Table 1. Experimental parameters

Parameter	Method	
	Chemical precipitation with high-temperature reduction	Sodium borohydride reduction
$FeCl_3$, %	5.0	3.0
NaOH, %	5.0	-
$NaBH_4$, %	-	2.0
SDS, %	0; 0.1	0; 0.1
Precipitation temperature, °C	25	-
Reduction temperature, °C	405	25
Drying temperature, °C	60	60
Passivation temperature, °C	25	25
Amplitude of ultrasonic vibrations, %	0; 50; 100	50; 100

A sample party of nZVI was prepared following the experimental procedures and parameters presented earlier (figure 1 and table 1 respectively). First 3 samples were obtained by chemical precipitation with high-temperature reduction (without surfactant and 50 % amplitude, with 0.1 % of surfactant and 50 % or 100 % amplitude) and other 3 samples were obtained by sodium borohydride reduction with same parameters. Besides that, 1 reference sample of nZVI was synthesized by chemical precipitation with high-temperature reduction with same experimental

IOP Conf. Series: Materials Science and Engineering **112** (2016) 012012 doi:10.1088/1757-899X/112/1/012012

parameters (figure 1B, table 1) without ultrasonication (500 RPM, mechanical stirrer made by Heidolph, Germany, model RZR-1).

All synthesis experiments were performed using an ultrasonic unit consisting of ultrasonic generator, piezoceramic transducer and titanium sonotrode (industrial grade ultrasonic device made by Hielscher, Germany, model UIP2000hdV); flow cell with a working volume of 150 ml with a cooling jacket; peristaltic pump (made by Heidolph, Germany, model PD 5201), worm pump, thermostat (made by Lauda, Germany, model RS232 Eco Silver), sliding stand, hoses, fittings and other peripherals. The experimental set-up is shown in figure 2.

Figure 2. Ultrasonic unit with Hielscher UIP2000hdV

After synthesis, all nZVI samples were characterized by low-temperature adsorption of nitrogen according to the Brunauer-Emmett-Teller theory of multimolecular adsorption (BET, device made by Quantachrome Instruments, USA, model Nova 1200e), X-ray diffraction analysis (XRD, powder diffraction with Bregg-Brentano geometry, device made by Scientific Instruments, Russia, model Diffrei 401), scanning electron microscopy (SEM, with device made by JEOL, Japan, model JSM-6700F, 15 kV of accelerating voltage) for samples synthesized under ultrasonic irradiation and transmission electron microscopy (TEM) for reference sample (JEOL, Japan, model JEM-2000 EX, 100 kV of accelerating voltage).

Particle size was calculated (assuming that all particles have spherical morphology) as follows:

$$d_{SSA} = \frac{\rho}{6 \cdot S} \quad (1)$$

where ρ – density, kg/m^3 and S – specific surface area, m^2/kg.

SEM and TEM images are presented in figure 1.

Mean crystalline size of nZVI was calculated from XRD data using well-known Scherrer equation as follows:

$$d_{Sher} = \frac{K_{hkl} \cdot \lambda}{\beta \cdot \cos\theta} \quad (2)$$

where d_{Sher} – mean size of the crystalline domains, nm; K_{hkl} – dimensionless shape factor; λ – X-ray wavelength, 10^{-10} m; β – line broadening at half the maximum intensity, radians; θ – Bragg's angle.

3. Results and discussions

Results for specific surface area measurement of nZVI samples are shown in table 2. XRD patterns for obtained samples are not presented due to irrelevance.

It should be noted that results of calculations from specific surface area and XRD data using (1) and (2) have good correlation for almost all samples, which corresponds that iron nanoparticles have monocrystalline or close to monocrystalline structure. That also means that most of presented agglomerates should be formed from separate nanoparticles.

Table 2. Specific surface area, calculated particle and crystalline sizes of synthesized nZVI samples

Method	SDS, %	Amplitude of ultrasonic vibrations, %	S, m^2/g	d_{SSA}, nm	d_{sher}, nm
Chemical precipitation with high-temperature reduction (reference)	0.1	0 %	12.0	63	59
Chemical precipitation with high-temperature reduction	0	50 %	8.6	90	72
	0.1	50 %	12.2	75	69
	0.1	100 %	12.9	70	64
Sodium borohydride reduction	0	50 %	11.3	67	59
	0.1	50 %	18.2	42	34
	0.1	100 %	21.6	35	23

SEM images of the obtained nZVi samples are shown in figures 3 and 4. As it could be seen, size of iron nanoparticles sufficiently correlates with calculated values from specific surface area measurement and XRD structure analysis.

Morphology of nZVI samples, obtained by chemical precipitation with high-temperature reduction, varies from quazi-regular shape for the reference sample (figure 3A) to dumbbell shape for 50 % amplitude without surfactant and finally to chain (figure 3C) and agglomerated (figure 3D). It is clear that presence of ultrasonic irradiation, as well as SDS additions, led to more dispersed structure of nZVI and therefore to higher values of specific surface area and dispersity.

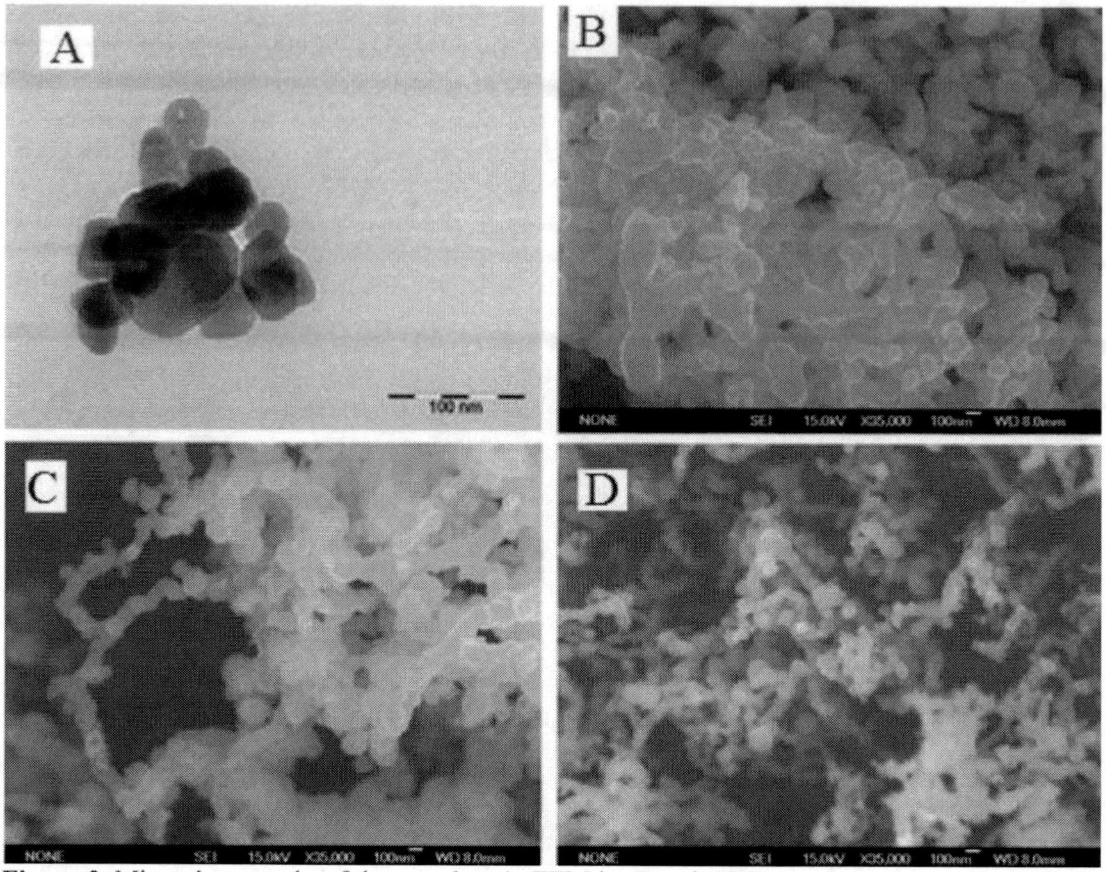

Figure 3. Microphotographs of the samples: A: TEM image of nZVI sample (reference, no US), B: SEM image of nZVI (0% SDS, 50 % amplitude), C: SEM image of nZVI (0.1% SDS, 50 % amplitude), D: SEM image of nZVI (0.1% SDS, 100 % amplitude)

Morphology of samples obtained by sodium borohydride reduction is mostly chain-shaped (figures 4A and 4B), but for 100% amplitude of ultrasonic irradiation morphology changes to agglomerated. It is curious, that in spite of quite coarse size of agglomerates (approximately), measures values of specific surface area and crystalline size are best of all experiments. This could be explained by taking into account the assumption, that all of the particles have relatively small size (20-30 nm), but as in order to compensate the excess surface energy they formed an agglomerated structure.

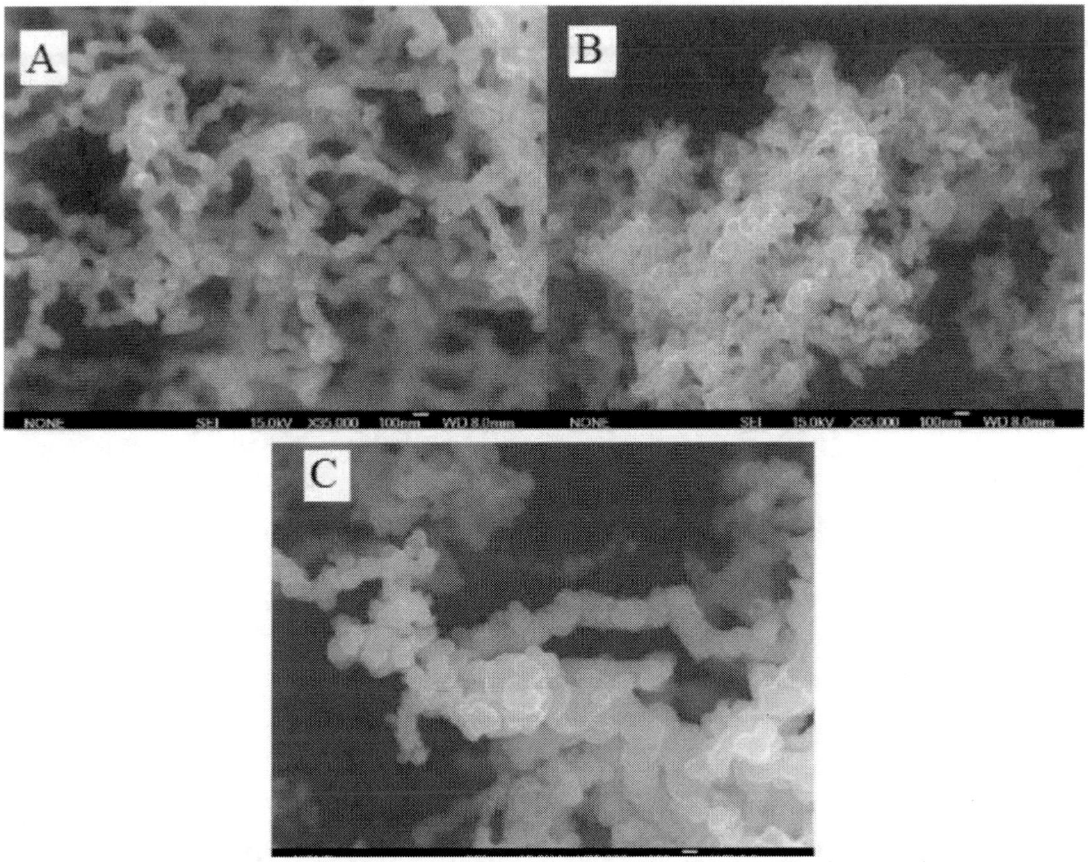

Figure 4. A: SEM image of nZVI (0% SDS, 50 % amplitude), B: SEM image of nZVI (0.1% SDS, 50 % amplitude), C: SEM image of nZVI (0.1% SDS, 100 % amplitude

4. Conclusions

Sample party of nZVI with different particle size and morphology have been prepared under ultrasonic cavitation via chemical precipitation with high-temperature reduction and sodium borohydride reduction methods.

During the study an impact of ultrasonic irradiation on specific surface area, particle size and morphology of nZVI has been observed. Results showed that compared to the sample obtained via synthesis without ultrasonic irradiation, presence of ultrasonic cavitation provides decreasing in particle size without significant changes in morphology.

Achieved values of nZVI particle size make it appropriate for use in environmental applications (wastewater treatment) as well as in development of novel processes for rare-earth metals extraction, requiring higher values of oxidation-reduction potential or efficient separation via sorption on the nanoparticles surface.

Acknowledgements

Authors gratefully acknowledge the Federal Program for Research and Development of the Russian Federation (Ministry of Science and Education, project # RFMEFI57514X0102) for financial support.

References

[1] Binnemans K, Jones P T, Blanpain B, Van Gerven T and Pontikes Y 2015 *J. Clean. Prod.* **99** 17
[2] Funari V, Braga R, Bokhari S N H, Dinelli E and Meisel T 2015 *Waste Manag.* **45** 201

[3] Tunsu C, Ekberg C and Retegan T 2014 *Hydrometallurgy* **144** 91
[4] Binnemans K, Jones P T, Blanpain B, Van Gerven T, Yang Y, Walton A and Buchert M 2013 *J. Clean. Prod.* **51** 1
[5] Su C, Pulsa R W and Krug T A 2012 *Water Res.* **46** 5071
[6] Taha M R and Ibrahim A H 2014 *J. Env. Chem. Eng.* **2** 1
[7] Yan W, Herzing A A, Kiely C J and Zhang W X 2010 *Journal Contam. Hydrol.* **118** 96
[8] Crane R and Scott T B 2012 *J. Hazard. Mater.* **211-212** 112
[9] Li X and Zhang W 2012 *J. Phys. Chem. C* **111** 6939
[10] Yang G C C and Chang Y-I 2011 *Sep. Purif. Technol.* **79** 278
[11] Gan S and Ng H K 2012 *Chem. Eng. J.* **213** 295

IOP Conf. Series: Materials Science and Engineering **112** (2016) 012013 doi:10.1088/1757-899X/112/1/012013

Chromium-based catalyst for HFC-125 synthesis: promoters effect

S I Reshetnikov[1], L G Simonova, A A Zirka and R V Petrov
Boreskov Institute of Catalysis, Siberian Branch, Russian Academy of Sciences, Pr. Lavrentieva, 5, Novosibirsk, 630090, Russia

[1]E-mail: reshet@catalysis.ru

Abstract. A variation of promoters, including rare-earth elements, allows to control of the specific surface and, perhaps, oxidation state of chromium and, consequently, catalyst activity and selectivity of chromium-based catalyst for HFC-125. To improve the catalytic properties of the 15% Cr_2O_3 - 85% γ-Al_2O_3 oxide system were added promoters (Ni or Cu or Co) in an amount of 5 wt% in terms of oxides. It was found that additional promotion Cr-Al samples by nickel and copper allow to increase the specific surface area of about 25-40% and the activity increased about 2 times. Modification of cobalt resulted in a decrease of the surface by 20% and the activity decreased by about 2 times.

1. Introduction

Various rare-earth metal oxides were widely investigated as promoters of heterogeneous catalysts for enhancing their activity, selectivity, and stability [1]. In particular, the ceria-containing catalysts use for diesel soot combustion and the reduction NO_x [2,3] hydrogen purification reactions, namely the water gas shift reaction [4] and etc. It is known, that the promoters affect the textural characteristics: specific surface area, pore diameter, and on the electronic state of the active component. Despite the large number of works, studying the role of promoters in the formation of the main characteristics of catalysts and their role in the formation of the active sites of the catalyst are still unclear.

In recent decades, considerable efforts have been focused on the synthesis of chladones, having zero ozone depletion potential. The most promising in this regard are chladones of the ethane row, namely, pentafluoroethane CHF_2–CF_3 (HFC-125), which is produced by the gas-phase catalytic hydrofluorination of perloroethylene Cl_2C=CCl_2. A fluorination catalyst based on of an amorphous Cr (III) compounds in combination with promoting and stabilizing admixtures of aluminum, magnesium, nickel, zinc and others [5,6]. Rare-earth metal oxides are used as promoters also, namely, lanthanum [7], scandium [8,9], and yttrium [10,11]. However, systematic studies of the mechanism of influence of promoters on the catalytic properties of chromium-containing fluorination catalysts are absent. These issues are of great scientific and practical interest.

In this work, we studied chromium-containing catalyst (~15% Cr_2O_3 - 85% γ-Al_2O_3) and the effect of promoters (~5% Ni, Cu and Co) on physicochemical and catalytic properties of the catalysts for the gas-phase hydrofluorination of perchloroethylene to pentafluoroethane (chladone HFC-125). Kinetic regularities of the reaction were also considered.

Content from this work may be used under the terms of the Creative Commons Attribution 3.0 licence. Any further distribution of this work must maintain attribution to the author(s) and the title of the work, journal citation and DOI.
Published under licence by IOP Publishing Ltd

2. Experimental Part

Catalysts obtained by precipitation of NH_4OH solution from solutions of chlorides of Cr, Al, Ni, Cu and Co. The precipitat was washed, filtered, dried in air at calcined in nitrogen at 330-400°C. The size and volume of pores, the catalyst activity in the reaction of hydrofluorination of perchloroethylene were measured for the samples.

The catalytic activity of samples in the gas-phase fluorination of perchloroethylene (PCE) was performed using the flow-type reactor. A mixture of reagents PCE and HF was fed in U-shaped nickel reactor with a diameter of 3 mm, placed in an oven. After the reactor the mixture entered absorbers, filled with alkali metal carbonate, where formed during the synthesis HCl and unreacted HF were neutralized, thus isolating the water and carbon dioxide After absorbers sampled gaseous reaction mixture, which mode *on-line* analyzed by gas chromatography Tsvet-500M (Russia) equipped with a thermal conductivity detector.

The reaction products were identified by CG-MS using a Saturn-2000 system (Varian, United States. According to the results of the chromatographic analysis was calculated mole fraction of PCE and reaction products. The conversion of perchloroethylene X was calculated as the difference between the molar fractions at the reactor inlet and outlet (corrected for the volume change during the reaction) to the mole fraction of PCE in the feed mixture. The selectivity of the reaction products was determined as the ratio of the mole fraction of each product to the sum of molar fractions of all the products in the reaction mixture. The rate constant expenditure of PCE was estimated by the equation for first order reactions

$$k_{PCE} = -\ln(1 - X)/\tau \qquad (1)$$

where τ is contact time and X is PCE conversion.

Experiments were performed in the kinetically controlled regime, which has been verified experimentally. The absence of external diffusion limitation was confirmed by varying the feed rate of the reaction.The absence of internal diffusion was determined by performing the experiment at different grain size.

3. Results and Discussion

It is known that individual Cr_2O_3 thermally not stable and at a relatively low temperature (~400°C) crystallizes in the α-Cr_2O_3, which is accompanied by a decrease of the specific surface. To increase the thermal stability of all investigated samples to Cr_2O_3 was add ~85% Al_2O_3. To improve the catalytic properties (activity and selectivity) in the Cr-Al samples were also added promoter (Ni or Cu or Co) in an amount of 5 wt% in terms of oxides.

It is found that additional promotion Cr-Al samples of nickel and copper increased the specific surface area of about 25-40% (table 1). Moreover, the activity increased about 2 times. Modification of cobalt resulted in a decrease of the surface by 20% and the activity decreased by about 2 times, while maintaining high selectivity. Thus, a direct relationship of the specific surface area of chromium-containing fluorination catalyst with the activity was not observed.

In the literature it is noted that the activity of Cr-catalysts of the gas-phase fluorination depends on chromium oxidation state [6]. We have shown [12] that in Cr-Al catalysts for the fluorination of chromium oxidation state of Cr depend essentially on the conditions of heat treatment: temperature and composition of the gas medium. We have investigated the physicochemical properties of coprecipitated oxide systems 95% of Cr_2O_3–5% Al_2O_3, obtained by heat treatment in nitrogen and air in the temperature range 110-600°C, and their influence on the catalytic activity in the reaction of hydrocorisone perchloroethylene.

Based on the methods of thermal and X-ray difraction analysis, diffuse reflectance spectroscopy, and specific surface measurements was found that $CrO_{1.9}$ compounds more active than $CrO_{1.5}$. However, getting them in superfine condition is difficult due to the high exothermicity of the oxidation stage, which leads to the decomposition of $CrO_{1.9}$ oxides to $CrO_{1.5}$ with subsequent crystallization and

IOP Conf. Series: Materials Science and Engineering **112** (2016) 012013 doi:10.1088/1757-899X/112/1/012013

sintering of decomposition products. The formation of highly dispersed and highly active chromium-containing oxides with stoichiometry $CrO_{1.9}$ possible in the two-stage heating of the hydroxide samples, first in nitrogen and then in air at relatively low temperatures:

$$CrO_{1.5}(am.) \xrightarrow{200\text{-}370^{\circ}C} CrO_{1.9}(am.) \xleftrightarrow{370\text{-}403^{\circ}C} CrO_{1.5}(am.) \xleftrightarrow{412\text{-}420^{\circ}C} \alpha\text{-}Cr_2O_3(cr.) \quad (2)$$

We can assume that the promoter can significantly effect on the electronic state of chromium. Perhaps this is the reason the difference in the activity of samples with different promoters. To clarify these issues, it is advisable carry out studies with a large set of promoters, including rare-earth elements.

Table 1. Effect of promoters on the BET specific surface area (S_{sp}), total pore volume (V_{pore}), mean pore diameter (D_{pore}) and catalyst activity (k at 370°C), PCE conversion (X) and selectivity towards main products (S_{sum}).

Sample	Catalyst composition	S_{sp} (m^2/g)	V_{pore} (cm^3/g)	D_{pore} (nm)	k_{370} (s^{-1})	X (%)	S_{sum} (%)
1	15%Cr_2O_3 85%Al_2O_3	245	0.6	35	0.15	33	96
2	5%NiO 10%Cr_2O_3 85%Al_2O_3	310	0.5	30	0.28	57	95
3	5%CuO 10%Cr_2O_3 85%Al_2O_3	340	0.45	30	0.27	54	100
4	5%CoO 10%Cr_2O_3 85%Al_2O_3	200	0.45	40	0.08	21	100

The overall reaction of the perchloroethylene hydrofluorination to pentafluoroethane is written as follows:

$$C_2Cl_4 + 5\,HF = C_2HF_5 + 4\,HCl \quad (3)$$

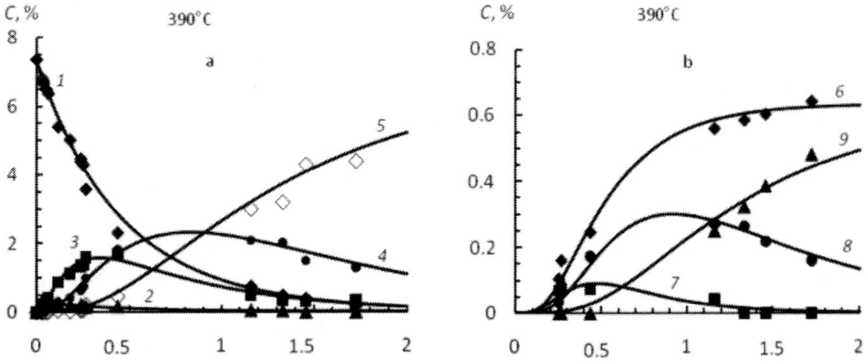

Figure 1. Concentrations of main (a) and by-products (b) along the reactor high. *1* – $CCl_2{=}CCl_2$, *2* – $CF_2Cl\text{–}CHCl_2$, *3* – $CF_3\text{–}CHCl_2$, *4* – $CF_3\text{–}CHFCl$, *5* – $CF_3\text{–}CHF_2$, *6* – $CF_3\text{–}CH_2Cl$, *7* – $CF_3\text{–}CCl_3$, *8* – $CF_3\text{–}CFCl_2$, *9* – $CF_3\text{–}CF_2Cl$.

However, the process includes a large number of possible reaction routes and, apart from its main products, might yield the products of side-reactions like isomerization, elimination and disproportionation. The reaction mixture contained the following main products: $CFCl{=}CCl_2$ difluorodichloroethylene isomers $CFCl{=}CFCl$ and $CF_2{=}CCl_2$, difluorotrichloroethane $CF_2Cl{-}CHCl_2$, trifluorodichloroethane isomers $CHCl_2{-}CF3$ and $CHFCl{-}CF_2Cl$, tetrafluorochloroethane isomers $CF_3{-}CHFCl$ and $CHF_2{-}CF_2Cl$, pentafluorochloroethane $CF_3{-}CF_2Cl$, and pentafluoroethane $CF_3{-}CHF_2$.

Besides, the following by-products were formed: trifluorotrichloroethane $CF_2Cl{-}CFCl_2$, trifluorochloroethane $CH_2Cl{-}CF_3$, tetrafluorodichloro ethane $CF_2Cl{-}CF_2Cl$, tetrafluoroethane $CHF_2{-}CHF_2$ and its isomer $CF_3{-}CH_2F$.

Typical distribution of the reaction products is shown in Figure 1. Analysis of experimental data obtained on samples with different promoters (Table1) testified that despite the difference in the activity and selectivity the scheme the products formation is identical. The observed dependence shows that the process of perchlorethylene hydrofluorination proceeds via complex parallel-series scheme, which can be represented in a form similar to that given in [13, 14] for the Cr-Mg catalyst:

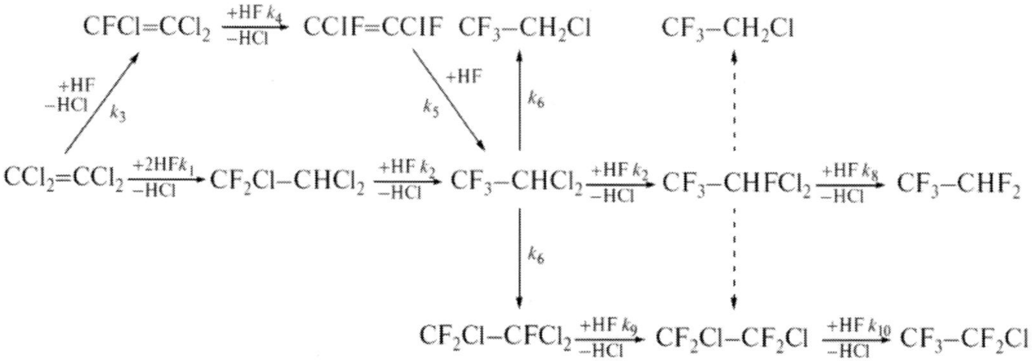

Figure 2. Simplified reaction pathway of PCE hydrofluorination.

This scheme consists of ten reactions. It includes consecutive substitution of fluorine for chlorine, and by-product formation, as a result of reactions of the trifluorodichloroethane $CHCl_2{-}CF_3$ and tetrafluorochloroethane $CF_3{-}CHFCl$ disproportionation.

4. Conclusion

A variation of promoters, including rare-earth elements, allows to control of the specific surface and, perhaps, oxidation state of chromium and, consequently, catalyst activity and selectivity.

To improve the catalytic properties of the 15% Cr_2O_3 - 85% γ-Al_2O_3 oxide system were added promoter (Ni or Cu or Co) in an amount of 5 wt% in terms of oxides. It was found that additional promotion Cr-Al samples of nickel and copper increased the specific surface area of about 25-40% (table 1) and the activity increased about 2 times. Modification of cobalt resulted in a decrease of the surface by 20% and the activity decreased by about 2 times.

Analysis of experimental data obtained on samples with different promoters testified that despite the difference in the activity and selectivity the scheme the products formation is identical.

Acknowledgements

This work was supported by the Russian Foundation for Basic Research, project No. 15-08-04789.

References

[1] Catalysis by Ceria and Related Materials. Catalytic science series. 2012 Edited by A.Trovarelli, P.Fornasiero. 2nd Editions. Imperial Collage Press. **12** 888

[2] Trovarelli A 1996 *Catal. Rev. Sci. Eng.* **38** 439

[3] Herz R K, Sell J A 1985 *J. Catal.* **94** 166
[4] Fu Q, Saltsburg H, Flytzani-Stephanopoulos M 2000 *Science* **301** 935
[5] Kemnitz E, Winfield J M 2000 *Adv. Inorg. Fluorides* 367
[6] Brunet S, Boussand B, Martin D. *J. Catal.* 1997 **171** 287
[7] Pat. US5571770 (A) — 1996-11-05
[8] Pat. CN85105080 (A) — 1986-12-31
[9] Pat. US5919728 (A) — 1999-07-06
[10] Pat. CN103143344 (A) — 2013-06-12
[11] Pat. CN101269323 (A) — 2008-09-24
[12] Simonova L G, Zirka A A, Reshetnikov S I, Larina T V, Litvak G S, Pinaeva L G and Isupova L A 2011 *Kinetics and Catalysis* **52** 418
[13] Zirka A A and Reshetnikov S I 2008 *Kinet. Catal.* **49** 663
[14] Reshetnikov S I, Zirka A A, Petrov R V, Ivanov E A 2011 *Chem. Eng. Journal* **176–177** 22

IOP Conf. Series: Materials Science and Engineering **112** (2016) 012014 doi:10.1088/1757-899X/112/1/012014

Separation of rare earth elements by zone recrystallization

D V Akimov[1], A N Dyachenko[1], N B Egorov[1] and N A Zhuravlev[1]

1 Physical technical institute, National Research Tomsk Polytechnic University, 634050, Russia, Tomsk, Lenina street, 30.

E-mail: egorov@tpu.ru

Abstract. The separation of holmium and cerium by zone recrystallization using a mixture of $HoCl_3 \cdot 6H_2O$ and $CeCl_3 \cdot 6H_2O$ was investigated. It is shown that holmium is enriched at the end of the crystal that the recrystallization zone moves to, while cerium is concentrated in the primary solidification zone. The possible reasons for the experimentally observed distribution of hydrated ions of cerium and holmium along the length of the ingot are discussed. Also the coefficients of enrichment and separation are calculated.

1. Introduction

The demand for rare earth elements (REEs) and also scandium and yttrium is ever increasing due to their wide range of applications, such as in radio electronics, instrumentation, nuclear engineering, mechanical engineering, chemical industry and metallurgy [1-12]. For this reason, there is a steady increase worldwide in the consumption and production of REEs. In order to reduce the cost of processing raw materials both the traditional and new technological schemes for processing monazite and other minerals which contain the REEs are developing now [13]. The separation of REEs individually has also always attracted the attention of researchers [14, 15].

The processes of ion exchange and extraction are the main methods in modern industry for separation of REEs into individual components. However, these methods have some disadvantages. If we use liquid extraction, it is necessary to consider the mutual solubility of the organic and aqueous phases. In this case we will lose the extracting agent, the cost of which is comparable with REEs. The method of the ion exchange provides low separation coefficients of REEs, and therefore this method may be just additional in the technological scheme of the extraction separation of REEs. For these reasons, there is today a search for new methods for separating REEs [16].

Zone recrystallization, often called zone melting, is widely used for purification of the different substances, to obtain them in single-crystal form, and also for separating those elements with similar properties [17].

The literature contains information on the distribution of REEs in melts of NH_4NO_3, NH_4SCN, $MgCl_2$ and $BaCl_2$ in which it is shown that the ratio of the concentrations of REEs of the yttrium and cerium group in the zone recrystallization varies slightly [18-20].

Here we report our investigations into the process of separating the chlorides of REEs by using zone recrystallization of their chloride hexahydrates. This choice was influenced by two factors. The first one is the full melting of REEs chloride hexahydrates at low temperatures. Secondly, compounds also contain water molecules, which can affect the distribution of the REEs during the process of zone recrystallization.

Content from this work may be used under the terms of the Creative Commons Attribution 3.0 licence. Any further distribution of this work must maintain attribution to the author(s) and the title of the work, journal citation and DOI.
Published under licence by IOP Publishing Ltd

In this way, the aim of this work was to investigate the possibility of separating REEs by using zone recrystallization of their chloride hexahydrates.

2. Experimental

2.1 Reagents

$CeCl_3 \cdot 6H_2O$ (analytical grade) and $HoCl_3 \cdot 6H_2O$ (analytical grade) were used. Their mixture was prepared by melting in the percentage ratio of $CeCl_3 \cdot 6H_2O$: $HoCl_3 \cdot 6H_2O$ = 59.5:40.5.

2.2 Instrumentation

Investigations were conducted on samples with a length of 90 mm and a diameter of 10 mm. For preparing samples, the chloride hexahydrates of REEs were heated until they melted. After that, part of the melt was placed into a glass tube, which was sealed at one end.

All experiments were carried out on the installation, which has five heating elements and five cooling elements (Figure 1). Nichrome wire was used as the heating element. The temperature of the melt zone was equal to 160±1 °C. The length of the melt of the ingot was from 1 to 1.5 cm.

The main part of the installation was a platform with the heating elements placed at equal distance in the form of a ring, the inner diameter of which corresponds to the outer diameter of the glass tube. Power supply GW INSTEK PSH -10100 was used for regulating the temperature of the heating elements. Temperature was measured by using a chromel-alumel thermocouple. Between the heating elements the cooling elements were installed, which were cooled by tap water. The temperature of cooling elements was equal to +23±1 °C.

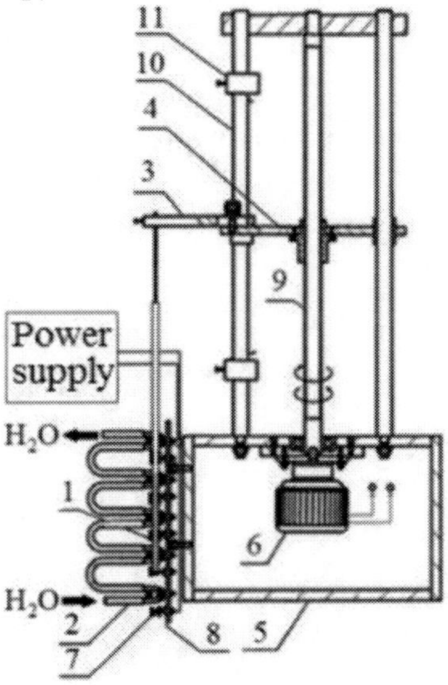

Figure 1. Installation for zone recrystallization (1 – mixture of REEs; 2 – cooling element; 3 – rod; 4 – carriage; 5 – housing; 6 – stepper motor; 7 – heating element; 8 – platform; 9 – shaft; 10 – guides; 11 – sensor).

The glass tube with the mixture of REEs was placed on the platform and fixed by using the rod on the carriage. The speed of the tube was 3 cm/hour. Our process was finished when the number of zone processes was equal to 40. After this, the glass tube was cut into samples, 10 mm in length. Samples were removed, weighed and dissolved in distilled water. In the obtained solutions we determined the content of cerium and holmium; these solutions were also investigated by spectrophotometry. Quantitative analysis was carried out using an atomic-emission spectrometer iCAP 6300 with inductively coupled plasma. The electronic absorption spectra of solutions were obtained by using a spectrophotometer Evolution 600 in quartz cuvettes 10 mm thick.

3. Results and discussion

Figure 2 shows both the electronic absorption spectra of soluble samples of the REE mixture, which were obtained after the process of zone crystallization, and the spectra of the original sample

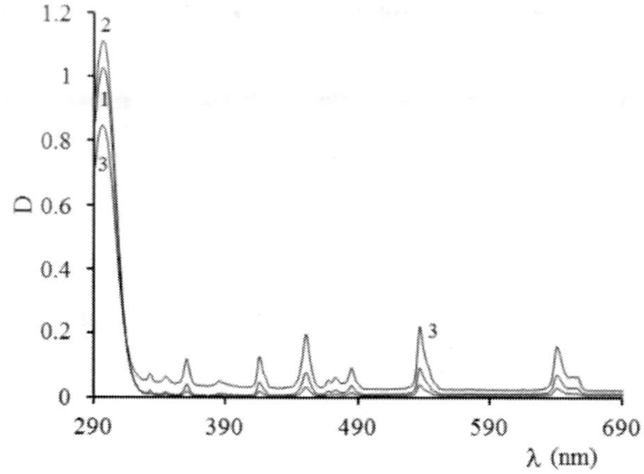

Figure 2. The electronic absorption spectra of solutions of cerium and holmium chlorides mixture: 1 – the original; 2 – the sample which was obtained on the 2 cm length of the ingot; 3 – the sample which was obtained on the 8 cm length of the ingot.

The maximum in the electronic absorption spectra at 296 nm belongs to the absorption by the aqua complexes of cerium chloride, all the rest belong to the aqua complexes of holmium chloride [21]. As we can see from the figure, the maximums related to the ions of cerium are increased in the initial crystallization zone of the samples, and the maximums related to the holmium ions are reduced. This means that the concentration of cerium ions is increased in the initial crystallization zone of the sample, while holmium moves with the melt zone and concentrates at the edge of the sample.

From the data that were obtained by atomic-emission analysis, the distribution of cerium and holmium on the length of the ingot after the process of zone recrystallization of chloride hexahydrates was plotted in Figure 3.

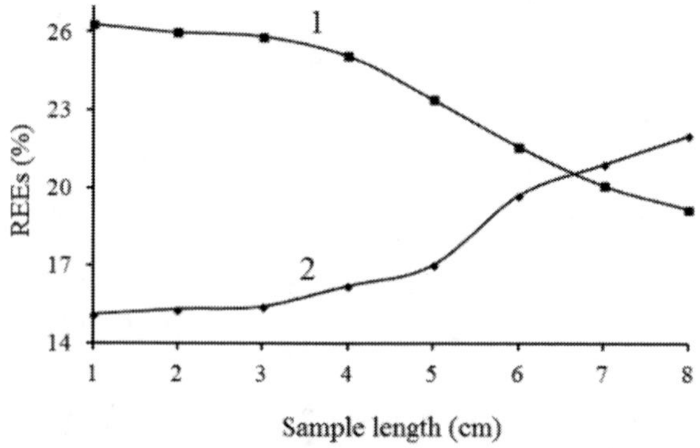

Figure 3. Distribution of cerium (1) and holmium (2) on the length of the ingot after the process of zone recrystallization

The results confirm that cerium, which is lighter, was concentrated in the initial zone of recrystallization, while the heavier holmium was concentrated in the end zone of the ingot. In addition, we can say that 40 recrystallization zones are insufficient for a full distribution of REEs in the sample.

For the determination of the separation coefficient we compared the initial concentrations of REEs with their concentrations after the process of zone recrystallization. The values of the separation and enrichment coefficients for both elements were calculated. The obtained data are presented in Table 1.

The values of the enrichment coefficients were calculated for the extreme zones of the ingot after zone recrystallization using the following formula:

$$\beta = \frac{x_e}{1-x_e} \cdot \frac{1-x_0}{x_0}, \quad (1)$$

where x_e – the contents of the element after the process of zone recrystallization in the maximum enriched zone, %;

x_0 – the initial contents of the element, %.

$$\beta_{Ho} = \frac{x_e}{1-x_e} \cdot \frac{1-x_0}{x_0} = \frac{22}{78} \cdot \frac{82.3}{17.7} = 1.311;$$

$$\beta_{Ce} = \frac{x_e}{1-x_e} \cdot \frac{1-x_0}{x_0} = \frac{26.3}{73.7} \cdot \frac{76.4}{23.6} = 1.155.$$

The separation coefficients (α) were found from the following relation:

$$lg\alpha = \frac{lg\beta}{n-1}, \quad (2)$$

where n – the number of passes of the zone.

$$lg\alpha_{\mathrm{Ho}} = \frac{lg\beta}{n-1} = \frac{0.118}{39} = 0.0030;$$

$$\alpha_{Ho} = 10^{0.0030} = 1.007.$$

$$lg\alpha_{Ce} = \frac{lg\beta}{n-1} = \frac{0.063}{39} = 0.0016;$$

$$\alpha_{Ce} = 10^{0.0016} = 1.004.$$

Table 1. The coefficients of separation and enrichment of cerium and holmium after the process of zone recrystallization.

Characteristics of the process of zone recrystallization	REEs	
	Ho	Ce
Coefficient of enrichment (β)	1.311	1.155
Coefficient of separation (α)	1.007	1.004

It is known that when the impurity increases the melting point, the concentration of this impurity will decrease upon crystallization in the liquid phase. Conversely when the impurity decreases the melting point, upon crystallization we will observe the reverse effect [17]. According to the data in [22, 23], the melting point of $CeCl_3 \cdot 6H_2O$ is 403 K, and that of $HoCl_3 \cdot 6H_2O$ is 437 K. Thus, changes in the concentrations of cerium and holmium during the process of zone recrystallization of the chloride hexahydrates cannot be explained by the difference in their melting temperatures. It is known that the solubility of the light lanthanide chlorides is higher than that of the heavy ones, and therefore the cerium ions should be concentrated in the region of the melt [22-24]. However, the experimental results show the inverse movement of REEs in the sample.

Burton, Prim and Slichter obtained the formula for the effective distribution coefficient [25]:

$$K = \frac{1}{1+\left[\left(\frac{1}{K_0}\right)-1\right]e^{-f\delta/D}}, \quad (3)$$

where K_0 – the equilibrium distribution coefficient, f – the crystal growth rate, δ – the thickness of the diffusion layer in front of the crystallization front, D – the diffusion coefficient.

From this equation we can conclude that differences between the effective distribution coefficient K and the equilibrium distribution coefficient K_0 will be more significant if the value of the diffusion coefficient D is higher and the thickness of the diffusion layer in front of the crystallization front is smaller. The chlorides of light REEs have a large diffusion coefficient and, correspondingly, a large mobility. For this reason they should move with the melting zone, but this fact also does not correspond with the observed experimental data.

Apparently, these results may be explained only by assuming that the distribution of REEs by the recrystallization process is superimposed by ion solvation (hydration), which can inhibit the diffusion of the hydrated cerium ions and thus contribute to the cerium–holmium separation.

It is well known that the solvation of ions in the solution is determined by the nature and ratio of different types of interactions in solution – the ion–ion, ion–molecular and intermolecular interactions [26]. For ionic systems, the solvation contributions can be estimated by considering the following processes:

1. The process of the formation of a cavity in a solvent (the contribution of interactions that depend on the size of the ion);
2. The process of the transition of uncharged particles, which are isoelectronic with respect to the ion, into a cavity;
3. The process of the redistribution of the electron density between the ion and solvent molecules (the contribution from other interactions that depend on the ion charge).

Apparently, the greatest contribution to the suppression of diffusion of the cerium ions is due to their interaction with the water molecules, i.e. the chemical hydration of the components, and to the fact that cerium has a smaller ion own size than the holmium ion. The value of the hydration shell will depend on the density of the ion charge on this surface. Therefore, ions that have a larger size also have a smaller hydration shell [27].

The molecule of H_2O is an electrodonor and has the donor–acceptor interaction with both ions of cerium and ions of holmium through the cation–solvent mechanism. If we have a low content of water molecules, the most part of these molecules will be coordinated by the cerium ion, which has the ability to have a large hydration shell. This fact will lead to a decrease in the diffusion of the hydrated ions of cerium and will also lead to an increase in the diffusion of the hydrated ions of holmium. This was confirmed experimentally.

According to this fact we can conclude that after the process of zone recrystallization the molecules of water will concentrate in the initial zone of crystallization. In our the next work we will determine the behaviour of the water molecules during the process of zone recrystallization of the chloride hexahydrates of REEs. We will find the optimal conditions for the separation of REEs by using the process of zone recrystallization. In addition we will conduct an experimental investigation of the process of separation of different multi-component mixtures of REEs.

4. Conclusions

For separation of salts REEs was applied the method of zone recrystallization. Also was studied the distribution of hexahydrates chlorides of cerium–holmium during the process of zone recrystallization. The coefficients of enrichment and separation were calculated. Results show that process of zone recrystallization can be used for separation of hexahydrates chlorides of REEs. If the salts of REEs will use as working substances which have low melting points (90–165 °C) it will allow obtain the individual high purity REEs with two advantages: expensive organic compounds will not use (the extractants and ion exchangers); the amount of water and water-organic waste will decrease. The both of these advantages will reduce the cost of REEs. Also the method of zone recrystallization can be used for effective separation of mixtures REEs.

Acknowledgements

This work is financially supported by Tomsk Polytechnic University within the Program of Improving the Competitiveness.

References

[1] Gschneidner K A and Pecharsky V K 2006 Rare Earths and Magnetic Refrigeration *Journal of Rare Earths* **24** 641

[2] Chen Y, Jin L, Fang D, Song Y and Ye R 2015 Effect of calcium, samarium addition on microstructure and mechanical properties of AZ61 magnesium alloy *Journal of Rare Earths* **33** 86
[3] Wong P S, Wan M H, Hussin R, Lintang H O and Endud S 2014 Structural and luminescence studies of europium ions in lithium aluminium borophosphate glasses *Journal of Rare Earths* **32** 585
[4] Miao Y, Yaohui L, Jiaan L and Song Y 2014 Corrosion and mechanical properties of AM50 magnesium alloy after being modified by 1 wt.% rare earth element gadolinium *Journal of Rare Earths* **32** 558
[5] Janos P, Kuran P, Kormunda M, Stengl V, Grygar T M, Dosek M, Stastny M, Ederer J, Pilarova V and Vrtoch L 2014 Cerium dioxide as a new reactive sorbent for fast degradation of parathion methyl and some other organophosphates *Journal of Rare Earths* **32** 360
[6] Balachandran G 2014 Extraction of Rare Earths for Advanced Applications *Treatise on Process Metallurgy* **3** 1291
[7] Visseaux M and Bonnet F 2011 Borohydride complex of rare earths, and their applications in various organic transformations *Coordination Chemistry Reviews* **255** 374
[8] Guanming Q, Xikum L, Tai Q., Haitao Z, Honghao Y and Ruiting M 2007 Application of Rare Earths in Advanced Ceramic Materials *Journal of Rare Earths* **25** 281
[9] Cao X 2007 Application of rare earths in thermal barrier coating materials *Journal of Materials Science and Technology* **23** 15
[10] Goonan T G 2011 *Rare earth elements – End use and recyclability* (Virginia: U.S. Geological Survey)
[11] Srinivasan R, Yogamalar N R, Elanchezhiyan J, Joseyphus R J and Bose A C 2010 Structural and optical properties of europium doped yttrium oxide nanoparticles for phosphor applications *Journal of Alloys and Compounds* **496** 472
[12] Ahmad H, Thambiratnam K, Paul M C, Zulkifli A Z, Ghani Z A and Harun S W 2012 Fabrication and application of zirconia-erbium doped fibers *Optical materials express* **2** 1690
[13] Egorov N B, Djyachenko A N, Akimov D V, Kiselyov A D, Obmuch K V and Chalov S A 2014 Influence of adding ammonium bifluoride when leaching monazite using sulphur acid *Procedia Chemistry* **10** 168
[14] Karve M and Vaidya B 2008 Selective Separation of Scandium (III) and Yttrium (III) from other Rare Earth Elements using Cyanex302 as an Extractant *Separation Science and Technology* **43** 1111
[15] Kim J S, Kuman N, Lee J Y Kantam M L and Reddy B R 2012 Separation and recovery of Light Rare-Earths from Chloride solutions using Organophosphorus based Extractants *Separation Science and technology* **47** 1644
[16] Liao Ch, Wu Sh, Cheng F, Wang S, Liu Y, Zhang B and Yan C 2013 Clean separation technologies of rare earth resources in China *Journal of Rare Earths* **31** 331
[17] Pfann W G 1967 *Zone melting* (New York: John Wiley & Sons)
[18] Valtsev K V, Avvakumov K V and Pyref M F 1961 Distribution of lanthanides in the melt of ammonium nitrate during the process of zone recrystallization *Bulletin of Siberian Branch of the USSR Academy of Sciences* **6** 71
[19] Valtsev K V, Oziashvili E D and Solovev L K 1962 The process of zone recrystallization of compounds of rare-earth elements from some of molten salts *Bulletin of Siberian Branch of the USSR Academy of Sciences* **2** 53
[20] Valtsev K V, Avvakumov K V, Pyrev M F and Kravchenko L Ch 1963 Separation of lanthanides in ammonium nitrate using the process of zone recrystallization *Bulletin of Siberian Branch of the USSR Academy of Sciences* **3** 152
[21] Ryabchikov D I and Pyabuchin V A 1966 *Analytical chemistry of rare earth elements and yttrium* (Moscow: Nauka)

[22] Mioduski T, Guminski C and Zeng D 2009 IUPAC-NIST Solubility Data Series 87 Rare Earth Metal Chlorides in Water and Aqueous Systems Part 2 Light Lanthanides (Ce–Eu) *Journal of Physical and Chemical Reference Data* **2** 441
[23] Mioduski T, Guminski C and Zeng D 2009 IUPAC-NIST Solubility Data Series 87 Rare Earth Metal Chlorides in Water and Aqueous Systems Part 3 Heavy Lanthanides (Gd–Lu) *Journal of Physical and Chemical Reference Data* **4** 925
[24] Mioduski T, Guminski C and Zeng D 2008 IUPAC-NIST Solubility Data Series 87 Rare Earth Metal Chlorides in Water and Aqueous Systems Part 1 Scandium Group (Sc, Y, La) *Journal of Physical and Chemical Reference Data* **4** 1765
[25] Wilson L 1978 On interpreting a quantity in the Burton, Prim and Slitchter equation as a diffusion boundary layer thichness *Journal of Crystal Grown* **44** 274
[26] Krestov G A, Novoselov N P, Perelygin I S, Kolker A M, Safonova L P, Ovchinnikova V D and Trostin V N 1987 *Ion solvation* (Moscow: Nauka)
[27] Turaev N S and Gerin I I 2006 *Chemistry and technology of uranium* (Moscow: Ore and Metals)

Surface layer composition of titania produced by various methods: the change of layer state under illumination

V Zakharenko[1,*], E Daibova[2], N Kosova[3], O Zmeeva[2]
[1] Boreskov Institute of Catalysis SB RAS, Novosibirsk, Russia
[2] Siberian Research Institute of Agriculture and Peat, Tomsk, Russia
[3]Tomsk State University, Tomsk, 634050, Russia

*E-mail: zakh@catalysis.ru

Abstract. The comparative analysis of experimental data over titanium dioxide powders prepared by various ways under ambient air is carried out. The results over TiO_2 prepared by high-temperature heating of anatase, produced by burning of titanium micro particles and grinding of rutile crystal are used for that comparison. Water and carbon dioxide were the main products released from the surface of the titania powders. It was found that under UV irradiation absorbed by titania, in absent oxygen, water effectively reacts with lattice oxygen of titanium dioxide.

1. Introduction

The composition of solid surface layer influences over adhesive, corrosion-resistant and mechanical properties of solids. The corrosion rate of some metals depends on metal oxide film formed on metal surface under ambient air. Such films are created on surfaces zinc (ZnO), aluminum (Al_2O_3) and titanium (TiO_2). As subject of inquiry was selected titanium dioxide. Titania is used as paintwork pigment, plastic filler, photocatalyst for cleaning of water and air [1].

The oxides are affected by the gas components of the atmosphere, and primarily by the components having the higher partial pressure. The latter are water, carbon dioxide and nitrogen oxides. So, according to recent studies [2] for magnesia obtained from $Mg(OH)_2$ and exposed by the atmosphere for a long time, the amount of adsorbed compounds attained are: NO and N_2 (from nitrogen oxides) ca.1% of monolayer, CO_2 ca.10% of monolayer, H_2O ca. one monolayer. The total amount of the adsorbed compounds in this situation equals to 5 weight % (according to the data of derivatographic analysis).

Traditionally before performing experiments or using in various chemical processes, the surfaces of adsorbents and catalysts (including photoadsorbents and photocatalysts) are exposed to a rigid pretreatment of various kinds. The main goal of this pretreatment (e.g., in vacuum, in an appropriate gas environment, at a certain pressure and temperature, etc.) is usually to obtain either reproducible results, which are consistent with those of other researchers, or the maximum of adsorption or catalytic (accordingly, photoadsorption or photocatalytic) activity. However, it is essential that the rigid pretreatment of a surface may result in some sufficient changes of chemical composition of the adsorption layer, stoichiometry of the surface and near-surface layers of an adsorbent or a catalyst. Our purpose has been to study the surface chemical properties of titania samples produced by various ways under ambient air.

Content from this work may be used under the terms of the Creative Commons Attribution 3.0 licence. Any further distribution of this work must maintain attribution to the author(s) and the title of the work, journal citation and DOI.
Published under licence by IOP Publishing Ltd

IOP Conf. Series: Materials Science and Engineering **112** (2016) 012015 doi:10.1088/1757-899X/112/1/012015

2. Experimental Part

A home-built static installation was used for the measurement of the rates of adsorption, desorption, catalytic oxidation in darkness and under irradiation. The experiments were performed in a high vacuum apparatus equipped with facilities of mass-spectrometer, Pirani gage and gas introduction. The base pressure of the apparatus was below about 10^{-7} Torr.

Water vapour did not get into the high-vacuum (measuring) part of the set-up, which allowed us to measure small (in comparison with water vapour) amount of adsorbates and products desorbed from the surface of oxides during the dark and photo processes.

The chemical composition during the course of the experiment was monitored with a mass spectrometer via their sampling through a leak valve. The mass spectrometer was made from a monopoly analyzer APDM-1. The kinetics of the observed processes was registered by measuring of the alteration in amplitude of the most intensive peak in the mass spectra of the corresponding substances.

The amounts of O_2, N_2O, NO, CO and other products of desorption from the titania surface were determined using a Pirani gauge, which was calibrated with oxygen and carbon monoxide. A necessary correction was introduced for the sensitivity with respect to other gases.

To expose the oxides under UV radiation, an OSL-1 illuminator with a high pressure mercury lamp (DRSH-250, power 250 W), including a thermal water filter and UV filter, were used. The total radiation flux density, reaching the reactor's surface and recorded by an RTN-20S thermopile, was equal to ~1 milliwatt · cm^{-2} for this filter. To separate monochromatic radiation, interference filters were used. The quantum yield of the photodesorption was found as the ratio of the quantity of photodesorbed molecules to the number of quanta passed through the reactor's frontal (transparent) wall.

2.1. Preparation of samples

The TiO_2 monocrystal of the crystalline structure of rutile produced by the Czochralski method [3] was broken and ground in a corundum mortar ("crushing" TiO_2). The powder specific surface was 1.6 $m^2 \cdot g^{-1}$.

The titania powder of anatase crystalline modification was heated at 1273 K for two hours in air in order to obtain from the anatase phase (basically) the rutile phase (basically). After that treatment the titanium dioxide ("heating" TiO_2) was a rutile modification with a specific surface of 8 $m^2 \cdot g^{-1}$.

Next sample of powdery titanium dioxide was prepared by combustion of pyrotechnic mixture consisting of ammonium perchlorate, hydroxy-terminated polybutadiene and titanium particles (particle size from 60 to 90 μm) under ambient air [4]. The highly dispersed titanium dioxide powder ("burning" TiO_2) had the specific surface area of 6 $m^2 \cdot g^{-1}$. After a treatment of the oxide surface at temperature 625 K for 30 min in air the specific surface area was decreased to 4 $m^2 \cdot g^{-1}$.

A suspension of TiO_2 in the distilled water was supported on the internal wall of a quartz glass reactor, which was transparent for the light with wavelength longer than 185 nm. The TiO_2 layer was dried at room temperature in air. Then, without any additional treatment, the reactor was sealed to the high vacuum system for the photoadsorption experiments. The reactor was evacuated to 10^{-5} Pa at room temperature for several hours.

2.2. Absorption spectra

Titania samples with stoichiometric chemical composition are white powders. The absorption of light with energies below the band gap of titania in rutile modification (Eg = 3.05 eV) is minimal by all such powders (Figure 1, curves 1 and 2). This absorption corresponds to the impurity absorption bands for single crystals or surface absorption for finely dispersed powders.

According to these data (Figure 1), there is an inflection of absorption edge curve for powdered titania samples consisting from a mixture of anatase and rutile modifications. Such an inflection of curve is not observed for pure anatase or rutile (Figure 1, curve 1) samples. The absorption band edge of anatase powder is shifted by 0.1-0.2 eV to shorter wavelengths compared to that of rutile powder (3.0 and 2.8-2.9 eV, respectively).

If the titania stoichiometry is changed due to partial reduction of Ti^{4+} to Ti^{3+} and formation of lattice oxygen vacancies during the oxide synthesis, the powders become colored and absorption in the surface band of the oxide appears.

The absorption properties of dispersed titania after storage in air for a long period of time and after activation by heating at 625 K in air for 30 min with respect to the light from different parts of the solar spectrum Figure 1, curve 3). The experiments with the samples contacted with the air for a long time are especially important for evaluating the prospects of titania prepared by this method for application in processes leading to purification of the environment during various man-caused catastrophes.

The absorption of titania powders prepared by combustion of pyrotechnic mixture containing titanium microparticles in the region before the TiO_2 intrinsic absorption edge (corresponding to the surface absorption region) depends on the synthesis method and after-treatment of the samples.

3. Results and their discussion

3.1. Composition of surface layer

The samples prepared under ambient air were deposited from water suspension in the form of thin layer on the walls of the reactor made from optical quartz. After soldering to high-vacuum set-up the sample was pumped out at room temperature to evacuate the most part of adsorbed CO_2.

Titanium dioxide ("heating") surface was cleaned from the adsorbed organic impurities, carbon dioxide, and water by long-time, high temperature, oxygen-vacuum treatment. The analysis of the products released upon heating from a rutile surface purified by on oxygen-vacuum treatment for 130 hrs shows the absence of CO_2 evolution and occurrence of water and oxygen evolution at about 450 K (H_2O alone) and above 800 K ($H_2O + O_2$). Increasing the time of oxygen-vacuum treatment to 300 hrs leads to the disappearance of the water thermodesorption peak at 450 K and, above 800 K, only small amount of oxygen evolve.

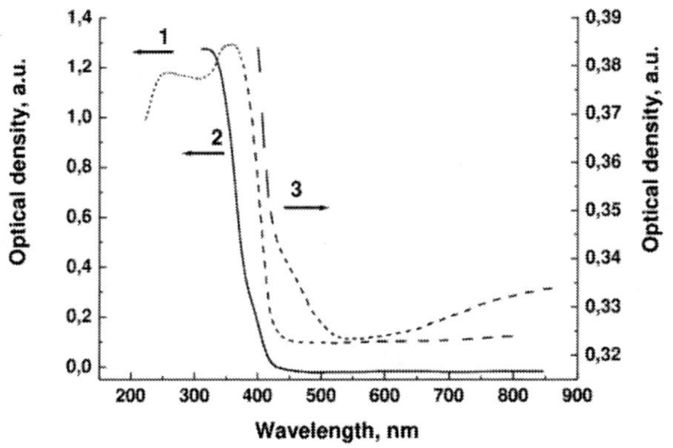

Figure 1. Diffuse reflection spectra of "crushing" (1), "heating" (2) and "burning" titania samples.

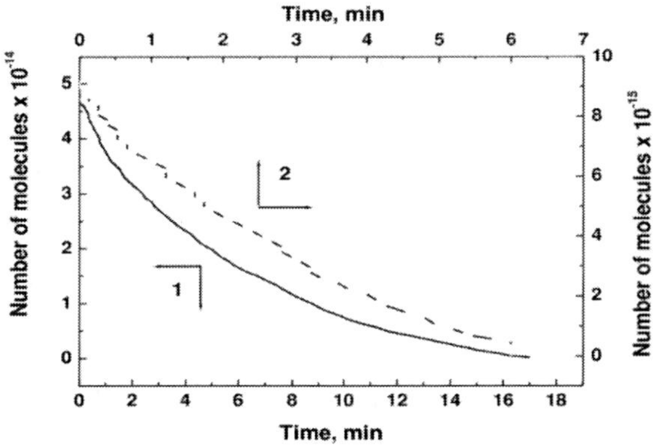

Figure 2. Kinetics of the desorption CO_2 from "crushing" TiO_2 surface. 1 – decrease of number of CO_2 molecules on TiO_2 surface (dark desorption); 2 - decrease of number of CO_2 (in addition to dark desorption) under illumination of TiO_2 surface through the UV filter (photodesorption).

3.2. Darkness processes

The presence of water vapour in the reactor was due to it's desorption from the "crushing" TiO_2 after the connection of the reactor to the set-up. During evacuation, water vapour was frozen in trap cooled down to 173 K. Thus, water vapour was always present in the reactor. After pumping-out of the reactor through this trap, mainly NO evolved from the TiO_2 surface (fourfold greater than CO_2 accumulated inside the reactor).

It can be assumed that nitrogen oxide (II) is produced in a sufficient amount when grinding the TiO_2 crystal in air due to molecular nitrogen oxidation at centres formed at the break of the Ti–O bonds of the titanium dioxide lattice. The NO generation is also possible through the well-known reaction of N_2O decomposition at the electron donor centre of the metal oxide surface [5, 6]. In this case, N_2O is adsorbed from air like the carbon dioxide.

Prolonged evacuation of the reactor volume for 1 h in high-vacuum installation through a trap with the cooling liquid resulted in partial removal of carbon dioxide adsorbed on the titania surface. As a result, quasi-equilibrium filling of the surface with CO_2 was established. The kinetics of approaching to the equilibrium CO_2 pressure is shown in figure. 2, curve 1. This kinetics is presented as the decrease of the amount of CO_2 molecules adsorbed on the titania surface (desorption). The kinetic curve follows the first-order equation with the time constant $\tau_1 \approx 6$ min. For comparison, the kinetics of CO_2 photodesorption, obtained after the attaining equilibrium of the dark adsorption, is also given in figure 2, curve 2.

No other gases except for CO_2 and H_2O were observed to be released in measurable amounts in the dark at room temperature from the surface of "burning" titania.

3.3. Processes under illumination

Surface photochemical reactions occur after absorption of radiation by compounds adsorbed on the surface of solid particles as they transform into the excited reactive state [7]. An example of such a reaction is photodecomposition of nitrous oxide adsorbed on the surface magnesia [8, 9]. The surface in this case plays the role of a "carrier" (substrate), changing the energy of excited state of the adsorbed compound, or role of stabilizer of the adsorbed compound in the state different from the state of molecules in the gas phase, for example, for such compounds as CO_3 complex, OH group, etc.

The photodesorption kinetics is characterized by the fast desorption of carbon dioxide in the initial time of illumination (up to 3–5 min), followed by slow desorption with a constant rate. In the first case, probably, desorption of carbon dioxide is connected with classic photodesorption under illumination of the surface under quanta with energy from the fundamental absorption band of TiO_2 [10, 11]. In the second case, desorption is attributed to the photocatalytic reaction of the oxidation oxygen of an oxide of the by adsorbed carbon-containing compounds (for example, carbon oxide from air) followed by reducing the titanium dioxide surface [12]. During this process, as it follows from mass-spectrometric data, the magnitude of mass peak 44 increases by several times.

Measurements of the CO_2 photodesorption rate under illumination by monochromatic light at different wavelengths (via the use of interference filters) and light flux intensities allowed for the estimation of the quantum yield of the CO_2 photodesorption. The calculation results are given in figure 3 (curve 1). Curve 2 shows the diffusion reflection spectrum of the powder-like TiO_2 measured relative to magnesium oxide as the reference standard.

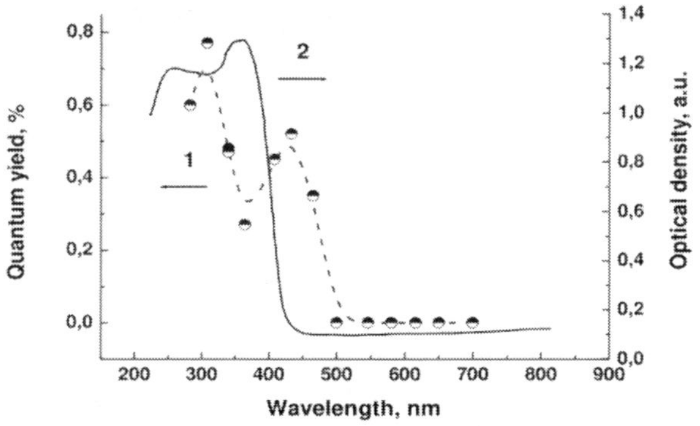

Figure 3. Spectral dependencies for the "crushing" TiO_2: (*1*) the quantum yield of CO_2 photodesorption and (*2*) the optical density relative to MgO.

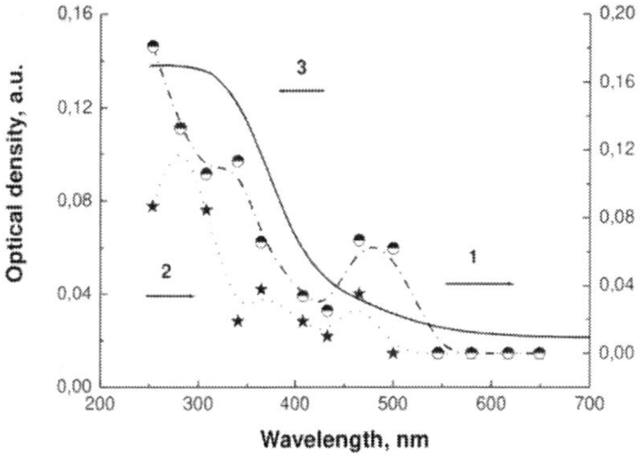

Figure 4. Spectral dependencies for "burning" TiO_2. 1 – photodesorption CO_2 after first illumination; 2 - photodesorption CO_2 after the long-time illumination through the UV filter; 3 – diffuse reflection spectrum for TiO_2 which is the same as located on the wall of a reactor.

Much more processes take place on the titania surface under visible light illumination than in the dark. In addition to the dark desorption of CO_2, its photodesorption was observed with the first-order kinetics similar to that of the dark desorption ($\tau_2 \approx 5$ min) by 20 times larger amount of desorbed CO_2 (Figure 2, curve 2). The photodesorption kinetics is presented as the decrease in the amount of CO_2 molecules adsorbed on the titania surface.

The measurement of the spectral dependencies of the CO_2 photodesorption quantum yield showed that the highest quantum yield was observed in the region of the TiO_2 intrinsic absorption (Figure 4, curves 1 and 2). A diffuse reflectance spectrum of the titania powder used in this study is shown for comparison (Figure 4, curve 3). If the time of the TiO_2 surface illumination was increased to several hours, the CO_2 photodesorption quantum yield decreased.

The CO_2 photodesorption in the region of the TiO_2 intrinsic absorption (including the 308 nm filter) is apparently caused by deep oxidation of adsorbed hydrocarbons with the oxygen from the air [13]. The processes predominating in the region of the surface absorption are related to the electron transfer from the oxide valence band to the surface and discharging of the surface compounds (e.g. CO_3^-, CO_2^-) with the electron transfer to the conduction band. In the former case, the CO_2 photodesorption results from discharging of the surface carbonate and carboxylate groups [14] with a mobile hole of the valence band. In the latter case, the CO_2 photodesorption to the gas phase is related to direct discharging of these compounds. The efficiency of the absorption with the formation of electron-hole pairs is substantially higher than that in the surface absorption band. The quantum yield of photodesorption in the region of the oxide intrinsic absorption is also much higher than in the surface absorption region (Figure 4).

It is known that the surface of the commercial (initial) TiO_2 material is reduced during heating in a vacuum, an inert gas, or hydrogen at high temperatures. This treatment can vary the color of the TiO_2 powder from slightly gray to dark blue depending on the degree of reduction of the sample [15, 16]. After high temperature oxygen–vacuum treatments of starting TiO_2, the degree of reduction of its surface by high temperature heating in a vacuum or illuminating decreases. The dark adsorption of oxygen on such a sample is not observed [17]. In light of the foregoing, it can be assumed that adsorbed water plays the key role in the reduction of the surface of hydrophilic titanium dioxide in a vacuum, inert gas, or hydrogen atmosphere, interacting with the lattice oxygen of the oxide.

Long-term illumination of the surface of TiO_2 particles in a water vapor atmosphere (in the reactor space) results in the reduction of the particle surface, which is detected both visually and by the diffuse reflectance spectrum of the sample in air. This is evidence for the effective photoreduction of titanium dioxide by water molecules, i.e., the oxidation of water by the titania lattice oxygen under the action of UV radiation. Under the given experimental conditions, as water, the hydrogen peroxide produced is frozen out into the trap with a cooling mixture. After thawing the trap to room temperature, oxygen is generated in the closed reactor volume; the amount of oxygen is measured with the Pirani gauge and the mass spectrometer. The oxygen produced is not adsorbed in the dark on the titanium dioxide surface. Thus, by the action of UV radiation and the subsequent storage, the titanium dioxide is reduced by the adsorbed water in the reactor space to produce an intermediate peroxide compound via the reaction:

$$(H_2O)_s + O^*_{Latt} \rightarrow (OH)_s + H\text{-}O_{Latt},$$

where O^*_{Latt} is the surface lattice oxygen activated by the free mobile "hole" of the valence band [13]. Some confirmation for this assumption is the STM data by Du et al. [18], who showed the formation of hydroxyl groups by the reaction of water with oxygen atoms formed during the adsorption of molecular oxygen on the (110) face of reduced TiO_2. The subsequent long-term holding of the sample in the reactor in the dark resulted in the release of molecular oxygen into the gas phase. For the TiO_2 sample prepared by grinding (dispersing) of the rutile crystal in air, the amount of oxygen formed after a 30-min exposure to light in water vapor and holding in air in the dark was greater than that corresponding to the monolayer coating of the surface of the titania sample.

Over a long period of illumination time by all radiation of OSL-1 illuminator in atmosphere of various gases (halogenorganic compounds, acetone) the titania surface is reducing as the result of lattice oxygen consumption for the carbon dioxide formation in the gas phase. At that, the formation of nonstoichiometric Ti^{3+} is detected by ESR method and by the absorption of light quanta with the wavelength more 400 nm and even by sight. Moreover, the darkness adsorption of oxygen is observed on reduced titania surface.

4. Conclusions

Water and carbon dioxide were the main products released from the surface of the titania powders. During exposure under UV light, which is absorbed by titania water interacts efficiency with the metal oxide lattice oxygen to subsequently release oxygen in darkness to gas phase, reduce titanium dioxide and form lattice oxygen vacancies.

References

[1] Chen X and Mao S S 2007 *Chem.Rev.* **107** 2891
[2] Zakharenko V S and Parmon V N 1999 *Russian J. Phys. Chem.* **73** 124
[3] Czochralski J 1918 *Zs. Phys. Chem.* **92** 219
[4] Khromova S A, Karasev V V, Onischuk A A, Glotov O G and Zarko V E 2005 *Nonequilibrium Processes*, ed G Roy, S Frolov and A Starik (Moskow: Torus Press)
[5] Lisachenko A A and Vilesov F I 1974 *Uspekhi. Fotoniki (Advances in Photonics)* **4** 18
[6] Volodin A M 1992 *Russian J. Chem. Phys.* **11** 1054
[7] Terenin A N 1967 *Fotonika molecul krasitelei (Photonics of dye molecules)* (Leningrad: Nauka)
[8] Basov L L, Kotelnikov V A and Lisachenko A A 1969 *Uspekhi. Fotoniki (Advances in Photonics)* **1** 76
[9] Kuznetsov V N and Lisachenko A A 1991 *Russian J. Phys. Chem.* **65** 1568
[10] Zakharenko V S and Moseichuk A N 2005 *Atmos. Oceanic Opt.* **18** 454
[11] Vanhieu N and Lichtman D 1981 *Surface Science.* **103** 535
[12] Zakharenko V S and Parmon V N 1996 *Kinetics Catal.* **37** 427

[13] Zakharenko V S 1997 *Catal. Today* **39** 243
[14] Jackson P and Parfitt G D 1972 *J. Chem. Soc. Faraday Trans. I.* **68** 896
[15] Mashchenko A I, Sharapov V M, Kazanskii V B and Kiselev V F 1965 *Teor. Eksp. Khim.* **1** 381
[16] Kazanskii V B, Nikisha V V and Shelimov B N 1969 *Dokl. Akad. Nauk* SSSR **188** 112
[17] Cherkashin A E, Volodin A M, Koshcheev S V and Zakharenko V S 1980 *Uspekhi. Fotoniki (Advances in Photonics)* **7** 86
[18] Du Y, Deskins N A, Zhang Z, Dohnalek Z, Dupuis M and Lyubinetsky I 2009 *Phys. Rev. Lett.* **102** 96102

IOP Conf. Series: Materials Science and Engineering **112** (2016) 012016 doi:10.1088/1757-899X/112/1/012016

Nanostructured microtubes based on TiO_2 doped by Zr and Hf oxides with the anatase structure

VV Zheleznov[1,*], EI Voit[1], YV Sushkov[1], SA Sarin[1], VG Kuryavyi[1], DP Opra[1], SV Gnedenkov[1], SL Sinebryukhov[1] and AA Sokolov[2]
[1]Institute of Chemistry Russian Academy of Sciences Far Eastern-Branch, 159, Prosp. 100-letya Vladivostoka, Vladivostok 690022, Russia
[2]Far Eastern Federal University, 8, Sukhanova Str., Vladivostok 690950, Russia

[*]E-mail: zheleznov_sergey@mail.ru

Abstract. The nanostructured microtubes based on TiO_2 have been prepared on the carbon fiber template using the sol-gel method. The microtubes consist of nanoparticles of metal oxides: TiO_2/ZrO_2 and TiO_2/HfO_2. The dependence of microtubes morphology and nanoparticles structure on the synthesis conditions has been studied using the methods of SEM, SAXS, and Raman spectroscopy. It has been demonstrated that at the stoichiometric ratio of up to 0.04 for Zr/Ti and up to 0.06 for Hf/Ti microtubes consist of uniform nanoparticles with the anatase structure. Along with further increase of the dopants content in the microtubes composition, nanoparticles acquire the core-shell structure. It has been suggested that nanoparticles have a core composed of the solid solutions $Ti_{1-x}Zr_xO_2$ or $Ti_{1-x}Hf_xO_2$ and a shell consisting of zirconium or hafnium titanate. The fabricated Zr- and Hf-doped TiO_2 materials were investigated in view of their possible use as anode materials for Li-ion batteries. Charge–discharge measurements showed that the doped samples manifested significantly higher reversibility in comparison with the undoped TiO_2. The method opens new prospects in synthesis of nanostructured materials for Li-ion batteries application.

1. Introduction

Recently times, a substantial attention has been paid to nanosized titanium and zirconium dioxides. The applications of these materials in the industry are diverse and include many branches: production of pigments for paint-and-varnish industry, manufacture of catalysts and their substrates, fabrication of glass, ceramics and refractory materials, electrical engineering, and photocatalysis [[1], [2], [3]]. The fabrication of modified forms of TiO_2 substantially extends the range of application of these materials [[4], [5]].

Up to present, many methods of synthesis of highly disperse materials based on titanium dioxides have been developed [[2], [6],[7],[8]]. One of the well-known methods of fabrication of nanosized oxides is the sol-gel technique including a number of consecutive stages: preparation of the precursor solution, sequential transformation of it into sol and, thereafter, into gel (as a result of hydrolysis and condensation processes), drying, and thermal treatment. The advantages of the sol-gel technique consist in the possibility to produce nanoparticles of multi-component oxides with high degree of chemical homogeneity and to change the nanoparticles phase composition, size, and shape. The main disadvantage of the synthesis method is related to the loss of control over the nanoparticles size and shape during gel drying and thermal treatment [[2]].

Content from this work may be used under the terms of the Creative Commons Attribution 3.0 licence. Any further distribution of this work must maintain attribution to the author(s) and the title of the work, journal citation and DOI.
Published under licence by IOP Publishing Ltd

In the present paper, the nanostructured oxide microtubes of the composition TiO_2/ZrO_2 and TiO_2/HfO_2 have been prepared by means of the of template sol-gel method. The relationship between the morphology, structure, and synthesis conditions of the materials was studied. The possibility of application of the syntehsized TiO_2-based microtubes as anodes for Li-batteries was discussed.

2. Synthesis of materials and methods of study

2.1. Synthesis

The scheme of synthesis is shown in Figure 1. The method of fabricating the microtubes is based on a slow hydrolysis in aqueous solutions of the inorganic precursors of titanium, zirconium, and hafnium accompanied with deposition of oxides on the template surface [[9]]. In the synthesis process, the following precursors were used: $TiCl_4$, $ZrOCl_2$, and $HfOCl_2$. Busofit T055 carbon fiber manufactured at the Belogorsk Chemical Fibers Plant (Belarus) was used as a template. The specific surface of the carbon fiber was determined using an ASAP 2020 V3.04 H (Micrometrics, USA) spectrometer from isotherms of low-molecular adsorption of nitrogen by the BET method: it was equal to 960 $m^2\ g^{-1}$. The porosity distribution was evaluated using the original density functional theory (DFT). The major part of the surface (780 $m^2 g^{-1}$) is formed d by micropores of a diameter of 2 nm. The template removal was carried out through slow annealing at a temperature of 500°C in air. The carbon template annealing was controlled using Raman spectroscopy. In the course of the extra thermal treatment at 550–850°C in air for 30 min, nanocrystallization of oxide tubes structural units was observed. Since the initial carbon fiber contains ~0.01 wt. % of silicon, some amount of SiO_2 is always present in the materials. The decrease of the SiO_2 concentration in the sample was achieved through autoclave pretreatment of the initial template by ammonium bifluoride. In the samples of as-prepared materials, the SiO_2 content was about 2-5 wt. %. The effect of the SiO_2 content in the above range on the microtube structure and morphology has not been identified. The proposed synthesis technique allows regulating over a broad range the phase composition, size, and shape of the nanostructural units by changing the synthesis conditions (solution pH, component concentrations, and annealing temperature).

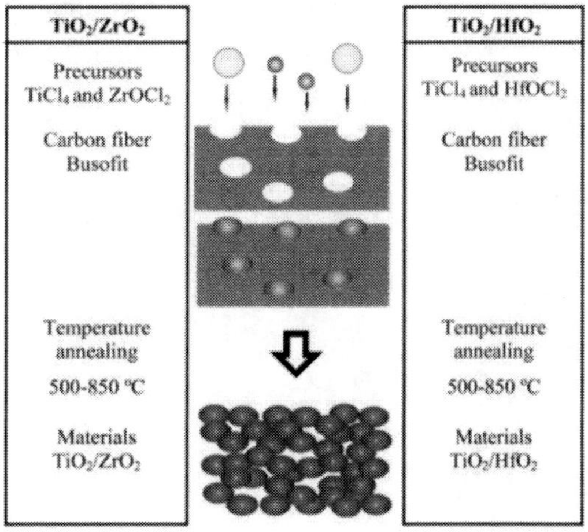

Figure 1. Synthesis scheme for nanosized oxide microtubes.

2.2. Methods of study

The content of titanium, zirconium, hafnium, and silicon oxides in the materials was determined by the energy-dispersive X-ray analysis on an EDX 700 (Shimadzu, Japan) spectrometer. The morphology was investigated by the method of scanning electron microscopy (SEM) using high-resolution S5500 and TEM-3000 (Hitachi, Japan) microscopes. The sizes of microtubes nanoparticles were determined by the method of small angel X-Ray scattering (SAXS). The measurements were carried out on an S3-MicroPIX (Hecus X-ray systems, Austria) spectrometer connected to a Xenocs GeniX generator (France) operated at 50 kV and 1 mA with using CuKα radiation (λ = 0.154 nm).

The Raman spectroscopy method was used to determine more accurately the structure and structural effects of separate components in the obtained material. The registration of the spectra was carried out on an RFS-100/S Fourier-Raman spectrometer (Bruker, Germany) with high-sensitive germanium detector. The Nd:YAG laser with the wavelength of 1064 nm was used as the excitation source.

The working electrode consisted of the active material, carbon black, and polyvinylidene fluoride binder at a weight ratio of 80 : 10 : 10. A lithium metal disc was applied as both counter and reference electrodes. 1 M solution of $LiBF_4$ salt in the mixture of propylene carbonate and dimethoxyethane at a volume ratio of 3:1 was used as an electrolyte. The design of the electrochemical half-cell is described in detail in [[10],[11],[12]]. The electrochemical performance was evaluated using a Solartron 1470E (UK) potentiostat/galvanostat. The parameters were measured by the galvanostatic discharge–charge at a current rate of 0.1*C* (C = 335 mAh g^{-1}) in the range from 3 to 1 V during 20 cycles. Within the scopes of the present work, due to half-cell applying, the lithiation process corresponds to discharge while the delithiation – to charge.

3. Discussion of results

3.1. Morphology

The morphology study by the SEM method demonstrated that all the fabricated materials consist of microtubes (Figure 2).

The diameters of oxide tubes vary from 1 to 5 μm, their lengths – from 10 to 100 μm, whereas the tube wall thicknesses change in the range of 100 to 500 nm. The microtubes are composed of nanoparticles of sizes from 15 to 50 nm.

The lengths, external and internal diameters, and porosity of microtubes as well as the nanoparticle sizes depend on the content of dopants and treatment temperature. Along with increase of the content of dopants and treatment temperature increase from 550C to 850°C, the decrease in the oxide microtubes lengths was observed and the degree of their destruction increased. One can see that TiO_2/ZrO_2 microtubes are characterized by a lesser diameter than TiO_2/HfO_2 ones.

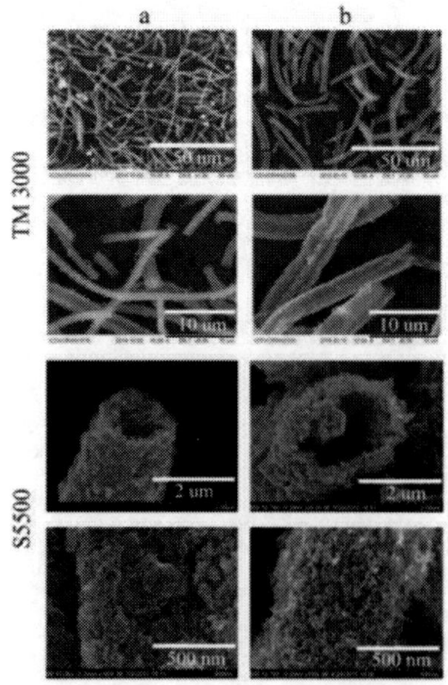

Figure 2. SEM micrographs of the TiO_2/ZrO_2 (a) and TiO_2/HfO_2 (b) obtained on S5500 and TM-3000 microscopes.

3.2. Nanoparticle sizes

To estimate the sizes of nanoparticles forming the tubes, the SAXS method was used. Based on the SAXS spectra, the pair distribution functions *P(r)* of the radii of gyration for scattering particles (R_g) were calculated, which allowed revealing the regularities in the arrangement of neighboring atoms, molecules, or atomic groups at a distance of *r*. The program GNOM was used [[13]]. The obtained *P(r)* and R_g for the materials are presented in Figures 3, 4.

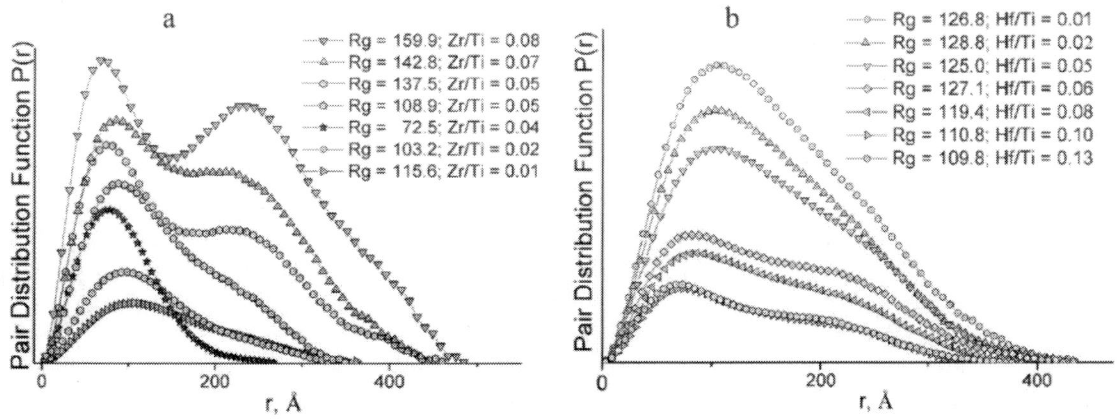

Figure 3. Dependence of pair distribution functions on the particles radius.

For Zr-doped microtubes, the type of *P(r)* curve corresponds to scattering from spheroids at Zr/Ti ratio of lower than 0.04 (doping level) (Figure 3a). In this concentration range, *P(r)* has one maximum. The dependence of R_g for nanoparticles on the Zr/Ti ratio shows that the average sizes of nanoparticles decrease nonlinearly (Figure 4). The decrease in average sizes of nanoparticles is probably related to the increase in heterogeneity of the nanoparticles' crystalline lattice and growing microstress in them due to dopant inclusion [[14]]. The calculated maximal diameters of the scattering particles vary from 25 to 35 nm (Figure 3a).

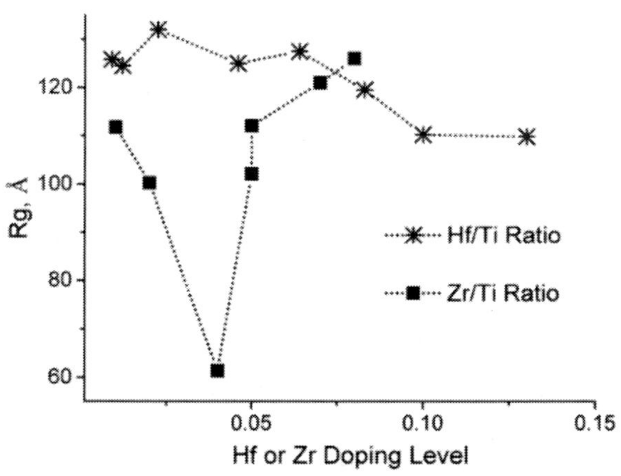

Figure 4. Dependence of the average inertia radius of the nanoparticles on the doping level.

If the Zr/Ti ratio is higher than 0.04, the second maximum increasing along with the increase of the ZrO_2 content emerges on the *P(r)* curve while R_g of nanoparticles increases. As was found earlier, the binary form of the *P(r)* curve corresponded to the formation of nanoparticles with the core-shell structure [[13], [15]]; in this case, the nanoparticle comprises a shelled nanocrystallite. The calculated maximal diameter of scattering particles varies in the range from 35 to 50 nm depending on the doping level (Figure 3b).

The character of changes in the sizes of nanoparticles forming the basis of Hf-doped microtubes differs from that of Zr-doped ones. At the annealing temperatures of 650°C and the doping level of up to 0.06, the average sizes of particles do not virtually change being equal to about 25 nm (Figure 4). Further increase of the Hf/Ti ratio results in the decrease in the nanoparticles sizes and the emergence of the second maximum on the *P(r)* curve (Figure 3b).

3.3. Structure of nanoparticles

To determine the structure of nanoparticles forming the microtubes and reveal the role of dopants in the synthesized materials, their Raman spectra were studied. Raman spectroscopy is known to enable one to identify changes in the matter crystalline state even if the particle sizes are small.

As was established earlier that in the Raman spectrum of the polycrystalline TiO_2 with the structure of anatase at the room temperature in accordance with the data of the factor-group analysis $\Gamma(D_{4h}^{19}) = 1A_{1g}(\text{Raman})+1A_{2u}(\text{IR})+2B_{1g}(\text{Raman})+1B_{2u}+3E_g(\text{Raman})+2E_u(\text{IR})$, six modes are active: $\nu_1+\nu_3=515$ cm^{-1} (A_{1g}, B_{1g}), $\nu_4=396$ cm^{-1} (B_{1g}), $\nu_6=638$ cm^{-1} (E_g), $\nu_7=196$ cm^{-1} (E_g), $\nu_8=143$ cm^{-1} (E_g) [[16]].

In the Raman spectra of the materials under study, the bands conforming to the structure of anatase appear at a Zr/Ti ratio of up to 0.04 and a Hf/Ti ratio of up to 0.06 (Figure 5). Figure 6 shows the dependence of the bands maxima ν_4, ν_6, ν_8 in the spectra on the Zr/Ti ratio (0 – 0.04) and Hf/Ti ratio (0 – 0.06). Along with the increase in the content of the dopant, a shift of the bands ν_4, ν_6 and ν_8 towards lower frequencies is observed.

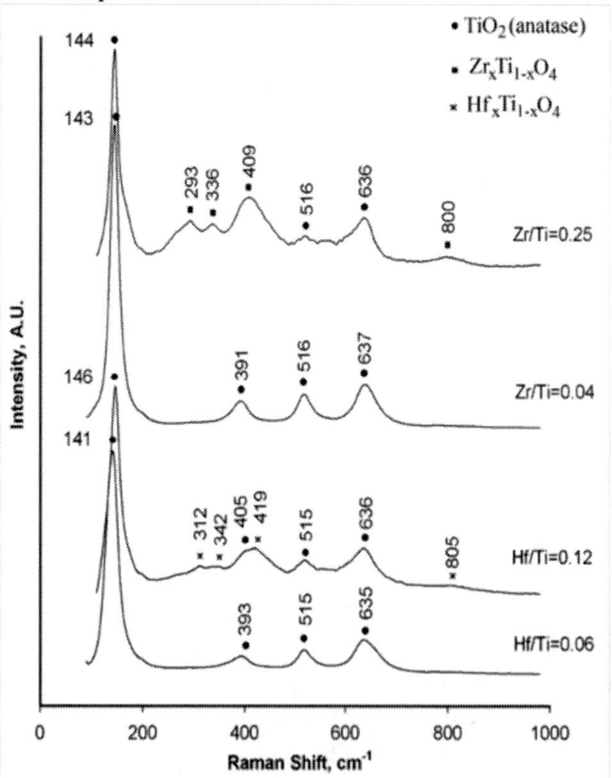

Figure 5. Raman spectra of the TiO_2/ZrO_2 (a) and TiO_2/HfO_2 (b) annealed at 650°C.

A number of studies of the reasons of the bands' location change in the Raman spectra of nanosized materials were performed [[17], [18], [19]]. The nanocrystallite size change was identified as the main reason of the above processes. As was shown on the example of nanodispersed TiO_2 (anatase), low-frequency shift of bands in the spectrum was observed along with the increase of the sizes of nanocrystallites [[17]]. According to the data of SAXS (Figure 4) and Raman spectroscopy, the above-mentioned dependence is not valid for materials examined in the present paper.

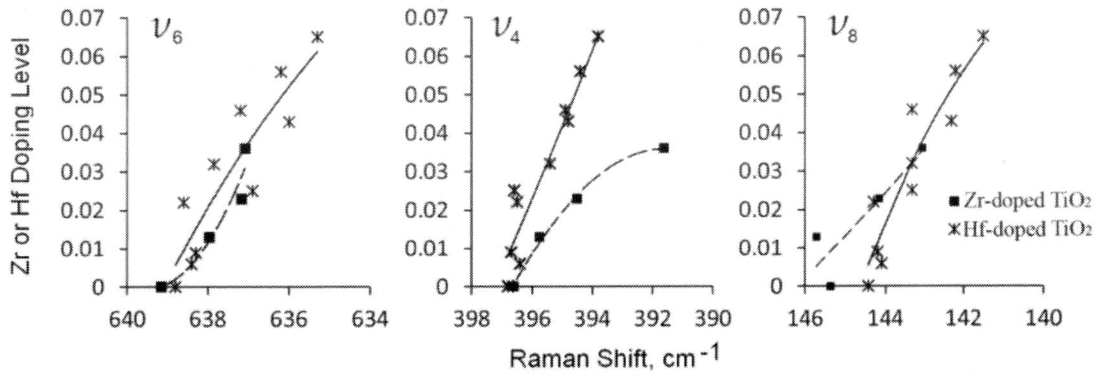

Figure 6. Dependence of bands maxima positions in the Raman spectra on the doping level.

It is generally accepted that the other cause of the low-frequency shift of the ν_4, ν_6, ν_8 bands in the spectra is related to the presence of structural defects [[18], [19]], which can include incorporation of the dopant atoms into the material structure that results in changes of lattice parameters. In the spectra of Zr-doped materials, the more noticeable shift of the ν_4 band (as compared to that of Hf-doped materials) was observed (Figure 6). Taking into account similarity of the Zr^{4+} and Hf^{4+} ionic radii, the different behavior of the bands in the spectra can be related to the difference of their atomic masses and the emergence of the relativistic effect for hafnium. The dependence of the bands' locations on the atomic masses of cations was demonstrated on the example of spectra of ZrO_2 and HfO_2 crystals [[20]].

Thus, according to the data of Raman spectroscopy and SAXS, the materials within the range of 0-0.04 for Zr/Ti ratio and 0-0.06 for Hf/Ti ratio consist of the $Ti_{1-x}Zr_xO_2$ or $Ti_{1-x}Hf_xO_2$ nanoparticles and have the structure of anatase. Further increase of the dopant contents in the studied materials results in changes in the Raman spectra. In addition to bands of anatase, the bands corresponding to the phase of zirconium or hafnium titanate emerge (Figure 5) [[21]]. Therefore, one can assume the formation, along with the increase of the doping level (Zr/Ti>0.04 and Hf/Ti>0.06), of nanoparticles having a core (nanocrystallite) composed of solid solutions of $Ti_{1-x}Zr_xO_2$ or $Ti_{1-x}Hf_xO_2$ with the anatase structure and a shell consisted of both anatase and zirconium or hafnium titanate. The emergence of the second maximum on the *P(r)* curve (Figure 3b) and changes in the Raman spectra (Figure 5) are in agreement with this assumption. According to the SAXS data, the shell growth takes place along with increase in the doping level. i.e., in changes of the maxima ratio on the *P(r)* curve changes. These conclusions agree well with different temperatures of crystallization of tentative structural components, since crystallization of the undoped TiO_2 can occur in the range 250-500°C while crystallization of $ZrTiO_4$ – in the range 600–1300°C depending on preparation conditions [[14], [22]].

3.4. Electrochemical investigations

To evaluate the electrochemical performance of nanostructured Zr- and Hf-doped anatase TiO_2 materials as anodes for Li-ion batteries, preliminary galvanostatic discharge–charge measurements were carried out. Figure 1 shows the results of 20-fold cycling for undoped, Zr-doped (Zr/Ti = 0.01; 0.02; 0.03), and Hf-doped (Hf/Ti = 0.01; 0.02; 0.04; 0.06) anatase TiO_2 electrodes at a 0.1*C*-rate.

The reversible capacities of nanostructured $Ti_{0.99}Zr_{0.01}O_2$, $Ti_{0.98}Zr_{0.02}O_2$, and $Ti_{0.97}Zr_{0.03}O_2$ are 80, 95, and 140 mAh g^{-1}, respectively, demonstrating better capacity retention of the doped electrodes in comparison with undoped TiO_2 that yielded only 65 mAh g^{-1}. It seems that enhanced capacities of doped samples are mainly related to facilitating of Li^+ ions diffusion due to the difference in the values of ionic radii of Zr^{4+} and Ti^{4+} ions leading to changes in lattice parameters of anatase TiO_2. In addition, according to electrochemical impedance spectroscopy, the charge redistribution in Zr-doped TiO_2 lattice results in the conductivity increase by two orders of magnitude from $4.4 \cdot 10^{-11}$ S cm^{-1} (undoped) to $2.2 \cdot 10^{-9}$ S cm^{-1} (Zr/Ti = 0.03) that promotes the electron transport [[23]].

The reversible capacities of nanostructured $Ti_{0.99}Hf_{0.01}O_2$, $Ti_{0.98}Hf_{0.02}O_2$, $Ti_{0.96}Hf_{0.04}O_2$, and $Ti_{0.94}Hf_{0.06}O_2$ are found to be equal to 50, 45, 85, and 55 mAh g^{-1}, respectively, which are some lower in comparison with Zr-doped TiO_2. Furthermore, the capacity dependence on the Hf concentration is not linear in contrast to Zr. The best result was observed for $Ti_{0.96}Hf_{0.04}O_2$. The electrochemical behavior of the doped TiO_2 is directly depends on the particle size distribution, parameters of the crystal lattice, the number of oxygen vacancies, etc. So we suggest that the inferior capacity enhancement for Hf-doped TiO_2 (as compared to Zr-doped one) is the effect of larger nanoparticles (Figure 7).

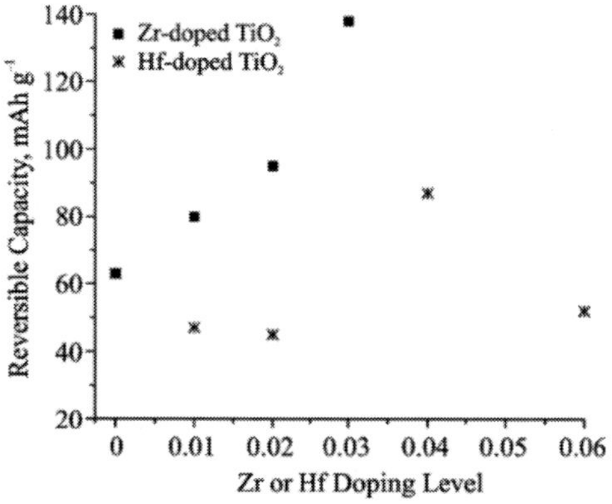

Figure 7. Dependence of reversible capacity after the 20-fold cycling on the doping level.

Note that $ZrTiO_4$ and $HfTiO_4$) are electric insulators [[24]]. Hence the formation of $Zr_{1-x}Ti_xO_4$ (or $Hf_{1-x}Ti_xO_4$) solid solution deteriorates the conductivity of the doped material leading to inferior electrochemical parameters of its application as Li-ion battery anode.

4. Conclusions

New TiO_2/ZrO_2 and TiO_2/HfO_2 materials of varying compositions have been fabricated by the template sol-gel method. The oxides consist of nanostructured microsized tubes formed by nanoparticles of sizes varying from 15 to 50 nm. TiO_2-based Hf-doped microtubes have larger diameters and consist of larger nanoparticles as compared to Zr-doped ones.

In the ranges of 0 to 0.04 for the Zr/Ti ratio and 0 to 0.06 for the Hf/Ti ratio, the materials are formed by nanoparticles composed of $Ti_{1-x}Zr_xO_2$ or $Ti_{1-x}Hf_xO_2$ and have the structure of anatase.

The increase in the material doping level above 0.04 (Zr/Ti) and 0.06 (Hf/Ti) yields changes in the structure of nanoparticles. Probably, the nanoparticles have a core (nanocrystallite) formed by solid solutions of $Ti_{1-x}Zr_xO_2$ or $Ti_{1-x}Hf_xO_2$ with the anatase structure and a shell consisted of both anatase and $Zr_xTi_{1-x}O_4$ or $Hf_xTi_{1-x}O_4$ at the above-mentioned content of dopants.

The promising electrochemical properties of both as-prepared Zr- and Hf-doped anatase TiO_2 samples as anodes for Li-ion batteries were identified. A good reversibility of the Li^+ intercalation into and deintercalation from the nanostructured Zr- and Hf-doped TiO_2 results from lattice parameters changes and conductivity increase and, thus, provides the improved dynamics for both electron and ion transport. To sum up, the method of synthesis is efficient in terms of the electrochemical performance of the battery systems.

Acknowledgements

The synthesis and study of the materials' morphology and structure were carried out under the program of fundamental scientific research of the Russian Academy of Sciences (project No. 0265-2014-0001, the Federal Agency of Scientific Organizations).

The electrochemical studies were carried out with the support of the Russian Science Foundation (project No. 14-33-00009) and the Government of the Russian Federation (the Federal Agency of Scientific Organizations).

References

[1] Kubacka A, Fernández-García M and Colón G 2012 *Chem. Rev.* **112** 1555–614

[2] Wang X, Li Z, Shi J and Yu Y 2014 *Chem. Rev.* **114** 9346-84

[3] Bai Y, Mora-Seró I, De Angelis F, Bisquert J and Wang P 2014 *Chem. Rev.* **114** 10095–130

[4] Reddy B M and Khan A 2005 *Catalysis Reviews* **47** 257–96

[5] Ma Y, Wang X, Jia Y, Chen X Han H and Li C 2014 *Chem. Rev.* **114** 9987–10043

[6] Macwan D P, Dave P N 2011 *J. Mater. Sci.* **46** 3669–86

[7] Cargnello M, Gordon T R and Murray C B 2014 *Chem. Rev.* **114** 9319–45

[8] Schneider J, Matsuoka M, Takeuchi M, Zhang J, Horiuchi Y, Anpo M and Bahnemann DW. 2014 *Chem. Rev.* **114** 9919–86

[9] Betenekov N D, Vasilevskij V A, Egorov Yu V and Nedobukh T A 1982 Radiokhimiya **24** 419–24 (Russ)

[10] Gnedenkov S V, Opra D P, Sinebryukhov S L, Kuryavyi V G, Ustinov A Yu and Sergienko V I 2015 *J. Alloy. Compd.* **621** 364

[11] Sinebryukhov S L. Opra D P, Gnedenkov S V, Minaev A N, Sokolov A A, Kuryavyi V G and Zheleznov V V 2015 *Int. Ocean and Polar Eng.* **1** 621

[12] Gnedenkov S V, Opra D P, Kuryavyi V G, Sinebryukhov S L, Ustinov A Yu and Sergienko V I 2015 *Nanotechnol.* **10** 353 (Russ)

[13] Semenyuk A V and Svergun D I 1991 *J. Appl. Cryst.* **24** 5 537–40

[14] Chang S M and Doong R M 2006 *J. Phys. Chem. B* **110** 20808–14

[15] Balmer J A, Mykhaylyk O O, Schmid A, Armes S P, Fairclough J P A and Ryan A J |2011 *Langmuir* **27** 8075–89

[16] Ohsaka T, Izumi F and Fujiki Y J 1978 *J. Raman spectrosc.* **7** 321–24

[17] Wang D, Chen B and Zhao J 2007 *J. Appl. Phys.* 101 113501-113501-5

[18] Gupta S K, Desai R, Jha P K, Sahoob S and Kirinc D 2010 *J. Raman Spectrosc.* **41** 350–55

[19] Parker J C and Siegel R W 1990 *J. Appl. Phys.* **57** 943

[20] Haines J and Le´ger J M, Hull S, Petitet J P, Pereira A S, Perottoni C A and da Jornada J A H 1997 *J. Am. Ceram. Soc.* **80** 1910–14

[21] Krebs M A and Condrate R A 1988 *J. Mater. Sci. Lett.* **7** 1327–30

[22] Dos Santos V, Zeni M, Hohemberger J M, and Bergmann C P 2010 *Rev. Adv. Mater. Sci.* **24** 44

[23] Opra D P, Gnedenkov S V, Sokolov A A, Zheleznov V V, Voit E I, Sushkov Yu V and Sinebryukhov S L 2015 *Scripta Mater.* **107** 136

[24] Vittayakorn N 2006 *J. Ceram. Process. Res.* **7** 288–91

IOP Conf. Series: Materials Science and Engineering **112** (2016) 012017 doi:10.1088/1757-899X/112/1/012017

Investigation of supramolecular structure of the rare and rare-earth elements nanoparticles carrier when modified using microwave irradiation

I A Lysak[1,2*], G V Lysak[1], T D Malinovskaya[2] and L N Skvortsova[3]
[1]EGID department, National Research Tomsk Polytechnic University, 30, Lenina ave., Tomsk, 634050, Russia
[2]Siberian Physical-Technical Institute, National Research Tomsk State University, 1, Novosobornaya sq., Tomsk, 634050, Russia
[3]AC department, National Research Tomsk State University, 36, Lenina ave., Tomsk, 634050, Russia

[*]E-mail: lysakilya@gmail.com

Abstract. In this paper, we present an IR and XRD study of the polypropylene fibrous nanoparticles carrier when its surface modified with rare and rare-earth elements nanoparticles using microwave irradiation, by the example of SnO_2/TiO_2 hetero-nanoparticles. The paper shows that the smectic mesomorph structure of the non-modified polypropylene fibrous carrier transforms into the monoclinic α-crystalline due to microwave irradiation. At the same time, the carrier material remains stereoregular and keeps its helical structure.

1. Introduction

Nowadays, polymer melt-blown fibres are widely used for the manufacture of textile goods, insulation materials and fibre composites, absorbents of oil, petroleum and heavy metals, filtering materials for water and air treatment or as nanoparticles carriers [1-3]. Spheres of its application are determined by physical properties, chemical constitution of polymers, structure and redox properties of surface. Among these materials, the polypropylene (PP) melt-blown webs have substantial advantages [4]. An extensive base of raw materials, constant developing of new methods of modification, and a valuable set of physical properties create supportive environment for the manufacture development and application of products produced on the basis of these webs. Application of these materials, as the filtering and absorbing elements in the water purifying systems, is especially relevant [5-7].

Nanosized particles attached to the polymer fibre surface can be used for the transformation of organic water pollutants, such as phenols, aromatics, surface-active reagents, and nitro compounds, into the compounds safe for humans [3, 8 and 9]. At recent time, a lot of success is achieved in the developing the filtering materials – the polypropylene fibres, whose surface is modified with active nanoparticles, such as the systems named as "nanoparticles (TiO_2, SnO_2, TiO_2/SnO_2) – polypropylene fibrous carrier".

Attached to the carrier surface hetero-nanoparticles are most advantageous in case of developing effective photocatalytic systems on account of generating more free radicals due to spatially separation of electron-hole pairs [10, 11]. Structure and properties of modified materials and, consequently, the functional features and application sphere largely depend on methods of their obtaining. Thus, arises a

Content from this work may be used under the terms of the Creative Commons Attribution 3.0 licence. Any further distribution of this work must maintain attribution to the author(s) and the title of the work, journal citation and DOI.
Published under licence by IOP Publishing Ltd

need of thorough study all the materials which were obtained using new methods. This work is aimed at using the IR spectroscopy technique to investigate the changes in the chemical structure of the polypropylene melt-blown web, when modified with SnO_2/TiO_2 hetero-nanoparticles.

2. Materials and methods

The PP melt-blown webs as a carrier for SnO_2/TiO_2 nanoparticles were obtained by the direct aerodynamic formation, a kind of the melt-blowing process, from melting of isotactic polypropylene (iPP) [12]. Such raw materials as the commodity polypropylene 21060-16 and 21080-16 grades issued in accordance with TU 2211-016-05796653-95 were used in the research. The selection of samples was provided according to the GOST 10213.0-2002. "Staple fibre and tow chemical. Acceptance rules and the method of sampling." The phase composition, crystallite size, and internal elastic strain ($\Delta d/d$) of the resulting materials were determined by X-ray diffraction (XRD) on an XRD-6000 diffractometer with CuK_α radiation. The surface microstructure of the modified fibers was examined by transmission electron microscopy (TEM) on a JEOL JEM_100CXII. Formation the SnO_2/TiO_2 hetero-nanoparticles – carrier systems was provided by a mutual precipitation of the stannic and the titan chlorides hydrolysis products to the polypropylene carrier surface.

The starting material for the synthesis of SnO_2/TiO_2 photocatalytic nanoparticles was water solutions of the stannic chloride (II) and the titan chloride (IV) with the chemical purity. 1 gram of the carrier was placed in a container with the prepared stannic chloride (II) and titan chloride (IV) solution, kept for 15 minutes at room temperature, then withdrawn from the solution, and subjected to the microwave radiation to produce the SnO_2/TiO_2 nanocrystals attached to the fibre surface. The irradiation was conducted in a laboratory microwave oven with an output power of 650 W and a frequency of 2.45 GHz. The working chamber was a rectangular resonator with a standing wave and a rotating sample holder continuously rotating by 360°. The exposure time to the microwave radiation was 3 minutes.

Before and after surface modification with metal nanoparticles, the samples of the carrier were submitted to an FTIR analysis to find out the transformation in their chemical structures [13]. All spectra were recorded with a Bruker Tensor 27 FT-IR spectrometer using the standard technique in the wave number range of 4200 - 400 cm^{-1}. The samples for IR spectroscopy were completely dried as a result of 21 day keeping in an air-tight glass weighing capsule in the presence of P_2O_5.

3. Results and discussion

Assuming that the spectroscopic properties of the nanoparticle carrier (NC) do not change in the process of the nanoparticles attachment, it is worthwhile to begin analysing the IR spectrograms of NC before modification and to check up the changes afterward. The spectrograms of a non-modified carrier in the wave number range of 1200–800 cm-1 show several strong and medium absorption bands that correspond to the stretching vibrations of CH group, symmetric bending vibration of CH2 group (ν=1435 cm-1), antisymmetric bending vibration of CH3 group (ν=1454 cm-1 and 1460 cm-1), and symmetric bending vibration of CH3 group (ν=1380 cm-1). The IR spectra of the non-modified NC also show the clearly pronounced absorption bands 1366, 1328, 1302, 1258, 1224, 1103, 1045, 998, 975, 899, 842, and 810 cm-1 that give evidence of isotactic structure of the carrier material. Optical micrograph of the fibrous carrier is presented in figure 1.

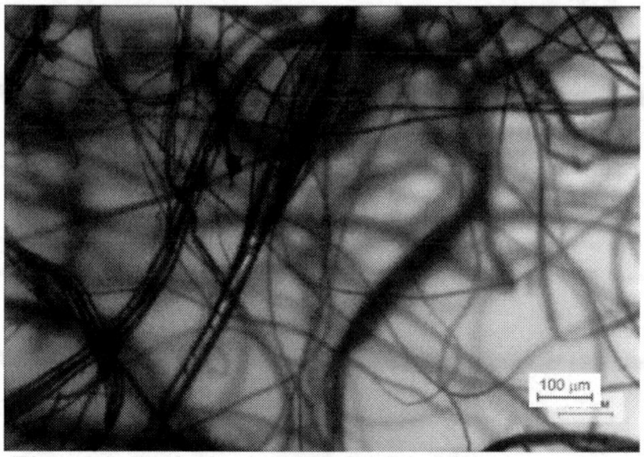

Figure 1. Optical micrograph of the fibrous carrier

The appearance of the 975 cm^{-1} band is caused by an interaction of the methyl group rocking vibration and the stretching vibration of the $-CH-CH_2-$ group. The band at 975 cm^{-1} is slightly related to the vibrations of the CH_2 group and $-CH-CH_2-$ groups adjacent to a molecular chain and does not depend on the conformation of macromolecules. The band at 998 cm^{-1} is caused by an interaction of the methylene group rocking vibration and stretching vibration of the $-CH-CH_3$ group. The power of this interaction depends on the conformation of macromolecules and, consequently, on the isotacticity degree. The degree of isotacticity is considered to mean the ratio of the intensities of the absorption bands 998 and 975 cm^{-1} in the IR spectra. The ratio of intensities of these bands is 0.86. According to [14], the band at 840 cm^{-1}, typical of the polypropylene spectra, defines the number of molecular chains which consist of the sequenced trans-gosh-conformers $(T-G)_n$ n>10. The band at 975 cm^{-1} defines a content of conformers $(T-G)_n$ n>4 in the amorphous areas of a high-molecular crystallising polymer. The presence of segments with helical conformation is proved by the band at 998 cm^{-1}. The coiled conformation, which can be identified by the band at 1155 cm^{-1}, is absent. The ratio of intensities of the absorption bands 975 and 840 cm^{-1} shows that the structure of NC is represented mainly by the long segments (n>10) in a helical trans-gosh-conformation, which lead to the formation of a long-length lamellar structure [15]. Hence, we can conclude that before the modification, the material of the nanoparticle carrier is isotactic polypropylene with high degree of isotacticity.

In addition, the IR spectrogram contains the following bands: a weak band at 1720 cm^{-1} belonging to the stretch vibration of C=O group that gives evidence of the oxygen-containing groups formation and absorption bands in the range of 2500–3700 cm^{-1} with the maximum of absorption at 3400–3460 cm^{-1}, which correspond to the stretch vibrations of OH groups in accordance with the results published in literature [16, 17]. See table 1. The appearance of these bands can be explained by the fact that when the melt of PP is blown, free radicals are formed due to the presence of the methylene groups in the PP macromolecular chains with low bond energy containing hydrogen at the tertiary carbon atom. In the presence of the air oxygen, thermal destruction goes by the radical-chain mechanism with the formation of the peroxide radicals and peroxides. It leads to the formation of the structure defects in the material including the unsaturated and oxygen-containing functional groups, not typical of initial polymer, such as hidroperoxides, carbonyl, ether, etc. [18]

Table 1. Absorption bands for initial polypropylene melt-blown web.

Wavenumber (sm^{-1})	Intensity	Absorbing group and type of vibration
809	weak	$\gamma_r (CH_2)$
842	average	$\gamma_r (CH_3)$
975	average	$\gamma_r (CH_3); \nu (C-C)-$

998	average	γ_r (CH_3) ; ν (C– CH_3)
1155	average	ν (C– CH_3)
1168	average	γ_r (CH_3); ν (C – C)
1380	strong	δ_s (CH_3)
1435	average	δ_s (CH_2)
1460	average	δ_a (CH_3)
1645	weak	ν (C = O)
1720	weak	ν (C = O)
2720	weak, narrow	ν (OH)
2838	average	ν_s (CH_2)
2879	average	ν_a (CH_2)
2921	very strong	ν (CH)
2950	very strong	ν_a (CH_3) symmetric to H – C– CH_3 plane
2959	very strong	ν_a (CH_3), antisymmetric to H – C– CH_3 plane
3400	weak, broad	ν (OH)

The presence of oxygen-containing groups on the NC surface makes it possible to carry out its modification by the method that includes the smearing of the carrier surface with an active component obtained as a result of the tin-salts hydrolysis. The active component is stabilized on the carrier surface due to the bonding of the metal cations by carboxylic groups coupled with the formation of hydrogen bonds between the carboxylic group hydroxyl hydrogen and the hydrolysed tin salt oxygen. These processes take place only at low pH. Formation and immobilisation of the tin dioxide nanoparticles on the NC surface occur due to the microwave irradiation. [19]. The results of an XRD study indicate that the supramolecular structure of the non-modified polypropylene fibrous carrier is represented by 33% of the smectic mesomorph phase, which can be identified by the presence of two wide peaks, situated at the $2\theta = 14.8°(5.99)$ и $21.4°(4.19)$ (figure 2.1). Furthermore, the supramolecular structure of the modified NC is represented by 42% of the monoclinic α-crystalline phase (figure 2.2).

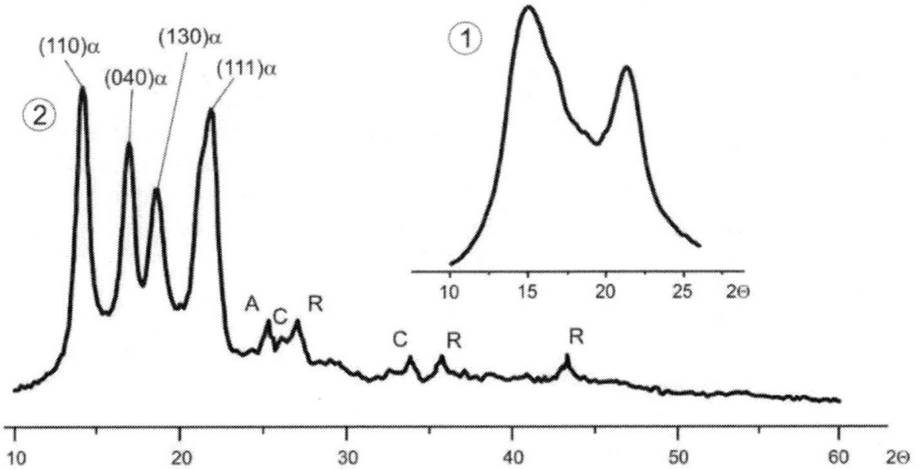

Figure 2. XRD spectra of polypropylene fibrous carrier before and after modification: 1 - non-modified NC; 2 - modified with SnO_2/TiO_2 nanoparticles NC

After modification, the IR spectra of NC samples demonstrate new absorption bands in the wave number range below 750 cm^{-1}. For the carrier material, weak bands located in this range are related to the skeleton bending vibration and internal chain rotation. In addition, the authors of [20, 21] identified the absorption bands in this range as related to the SnO_2 nanoparticles. With respect to these works, we can recognise the bands at the 427 cm^{-1}, 530 cm^{-1}, and 679 cm^{-1} as related to the metal – oxygen stretch vibrations (vO–Sn–O), which can be identified as the Sn–O bonds vibrations with various coordination saturability of tin. Moreover, new absorption bands appeared at the 417 cm^{-1} and 519 cm^{-1}, which correspond to the titan dioxide in the rutile structure. The TEM micrograph (figure 3) shows microfibers with (TiO_2/SnO_2) nanocrystals attached to their surface.

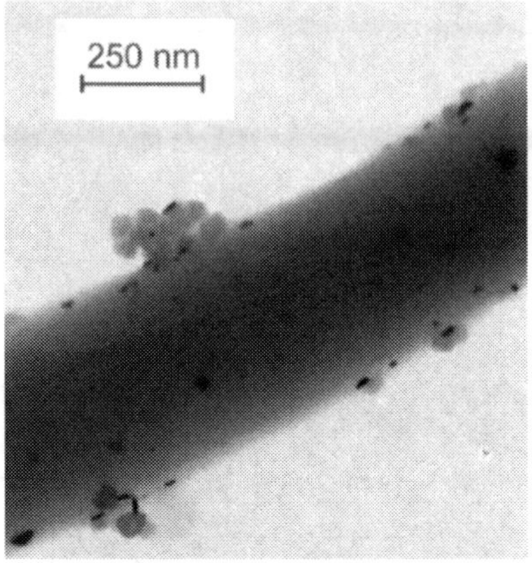

Figure 3. TEM micrograph of TiO_2/SnO_2 hetero-nanoparticles attached to the surface of a polypropylene fiber.

The spectra of the modified NC exhibit the clearly pronounced absorption bands 1366, 1328, 1302, 1258, 1224, 1103, 1045, 998, 975, 899, 842, and 810 cm^{-1} that indicate that the structure of the carrier material is isotactic. The ratio of intensities of the absorption bands 975 and 840 cm^{-1} remains unchanged. Consequently, after modification, the structure of NC is still represented mainly by long segments (n>10) in a helical trans-gosh-conformation. This is indicative of the fact that the steric configuration is largely unchanged. Furthermore, modification of the carrier surface does not lead to the destruction of chemical bonds. The ratio of intensities of the absorption bands 975 and 998 cm^{-1} in the spectra of the modified NC remained the same – 0.86. Thus, superficial modification of NC with the tin dioxide nanoparticles does not change its chemical structure.

4. Summary

Thus, this work has demonstrated the changes taken place as a result of the polypropylene melt-blown web superficial modification with the SnO_2/TiO_2 nanoparticles using microwave irradiation. These changes are manifested in an appearance of new absorption bands in the IR spectra. These bands do not overlap with the bands in the IR spectra of the unmodified carrier, and despite their weak intensity, can be surely identified as the bands of the SnO_2/TiO_2 nanoparticles. Having analysed the XRD analysis data, it has been concluded that the mesomorph structure of the carrier material transforms into the monoclinic α-crystalline due to microwave irradiation. Moreover, we have found out that the carrier material keeps its helical-chain conformation and remains stereoregular after modification. Therefore, the material of the carrier is not destructed under modification possessing the initial set of properties and attains new functional properties due to the presence of active nanoparticles attached to its surface.

Our findings lead us to the conclusion that a new opportunity appears of developing a new class of effective and safety filtering materials for the fresh and waste water purification using the polypropylene melt-blown webs modified with rare and rare-earth elements nanoparticles.

Acknowledgement

The work was financially supported by project RFMEFI57814X0026 of the Ministry of education and science of Russia.

References

[1] Pinchuk L S, Goldade V A, Makarevich A V and Kestelman V N 2002 *Melt-Blowing: equipment, technology, and polymer fibrous materials* (Berlin: Heidelberg, New York: Springer)
[2] Mohapatra H S, Chatterjee A and Kumar P 2013 *J. of Adv. and Nano Technol.* **1** 2347
[3] Perepelkin K E 2009 *Fibre Chem.*, **41(1)** 9
[4] 2003 *Handbook of Polypropylene and Polypropylene Composites, Revised and Expanded* Ed. Karian H G (Cleveland: CRC Press)
[5] Shi L, Kang W M, Zhuang X P and Cheng B W 2011 *Adv. Mater. Res.* **332, 334** 1287
[6] Kravtsov A G, Pinchuk L S and Goldade V A 2000 *Chem. Fibers* [in Russian] **6** 42
[7] Zhukovsky M S, Lysak I A, Lysak G V, Vazhenin S V, Malinovskaya T D, Beznosjuk S A 2011 *Russ. Phys. J.* **54(7)** 739
[8] Baoli T, Yu Q, Binbin H, Jiaruo S and Zuliang D 2012 *J. Mater. Res.* **27, 18** 2408
[9] Lysak G V, Lysak I A, Shabalina A V, Izaak T I, Malinovskaya T D 2010 *Rus. J. Appl. Chem.* **83(12)** 2193
[10] Lysak I A, Lysak G V, Malinovskaya T D, Tchaikovskaya O N and Artyushin V R 2015 *Adv. Mater. Res.* **1085** 107
[11] Ungur G and Hrůza J 2014 *Adv. Mat. Lett.* **5(8)** 422
[12] Lysak G V, Lysak I A and Malinovskaya T D 2014 *Polzunovsky Almanah* [in Russian] **1** 30
[13] 2012 *Advanced Aspects of Spectroscopy*, ed M A Farrukh (Rijeka: INTECH Open Access Publisher)
[14] Barbara H S 2004 *Infrared Spectroscopy: Fundamentals and Applications* (New York: John Wiley & Sons Ltd.)
[15] 1999 *Spectroscopy of Polymers*, 2nd edition, ed Koenig J L (New York: Elsevier Science Inc.)
[16] Novoselova L Yu and Bordunov V V 2002 *Plastics* [in Russian] **6** 46
[17] Gorohovatskii Yu A, Temnov D E, Chistyakova O V, Tazenkov B A, Aniskina L B, Victorovich A S and Kojevnikova N O 2005 *Proc. of RSPU.* Herzen: Scientific Journal: Natural and exact sciences [in Russian] **5(13)** 91
[18] Fridman M L 1982 *Technology for Processing Crystalline Polyolefins* (Moscow: Chemistry)
[19] Lysak G V, Lysak I A, Malinovskaya T D and Volokityn G G 2010 *Inorg. Mater.* **46(2)** 183
[20] Khan A F, Mehmood M, Aslam M and Ashraf M 2010 *Appl. Surf. Sci.* **256** 2252
[21] Chacko S, Philip N S, Gopchandran K G, Koshy P and Vaidyan V K 2008 *Appl. Surf. Sci.* **254** 2179

IOP Conf. Series: Materials Science and Engineering **112** (2016) 012018 doi:10.1088/1757-899X/112/1/012018

LiOH – H_2O_2 – H_2O trinary system study for the selection of optimal conditions of lithium peroxide synthesis

R A Nefedov[1,*], Yu A Ferapontov[1] and N P Kozlova[2]
[1]Tambov State Technical University, 106, Sovetskaya Str., Tambov, Russia
[2]JSC Corporation "Roskhimzashchita", 19, Morshanskoe Highway, Tambov, Russia

[*] E-mail: kliorik@mail.ru

Abstract. Using solubility method the decay kinetics of peroxide products contained in liquid phase of LiOH – H_2O_2 – H_2O trinary system with 2 to 6% by wt hydrogen peroxide content in liquid phase in 21 to 33 °C temperature range has been studied. Conducted studies have allowed to determine temperature and concentration limits of solid phase existence of $Li_2O_2{\cdot}H_2O$ content, distinctness of which has been confirmed using chemical and qualitative X-ray phase analysis. Stabilizing effect of solid phase of $Li_2O_2{\cdot}H_2O$ content on hydrogen peroxide decay contained in liquid phase of LiOH – H_2O_2 – H_2O trinary system under conditions of experiments conducted has been shown.

1. Introduction

Academic interest in peroxide compound chemistry of alkali and alkaline earths is connected with their basic application range – products for chemical recovery of gaseous environment in isolated volumes (regenerative products). These compounds are able to generate oxygen required for human breathing when reacting with water vapour and carbon dioxide [1-3]. The development of safe and compact isolating products for human respiratory organs protection in case of emergency situations is based on this property [4]. Data regarding such class most common substance properties, suitable for usage as basic regenerative products, are given in the Table 1.

Comparison of these data shows that due to such properties, important for regenerative products like stoichiometric capacity per carbon dioxide and oxygen, stability at temperatures close to ambient temperature, superoxide of sodium, potassium and lithium peroxide hold preferred position in peroxide compound series. However, unlike sodium and potassium superoxide, there is no industrial implemented technology of lithium peroxide obtainment in Russia as of today.

It is very important to find an optimal ratio between resulting product quality and minimal costs of source raw material and energy resources for implementation of particular production method of chemical product on industrial scale. Most of the known lithium peroxide obtainment methods, which can serve as prototype for industrial implementation, are based on the interaction of lithium hydroxide with hydrogen peroxide followed by separation, multiphase flushing and thermal dehydration under vacuum or under dryers of intermediate adducts – lithium peroxide solvates of different composition ($Li_2O_2{\cdot}H_2O$; $Li_2O_2{\cdot}H_2O_2{\cdot}3H_2O$; $Li_2O_2{\cdot}H_2O_2{\cdot}2H_2O$; $Li_2O_2{\cdot}2H_2O_2$) [19]. For a more comprehensive lithium compounds separation from mother liquor, organic compounds (alcohols, xylene, toluene, pyridine, etc.) are used as salting-out agents [19-26]. Though the usage of water-organic solvents reduces lithium compounds solubility in mother liquor, ultimately forbids any Li^+ ions removal. Due to this factor the problem regarding either regeneration of mother liquor, constituting heavy mixture of

Content from this work may be used under the terms of the Creative Commons Attribution 3.0 licence. Any further distribution of this work must maintain attribution to the author(s) and the title of the work, journal citation and DOI.
Published under licence by IOP Publishing Ltd

water, solvent and lithium compounds, or its disposal will arise hereinafter [22, 27, 28]. All of this is associated with resources and material additional costs. Also one should bring to notice that the necessity of reaction mixture heating, consisting of organic compounds and peroxide products, up to the temperature of 70-80 °C may cause explosion hazard.

Table 1. Physical - chemical properties of peroxides, superoxides, and ozonides of alkali and alkaline earth metals.

Compound	Content of active oxygen, l/kg	Capacity on CO_2, l/kg	Decomposition temperature, ^0C	References
Li_2O_2	244	488	345 - 360	[5,6]
Na_2O_2	143	287	636	[7]
K_2O_2	101	203	-	[8]
Rb_2O_2	55	111	600	[8]
Cs_2O_2	37	75	640	[9]
MgO_2	199	405	-	[10]
CaO_2	155	311	400	[11]
LiO_2	430	287	-73	[12]
NaO_2	305	203	100-120	[13]
KO_2	236	157	>390	[14]
RbO_2	143	95	540	[15]
CsO_2	102	68	450	[8]
NaO_3	394	158	- 10	[16]
KO_3	321	129	60	[17]
CsO_3	155	62	82	[18]

From our point of view, the most preferable reaction route of lithium peroxide obtainment from hydrogen peroxide and lithium hydroxide occurs via formation of lithium peroxide monohydrate ($Li_2O_2 \cdot H_2O$) from water medium of intermediate adduct. This choice is conditioned by the fact that during dehydration of this particular compound the minimum water amount from reaction region is required to be removed. This procedure is important not only for the obtainment of final product with high Li_2O_2 content, but it also significantly reduces process operations timing and allows the reduction of energy consumption and hydrogen peroxide specific flow rate per final product unit. The interaction of hydrogen peroxide and lithium hydroxide in water medium will reduce explosion hazard possibility.

A major contribution to $LiOH - H_2O_2 - H_2O$ trinary system study belongs to the authors [19, 29-31], who established crystallization regions of the majority of lithium peroxide solvates. However, in our opinion, the studies previously conducted by these authors due to which one can get information on concentration and temperature ranges of $Li_2O_2 \cdot H_2O$ crystallization from $LiOH - H_2O_2 - H_2O$ system are insufficiently complete (T. A. Dobrynina has determined that lithium peroxide monohydrate formation has been observed only within a narrow range of 1.63 to 6.17% by wt hydrogen peroxide concentration in liquid state at T=30.5 °C [31]).

Experimental studies, the results of which are set out in this paper, are intended for increase of knowledge about $Li_2O_2 \cdot H_2O$ crystallization conditions from $LiOH - H_2O_2 - H_2O$ trinary system in order to select optimal conditions for lithium peroxide obtainment [32].

2. Experimental part

Systems, where hydrogen peroxide is one of the components, differ from conventional water-salt systems in relative instability of liquid and solid phases. The interaction of alkali metal hydroxide and hydrogen peroxide rapidly occurs with significant thermal effect, hence with violation of study temperature condition. It should also be taken into account that solid phases of peroxy-hydrate type forming in $LiOH - H_2O_2 - H_2O$ system can affect decay speed of hydrogen peroxide contained in liquid phase of system under study. For lithium peroxide compounds having significant solubility [19],

IOP Conf. Series: Materials Science and Engineering **112** (2016) 012018 doi:10.1088/1757-899X/112/1/012018

solubility constant value can be obtained within short time interval, and any changes in liquid phase compound may cause a rapid change of content and composition of a second component of the system under study. Therefore, at the beginning of the experiments we have conducted the assessment of solubility method application possibility for $LiOH – H_2O_2 – H_2O$ trinary system study in temperature and concentration ranges which are of interest to us.

During experiments lithium hydroxide and 50% hydrogen peroxide of «High Pure» grade have been used. At the initial stage we conducted liquid phase decay study of $LiOH – H_2O_2 – H_2O$ system with the 6.09% by wt concentration of hydrogen peroxide in liquid phase at T=30.5 °C, i.e. when the solid phase according to data [31] is $Li_2O_2 \cdot H_2O$. For this purpose, required amount of lithium hydroxide has been added to hydrogen peroxide, thermostated on water bath up to ±0.2 °C, with the velocity due to which studied system temperature would not change more than 0.5 °C. So, two samples were being prepared simultaneously. One sample remained stable, from the second the solid phase was separated on Buchner funnel at experimental temperature (solid phase content corresponded to $Li_2O_2 \cdot H_2O$ formula, which was confirmed by chemical and X-ray fluorescence analysis). Test samples were placed into thermostat at T=30.5 °C and hydrogen peroxide decay kinetics contained in liquid phase has been studied. During the experiment thermostat was placed in dark place. Content in tested active oxygen solution was determined using permanganatometry [33]. During the experiment the weight of test solution containers was also recorded.

It has been known that hydrogen peroxide and peroxide products catalytic decay in liquid phase can be divided with a reasonable degree of accuracy into two components - homogeneous (ions impact contained in test solution) and heterogeneous (reaction vessel and solid phase walls impact which are present in system under study) [34-36]. Therefore, in order to achieve experimental conditions equivalence the solutions of the same size placed in pyrex glass containers with identical geometrical shapes and sizes allowing the reduction of heterogeneous component contribution of hydrogen peroxide decay on the walls of reaction vessel during all experiments to constant value have been studied. Prior to basic experiments all containers were passivated with test solution at 70 ÷ 75 °C temperature within 24 hours. Reproduced kinetic data has been obtained after three previously conducted passivating operations.

Conduction of experiments in order to study the possibility of solid phase crystallization of $Li_2O_2 \cdot H_2O$ composition from $LiOH – H_2O_2 – H_2O$ system at different temperatures and initial reagents concentrated ratio are similar to one described in this paper [29]. The system under study has been heat stabilized up to equilibration (1.5-2.5 hours) in water bath up to ±0.2 °C. Liquid phase and the solid residue have been analyzed for active oxygen [33] and lithium hydroxide content. Solid phase samples studied for active oxygen content have been preliminarily stabilized with boracic acid; after which permanganometric titration has been carried out. Samples studied for LiOHcontent have been water-boiled within 15-20 minutes in order to destroy peroxide compounds; after which 0.1N H_2SO_4 has been titrated per phenolphthalein. Therewith, studied sediments have been analyzed using thermogravimetric method and X-ray fluorescence analysis method (XRFA). Distinctness of $Li_2O_2 \cdot H_2O$ compound has been confirmed with qualitative X-ray phase analysis using DRON-6 X-ray diffraction meter with Cu K_α filtered radiation (λ = 0.154051 nm). Scanning step is 0.05°, scanning range is $20° \leq 2\Theta \leq 120°$, exposition - 3 seconds at each point.

3. Results and discussion

The results of experiments intended to solubility method qualification assessment for $LiOH – H_2O_2 – H_2O$ trinary system study at interested to us temperature and concentration ranges of hydrogen peroxide in liquid phase are given in figure 1.

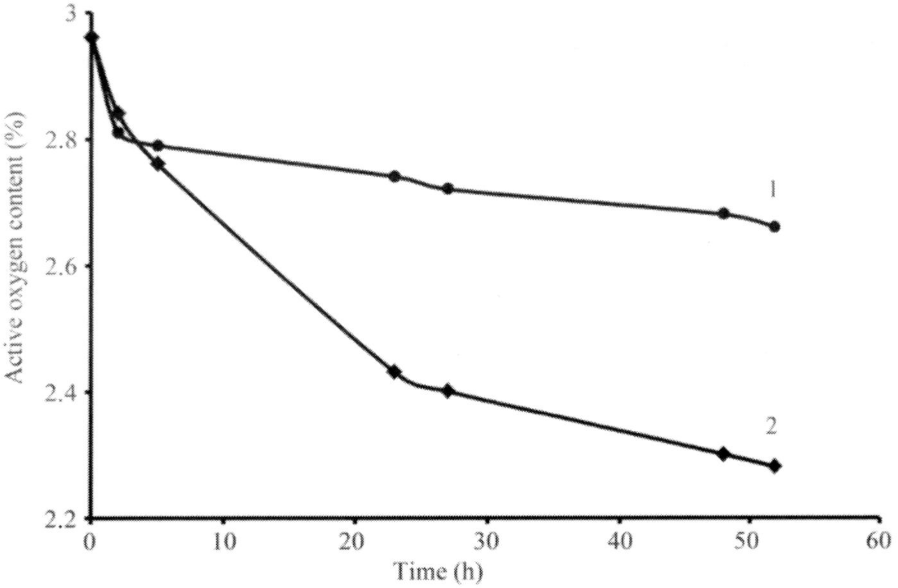

Figure 1. Kinetic curve decomposition of a liquid phase of LiOH system – H_2O_2 – H_2O in the presence of a firm phase of structure of $Li_2O_2{\cdot}H_2O$ and without it.

Curve 1 describes peroxide products decay of system liquid phase comprising $Li_2O_2{\cdot}H_2O$ composition solid phase. Curve 2 describes peroxide products decay in liquid phase after solid phase separation through filtration. Initial section at Curve 1, characterized by active oxygen content reduction in liquid phase of LiOH – H_2O_2 – H_2O trinary system with 2.96% to 2.81% at its weight constancy, corresponds to equilibrium condition reaching between phases. Equilibrium condition (constant composition of liquid and solid phases) may be considered as reached approximately 2 hours after primary components mixing. Fig. 1 shows that liquid phase in LiOH – H_2O_2 – H_2O system at concentration of hydrogen peroxide 6.09% by wt at T=30.5°C therein during long time preserves its composition (for 50 hours after equilibrium condition reaching the content of active oxygen in liquid phase decreased by 4.62%). Similar results have been obtained for system under study with hydrogen peroxide content in 2 to 6% by wt liquid phase at temperature range from 21 to 33 °C (temperature and concentration limits have been set based on conducted researches regarding crystallization rage determination of $Li_2O_2{\cdot}H_2O$ phase in LiOH – H_2O_2 – H_2O system, as will be discussed below).

On the basis of experiments conducted, one can conclude that solubility method is acceptable for LiOH – H_2O_2 – H_2O system study in the 21 to 33 °C temperature range and 2 to 6% by wt hydrogen peroxide content in liquid phase. From the rate of kinetic curves of liquid phase decay given in Fig. 1 one can see that in conducted experiment conditions the presence of solid phase of $Li_2O_2{\cdot}H_2O$ composition slows liquid phase decay of LiOH – H_2O_2 – H_2O trinary system.

Data on solid phase crystallization area of $Li_2O_2{\cdot}H_2O$ composition from LiOH – H_2O_2 – H_2O trinary system are shown in Table 2.

XRD method analysis results of solid phase sample extracted from LiOH – H_2O_2 – H_2O system at 33°C and hydrogen peroxide content in liquid phase 2.9% by wt are given in Figure 2 and Table 3.

Table 2. Area of crystallization of $Li_2O_2{\cdot}H_2O$ for system isotherms – LiOH – H_2O_2 – H_2O

The content of H_2O_2 in the liquid phase transition points at different temperatures, wt %						Structure of a solid phase
18 ^{0}C	21 ^{0}C	25 ^{0}C	29 ^{0}C	33 ^{0}C	35 ^{0}C	
0 – 4.96	0 – 1.56	0 – 1.59	0 – 1.63	0 – 1.82	0 – 5.62	$LiOH{\cdot}H_2O$
-	1.56 – 5.67	1.59 – 5.84	1.63 – 6.05	1.82 – 6.43	-	$Li_2O_2{\cdot}H_2O$
> 4.96	> 5.67	> 5.84	> 6.05	> 6.43	> 5.62	$Li_2O_2{\cdot}H_2O_2{\cdot}3H_2O$

Note. The upper bound of area of receiving $Li_2O_2{\cdot}H_2O_2{\cdot}3H_2O$ depending on the maintenance of H_2O_2 in LiOH system – H_2O_2 – H_2O at the considered temperatures wasn't defined by us.

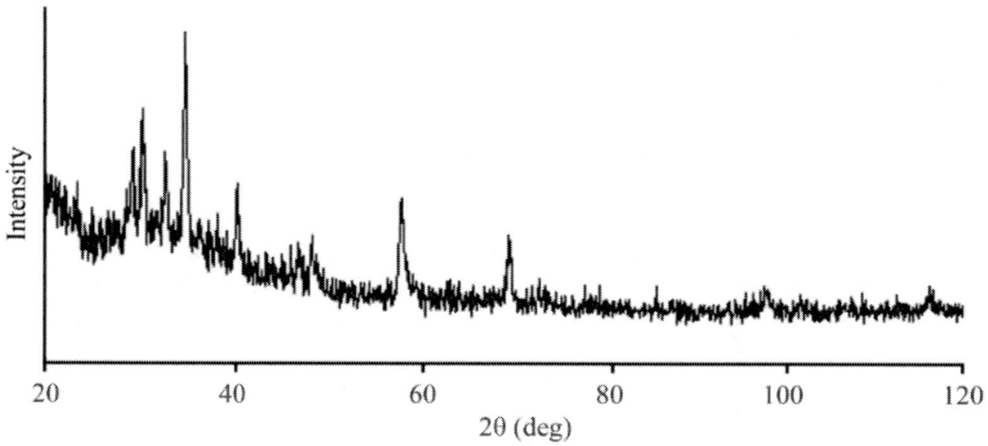

Figure 2. The diffractogram of the solid phase of the sample extracted from LiOH - H_2O_2 - H_2O system at 33 ^{0}C and hydrogen peroxide content in the liquid phase of 2.9 % by weight.

Table 3. The interplanar distance and the intensity of the main diffraction peaks of the compound $Li_2O_2{\cdot}H_2O$.

Option	Main diffraction peaks									
	1	2	3	4	5	6	7	8	9	10
Angle of slide 2Θ, gradian	23.26	29.30	30.47	32.98	35.07	40.71	47.67	49.01	58.89	70.65
Intensity , %	13	30	55	30	100	35	15	18	45	33
Interplanar distance d, A^0	3.820	3.04	2.93	2.71	2.557	2.21	1.91	1.86	1.57	1.33

As can be seen from the data given, $Li_2O_2{\cdot}H_2O$ crystallization range from LiOH – H_2O_2 – H_2O system is broader than it was stated in the works [19, 31]. We have managed to allocate lithium peroxide hydrate from LiOH – H_2O_2 – H_2O system in solid state in the temperature range from 21°C (hydrogen peroxide concentration in liquid phase is 1.56 to 5.67% by wt) to 33°C (hydrogen peroxide concentration in liquid phase is 1.82 to 6.43% by wt). Solid phase composition distinctness ($Li_2O_2{\cdot}H_2O$) has been confirmed by chemical and qualitative X-ray phase analysis data.

Based on the experiments data, we have added a polytermic diagram (Figure 3) of solid phase existed limits in saturated solution of LiOH – H_2O_2 – H_2O trinary system in H_2O_2 temperature/content plane projection in weight percentage in the liquid phase represented in paper [19].

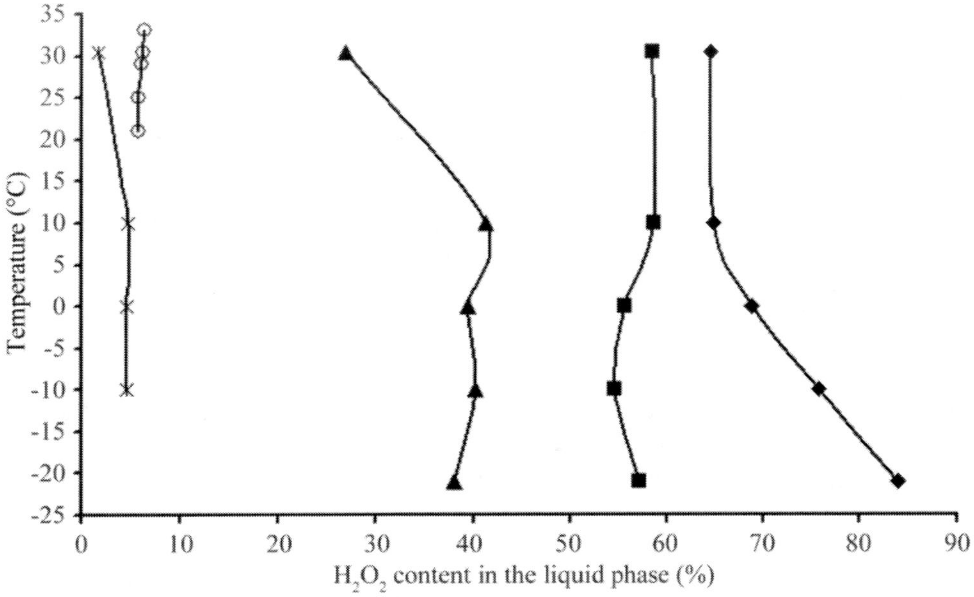

Figure 3. Polythermal diagram of limits of existence of solid phases of saturated solutions in the ternary system LiOH – H_2O_2 – H_2O.

Crystallization ranges from LiOH – H_2O_2 – H_2O system of solid phases of $LiOH{\cdot}H_2O$; $Li_2O_2{\cdot}H_2O_2{\cdot}3H_2O$; $Li_2O_2{\cdot}H_2O_2{\cdot}2H_2O$; $Li_2O_2{\cdot}2H_2O_2$ composition are applied as per data [19]. Solid phase crystallization range of $Li_2O_2{\cdot}H_2O$ composition, applied on diagram Fig.3 and marked with hatching, has been obtained based on experiments results conducted by the authors of this work.

4. Conclusion

Based on the study of kinetic decay of peroxide products, contained in liquid phase of LiOH – H_2O_2 – H_2O trinary system, application possibility of solubility method for its study with 2 to 6% by wt hydrogen peroxide content in liquid phase in 21 to 33 °C temperature range has been shown.

Stabilizing effect of solid phase of $Li_2O_2{\cdot}H_2O$ composition on hydrogen peroxide decay contained in liquid phase of LiOH – H_2O_2 – H_2O trinary system under the conditions of conducted experiments has been determined.

The study, using solubility method of LiOH – H_2O_2 – H_2O trinary system, has allowed to indicate more detailed temperature and concentration limits of solid phase of $Li_2O_2{\cdot}H_2O$ composition, distinctness of which has been confirmed by chemical and qualitative X-ray diffraction analysis.

Acknowledgement
This study was supported by the Russian Science Foundation (Agreement No. 15-19-10028).

References
[1] Presti J, Wallman H and Petroctlli A 1967 *Undersea Technol.* **8** 20
[2] Voronin G I and Polivoda A I 1967 *Sustenance Space Crews [*Zhizneobespechenie ekipazhey kosmicheskikh korabley – in russian] (Moscow: Mashinostroenie Publ.)
[3] Petroctlli A and Capolesto A 1964 *Aerospace Med.* **31** 442
[4] Didenko N S 1984 *Regenerative Respirators for Rescue Work [*Regenerativnye respiratory dlya gornospasatel'nykh rabot – in Russian] (Moscow: Nedra Publ.)
[5] Meyer M and Markowitz M 1964 *Ind. Engng. Chem.* **3** 321
[6] Rode T V, Dobrynina T A and Gol'der G A 1955 *Izv. AN SSSR OKhN [Rus. Academy of Sciences Library]* **4** 611
[7] Brewer L 1953 *Chem. Rev.* **52/1** 6
[8] Vol'nov I I 1964 *Peroxides, Superoxides, and Ozonides of Alkali and Alkaline Earth Metals* [Perekisi, nadperekisi i ozonidy shchelochnykh i shchelochnozemel'nykh metallov – in Russian] (Moscow: Nauka Publ.)
[9] Rengade E 1907 *Ann. Chem. Phys.* **11/6** 348
[10] Vol'nov I I 1983 *Peroxides of Alkaline Earth Metals* [Perekisnyye soyedineniya shchelochnozemel'nykh metallov in Russian] (Moscow: Nauka Publ.)
[11] Makarov S Z and Grigor'yeva N K 1954 *Izv. AN SSSR OKhN [Rus. Academy of Sciences Library]* **3** 611
[12] Thompson J K and Kleinberg J 1954 *J. Am. Chem. Soc.* **76** 3279
[13] Rode T V and Gol'der G A 1956 *Izv. AN SSSR OKhN [Rus. Academy of Sciences Library]* **3** 299
[14] Firsova T P *1965 Izv. AN SSSR OKhN [Russian Academy of Sciences Library]* **9** 1678
[15] Tsentsiper A B and Dobrolyubova M S 1975 *Inorganic peroxide compounds* [Neorganicheskiye perekisnyye soyedineniya - in Russian] (Moscow: Nauka Publ.)
[16] Tokareva S A and Pilipenko G P 1964 *Izv. AN SSSR OKhN [Russian Academy of Sciences Library]* **4** 740
[17] Sokovnin Ye I 1963 *Izv. AN SSSR OKhN [Rus. Academy of Sciences Library]* **1** 181
[18] Popova T I 1967 *Reactivity of lithium peroxide in the solid state [*Reaktsionnaya sposobnost' perekisi litiya v tverdom sostoyanii – in Russian] PhD Thesis Minsk
[19] Dobrynina T A 1964 *Lithium Peroxide* [Perekis' litiya - in Russian] (Moscow: Nauka Publ.)
[20] Strater H H and Heights M 1960 *US Pat. 2962358*
[21] Bach R O and Boardman W W 1964 *US Pat. 3147075*
[22] Klebba E L 1965 *US Pat. 3212850*
[23] Bach R O 1964 *US Pat. 3134646*
[24] Winternitz P 1949 *US Pat. 2488485*
[25] Smith W. N. 1969 *US Pat. 3446588*
[26] Brauer G 1985 *Guide to Inorganic Synthesis,* vol. 3 (Moscow: Mir Publ.)
[27] Ferapontov Yu A, Simanenkov S I, Gladyshev N F and Putin B V 2002 *RF Pat.2193522*
[28] Ferapontov Yu A, Simanenkov S I and Gladyshev N F 2004 Chemical Technology [Khimicheskaya Tekhnologiya – in Russian] **4**
[29] Makarov S Z and Dobrynina T A 1955 *Izv. AN SSSR OKhN [Russian Academy of Sciences Library]* **3** 411
[30] Makarov S Z and Dobrynina T A 1956 *Izv. AN SSSR OKhN [Russian Academy of Sciences Library]* **3** 294
[31] Dobrynina T A 1960 *Izv. AN SSSR OKhN [Russian Academy of Sciences Library]* **6** 961
[32] Ferapontov Yu A, Ul'yanov M A and Sazhneva T V 2007 Chemical Technology [Khimicheskaya Tekhnologiya – in Russian] **5**
[33] Seyb E and Kleinberg J 1951 *J. Am. Chem. Soc.* **73** 2308

[34] Seryshev G A 1984 *Chemistry and technology of hydrogen peroxide* [Khimiya i tekhnologiya perekisi vodoroda – in Russian] (Leningrad: Khimiya Publ.)
[35] Shamba V, Setterfield Ch and Ventvort P 1958 *Hydrogen Peroxide* [Peroksid vodoroda – in Russian] (Moscow: Inostrannaya literatura Publ.)
[36] Ferapontov Yu A, Ul'yanov M A and Zhdanov D V 2005 *Chemical Technology* [Khimicheskaya Tekhnologiya – in Russian] **1**

IOP Conf. Series: Materials Science and Engineering **112** (2016) 012019 doi:10.1088/1757-899X/112/1/012019

Investigation of mechanical properties of masterbatches and composites with small additions of CNTs

I N Burmistrov[1], T I Yudintseva, I A Ilinykh, B B Khaydarov, I N Mazov, S M Anshin and D V Kuznetsov
Department of Functional Nanosystems and High-Temperature Materials, National University of Science and Technology "MISiS", 4 Leninsky prospekt, Moscow, 119049, Russia

[1] Email: glas100@yandex.ru

Abstract. The present paper investigated physical and mechanical properties of the nanotube masterbatches and the polymer composites with low contents of carbon nanotubes (CNTs), which were obtained by diluting masterbatches. Ethylene-octene copolymer was used as the binder for the masterbatches, which provides the elasticity of the material at a content 20 wt% of CNT. Masterbatches were obtained with a 2-roller mixer, and their additive to polypropylene was carried out on a single screw injection molding machine. Strength properties of ethylene-octene copolymer increased when additing CNTs in an amount of 5-20 wt%. When the concentration of CNT in masterbatches is reduced to 0.01-0.1 wt% its strength characteristics increased up to 4-18%. The most effective strengthening of polypropylene was observed with the content of CNTs 0.1 wt%.

1. Introduction

Carbon nanotubes (CNTs) have a unique complex of physical and mechanical properties, which provides great perspectives of their use in the creation of functional and constructional materials [1-4]. The mechanical properties of CNTs are extremely high, but the problems associated with their dispersion into a polymer matrix are major challenges to realize these properties. Therefore the CNTs and CNTs-based composites are advanced materials for modern materials engineering. CNTs have been considered as a nanoreinforcing filler with a random orientation similar to conventional short fiber reinforced composites [5-8]. However, CNTs are characterized by an extremely high ratio of length to diameter, which contributes to tangling and significantly complicates the process of compounding makes it extremely difficult engineering challenge.

Many different dispersion methods have been applied for loading polymer matrices with CNTs. Sonication is the most commonly reported method [9-12]. It is a process where a transducer is used to agitate particles in solution with a high power ultrasound field with the cavities which implode with a violent and localized release of energy. In high viscosity systems, such as melt of thermoplastic sonication have many problems. Method of compounding on twin screw extruder was used usually for high-viscosity melt of thermoplastics [13, 14]. To improve the mechanical properties of the composites generally is enough small additions of CNTs, but the preparation of composites containing nano-dispersed fillers in an amount of less than 1 wt% is greatly complicated by the problem of effective compounding.

Content from this work may be used under the terms of the Creative Commons Attribution 3.0 licence. Any further distribution of this work must maintain attribution to the author(s) and the title of the work, journal citation and DOI.
Published under licence by IOP Publishing Ltd

Highly filled systems based on CNTs and polyolefin binder with the high physical and mechanical characteristics can be used as heat or electrically conductive elastic composites, as well as a masterbatch of CNTs to modify with small additives the properties of polymer composites during processing in industrial plants.

A highly promising method of dispersing nanofillers in a polymeric binder, especially in materials with a wide processing temperature range, such as thermoplastic and amorphous modified of polyethylene is rolling. This method allows obtaining highly filled systems with a good distribution of the CNTs for different applications for modern materials engineering. Therefore in this paper we present investigation of the properties of CNTs masterbatch obtained by rolling and evaluated their effectiveness in improving the mechanical properties of thermoplastics.

2. Materials and methods

The objects of study were chosen MWCNTs brand "Taunite-M", carbon black (CB) "Printex PX-2" and polygraph produced by Ltd. «NanoTechCenter», Tambov, Russia. For obtaining masterbatches we used ethylene-1-octene copolymer Lucene™ LC370 (by LG Chem. Ltd.) as polymer matrix. To synthesize the polymer-nanotube masterbatch was used 2-roller mixer (UB-6175), at a processing temperature of 120±3 °C with a time of treatment of nearly 30 min, to prepare composite samples for mechanical tests we used polypropylene (PP) brand PPG 1120-16, Russia. Obtaining of composites based on masterbatches was carried with Injection Molding Machine Haitian SA 900 with the parameters shown in Table 1. The choice of processing parameters and the formation of a polymer composite was based on the recommendations of the manufacturer for this brand (PPG 1120-16) polymer binder.

Table 1. Parameters of obtaining PP samples for mechanical testing

Parameter	Parameter Value
Mark of binder	PPG 1120-16
Zone temperature of the material cylinder, °C I / II / III / IV / nozzle	215/220/225/230/230
Mold temperature, °C	60±5
Screw speed, %	60
Rate of injection/ molding pressure, c.u./%	35/35
Shrinkage pressure, c.u.	43
Forming time, s	35±5
Cooling time, s	20±5

Breaking bending stress was determined according to ISO 178:210. The impact strength was determined according to ISO 179-1:2000. Tensile stress at break and elongation at break were determined in accordance with ISO 527-1.

3. Results and discussions

In this work we were investigated the mechanical properties of highly filled composites based on two modifications of CNT. Also PP materials were obtained with additions of these nanotubes in an amount of 0.01-0.1 wt%.

The results of investigating the mechanical properties of the highly filled systems based on the thermoplastic polymer matrix LC370 are presented in Table 2. Rigidity of obtained material, which was evaluated by the elastic modulus, regularly increases with increasing of filler content, wherein the properties of material based on CB Printex PX-2 inferior to those of materials based on the CNT at the same concentrations.

Table 2. Mechanical Properties of highly filled systems based on LC370

Composition	Modulus of elasticity, MPa			Stress at break, MPa		
	along the direction of rolling	transversely to the direction of rolling	average over all	along the direction of rolling	transversely to the direction of rolling	average over all
LC370	14	14	14	8	8	8
LC370 + CNTs, 5 wt%	21	19	20	9	9	9
LC370 + CNTs, 10 wt%	37	31	34	9	10	10
LC370 + CNTs, 20 wt%	25	70	47	12	12	12
CB Printex PX-2	32	29	30	8	9	8

Stress at break slightly higher for the samples to which the application of a load carried in the direction perpendicular to the linear displacement of the material on the rollers. There is also an increase in the stress at break when increasing the concentration of CNTs.

Investigation of mechanical characteristics of highly filled systems based on PP matrix with small additions of CNTs is presented in Table 3.

Table 3. The mechanical properties of PP with small additions of CNTs

Composition	Stress at break, MPa	Elongation at break,%	Modulus of elasticity, MPa	Charpy impact strength notched kJ/m^2
PP+LC370	18.1±0.5	551±8	1138±24	5.6±0.3
0.01 wt% of Taunite-M (masterbatch with 10 wt% of CNTs)	18.9±0.5	568±8	1314±24	5.8±0.3
0.1 wt% of Taunite-M (masterbatch with 10 wt% of CNTs)	20.4±0.5	566±8	1348±24	6.1±0.3
0.01 wt% of Taunite-M (masterbatch with 5 wt% of CNTs)	18.4±0.5	569±8	1318±24	6.0±0.3
0.01 wt% of CB Printex PX-2 (masterbatch with 10 wt% of carbon black)	19.1±0.5	539±8	1307±24	5.7±0.3
0.1 wt% of CB Printex-PX2 (masterbatch with 10 wt% of carbon black)	17.3±0.5	530±8	1279±24	6.2±0.3

Adding CNT masterbatches at extremely low concentrations enhances the stress at break by 13% and the modulus of elasticity of 18% for composites containing 0.1 wt% of Taunite-M with 10 wt% of CNT. Addition of the CNT in an amount of 0.01 wt% irrespective of the masterbatch types provides increased strength properties not more than 4-15% compared with the pure PP. The introduction of CB increases lower of mechanical properties compared to introduction of the CNTs. The concentration dependences in the case of introducing carbon black are the same as in the case introduction of CNTs.

4. Conclusions

Were investigated the physical and mechanical properties of highly filled systems based on CNTs and polyolefin binders (masterbatches) and polymer composites with low contents of CNTs obtained by dilution it with a screw injection molding machine.

It is found that by using ethylene-octene copolymers as a binder can be obtained the elastic materials even when CNT content of 20 wt%. When adding CNTs to ethylene-octene copolymer observed increase in its strength properties, which could allow using it as a final product, such as creating conductive articles.

When adding CNT masterbatches in polypropylene, with a final CNT concentration of 0.01-0.1 wt%, was an increase in the strength properties of up to 18%. The most efficient hardening is observed at CNT content of 0.1 wt%. The addition of CB under similar conditions provides a smaller increase strength properties, and in some cases, their decline.

Acknowledgements

This research was supported by the Ministry of Education and Science of the Russian Federation (project # 02.P25.31.0123).

References

[1] Baughman R H, Zakhidov A A and de Heer W A. 2002 *Science* **297** 787
[2] Cao J, Wang Q, Rolandi M and Dai H 2004 *Phys. Rev. Lett.* **93** 1
[3] Tibbetts G G and Beetz C P 1987 *J. Phys. D-Appl. Phys.* **20** 292
[4] Callister W D 2003 *Materials science and engineering, an introduction* (New York: Wiley)
[5] Hong S-K, Kim D, Lee S, Kim B-W, Theilmann P and Park S-H 2015 *Compos. Part A Appl. S* **77** 142
[6] Schilde C, Schlömann M, Overbeck A, Linke S and Kwade A 2015 *Compo.s Sci. Technol.* **117** 183
[7] Sadollah E and Hashem R-T 2015 *Comp. Mater. Sci.* **101** 189
[8] Nam T H, Goto K, Yamaguchi Y, Premalal E V A, Shimamura Y, Inoue Y, Arikawa S, Yoneyama S and Ogihara S 2016 *Compos. Part B Eng.* **85** 15
[9] Montazeri A and Chitsazzadeh M 2014 *Mater. Design* **56** 500
[10] Mazov I, Burmistrov I, Il'inykh1I, Stepashkin1A, Kuznetsov D and Issi J-P s 2015 *Polym. Compos.* **36** 1951
[11] Tkachev E N et. al. 2013 *Journal of Experimental and Theoretical Physics* **5** 860-5
[12] Mazov I N, Ilinykh I A, Kuznetsov V L, Stepashkin A A, Ergin K S, Muratov D S, Tcherdyntsev V V and Kuznetsov D V 2014 *J. Alloy Compd.* **586** 440
[13] Macutkevic J, Kuzhir P, Paddubskaya A, Shuba M, Banys J, Maksimenko S, Kuznetsov V L, Mazov I N and Krasnikov D V 2013 *Phys. Status Solidi A* **11** 2491
[14] Macutkevic J, Paddubskaya A, Kuzhir P, Banys J, Maksimenko S, Kuznetsov V L, Mazov I N and Krasnikov D V 2014 *J. Nanosci. Nanotechnol.* **14** 5430

IOP Conf. Series: Materials Science and Engineering **112** (2016) 012020 doi:10.1088/1757-899X/112/1/012020

Nanostructured $SmFeO_3$ electrophysical properties

K G Abdulvakhidov[1], S N Kallaev[2], M A Kazaryan[3], P S Plyaka[4], S A Sadikov[5], M A Sirota[1], S V Zubkov[1]
[1] Southern federal university, 105/42 B Sadovaya str., Rostov-on-Don, 344006, Russia
[2] Dagestan Scientific Center of Russian Academy of Sciences, 45 Hajiyev str., Makhachkala, Dagestan Republic, 367025, Russia
[3] P.N. Lebedev Physical Institute of Russian Academy of Sciences, 53 Lenin ave., Moscow, 119991, Russia
[4] Southern Scientific Center of Russian Academy of Sciences, 41 Chekhova str., Rostov-on-Don, 344006, Russia
[5] Dagestan state university, 43A Hajiyev str., Makhachkala, Dagestan Republic 367000, Russia

[1]E-mail: phys.kam@mail.ru

Abstract. Samarium orthoferrite conventional and nanostructured ceramic physical properties and structure investigations are reported in the paper. Ceramic samples have been prepared both by classical sintering method after solid phase synthesis and by Bridgman anvil mechanoactivation after synthesis. The authors have established that the mechanoactivation suppress direct current conductivity and increase spontaneous polarization. Polycrystalline samarium orthoferrite physical properties can be purposefully varied on fabrication stage without doping.

1. Introduction

Samarium orthoferrite ($SmFeO_3$) is representative of multiferroic, it have both ferroelectric and antiferromagnetic properties. From a practical point of view it is a promising material for gas sensors [1] and magnetic field sensors [2]. Ferroelectric and antiferromagnetic subsystems coupling allow one to drive material magnetic properties by electric field and vice versa to drive electric properties by magnetic field [3]. Ceramics and thin films multiferroics promising for memory, functional electronic elements, actuators and MEMS development [3, 4]. Electrical conductivity is the main problem, interfered to multiferroics practical application [3], especially for based on electrotransport properties gas sensors [1]. Authors [5] reports about samarium orthoferrite conductivity increasing with temperature rise, but measurements have been executed only on low temperatures. Gases are most active on high temperatures, thus conductivity dependence in wide temperature range is actual.

As known, main ceramics physical properties directly dependent from temperature-temporal manufacturing regimes, element composition and doping concentration [1, 3, 4]. But doping adding, strengthening some parameters can reduce other useful ceramics characteristics. The aim of presented work is to investigate the alternative ceramic properties control method, based on mechanoactivation. We have thirst studied ceramic samarium orthoferrite electro physics in particular electrotransport

Content from this work may be used under the terms of the Creative Commons Attribution 3.0 licence. Any further distribution of this work must maintain attribution to the author(s) and the title of the work, journal citation and DOI.
Published under licence by IOP Publishing Ltd

properties control possibilities by structure defects concentrations and crystallite sizes changing. Mechanical force action and shift deformation methods have been apply for this aim.

2. Experimental procedure

$SmFeO_3$ synthesis have been done from initial chemically pure oxides Sm_2O_3 and Fe_2O_3 in two hours under stabilized temperature T=1200 °C in closed platinum melting pot. The resulting composition phase analysis for impurity phases revealing have been performed on powder X-rays diffractometer HZG-4B in Cu $K\alpha_1,\alpha_2$ radiation, dedicated from common spectra by Ni filter. PCW-2.3 program packet [2a] have been used for Bragg angles corresponding (hkl) diffraction maximums indicating.

Sinthesed samarium orthoferrite mechanoactivation for nanostructuring have been performed by Bridgman anvils, bottom of them have rotated with slow angular velocity. Uniaxial pressure of 200 MPa magnitude has been applied on top anvil. Resulting shift deformation can be evaluate as

$$\zeta = \ln(v\, r/d),$$

where v is the whole anvil rotation angle in radians, equal for used regimes to 12, r and d– sample radius and thickness respectively.

Initial and activated powder samples micrographs have been made by FE-SEM Zeiss SUPRA 25 electron microscope. Activated under 200 MPa pressure sample is characterized by "loose coat" on crystallite surfaces, as clearly visible on SEM-images, presented on Figure 1. Nanostructured powder has increasing crystallites specific surface area and increasing broken bonds ions amount on surface. Activated powder sample crystallite sizes spread has multimodal character and in the range 10-600 nm. Low activation energy values and large diffusion coefficient values is characterized for ceramics sintering from such nanostructured crystallites.

a) b)

Figure 1. SEM-images of samarium orthoferrite powder samples, conventional (a) and mechanoactivated (b)

Disc form ceramic samples from both types of powder, 2 hours sintered on temperature T = 1100 °C have dimensions of 10 mm diameter and ~1 mm thickness. Silver paste has been brazed on 750 °C temperature for two hours on flat surface, formed electrodes. Conventional (etalon) and mechanoactivated ceramics properties have been investigated. Ceramic samples dielectric spectra for a lot of frequencies have been measured in wide positive temperature interval by automated setting, based on immitance measuring device E7-20. For specific dc conductivity measuring has been utilized classical voltmeter – ammeter scheme with 1.5 V batteries as dc power supply using. Ceramic samples spontaneous polarization Ps has been studies by classical Sawyer-Tower scheme using.

3. The experimental results and discussion

On X-ray diffraction investigations have been established as well as classical technology manufactured powders and mechanoactivated powders and sintered ceramic samples under room temperature are referring to *Pnma* space symmetry group. Any impurity phases have not been found within the X-ray diffraction analysis sensitivity as presented on Figure 2 diffraction pattern shows. Ceramics unit cell parameters have the following values: $a = 5.5935$ Å, $b = 7.7099$ Å, $c = 5.3984$ Å.

As have been established, powder mechanoactivation results to coherent scattering regions dimensions decreases from 1200 to 50 nm and microstrain $\Delta d/d$ slightly increases. Bragg maximums integral intensities are also increasing after activation as results of crystallite mosaic degree rising [3a]. As known, crystallite mosaic degree rising is followed by crystallites dislocation concentration increasing. Thus, after 200 MPa mechanoactivation dislocations concentration prevails on point defects concentration in $SmFeO_3$ powder generated structure defects. Polycrystalline point defects concentration can be estimated qualitatively only from X-ray diffraction profiles broadening and integrated intensities changing. A significant diffraction profiles broadening activated under 200 MPa pressure samarium orthoferrite have not been found. But, it should be understood, that Bridgman anvils powder mechanoactivation on lot of pressure can cause to dynamic recrystallization processes developing, accompanied by coherent scattering regions increasing and some microdeformations decreasing. The absorption of some crystallites to other is often took place in such situations [4a], but we have not study this phenomenon.

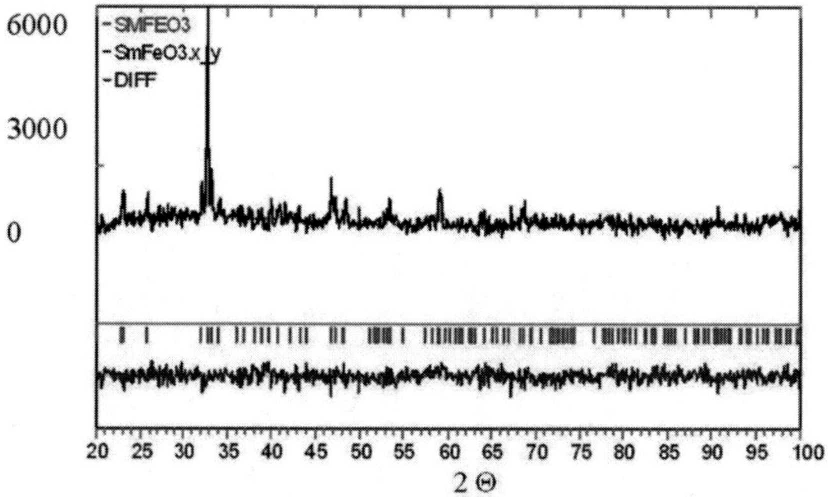

Figure 2. Samarium orthoferrite diffraction pattern

The dielectric constant ε, as ferroelectric fundamental properties, has been investigated in temperature range. The results for a number of frequencies are presented on Figure 3. As can be seen unactivated $SmFeO_3$ temperature dependences (a) are not corresponded to relaxor ferroelectrics. This is because that conventional sample has a sufficient amount of impurities, with energy, corresponded to located in the forbidden zone energy levels. During sample heating an impurity levels thermal devastation impurity levels and charge carriers transport in the conduction band took place. Enough large through conductivity of the sample masks its ferroelectric properties and it is not always possible to refer rare earth orthoferrites to relaxor ferroelectrics. This is confirmed by ceramic specific conductivity temperature dependence plots, presented on Figure 4. As have been found the activated sample specific conductivity in wide temperature range is two orders of magnitude less than that of the etalon sample.

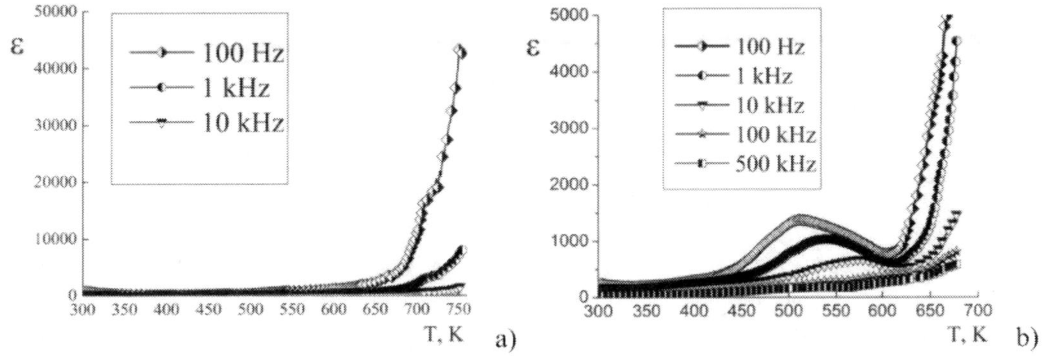

Figure 3. $SmFeO_3$ ceramic permittivity temperature dependence for a number of frequencies: a) etalon sample, b) mechanoactivated sample.

Mechanoactivation reduces the activation energy of impurities diffusion processes under ceramics sintering and facilitate their exit to the surface. Activated sample permittivity ε temperature and frequency dependence, presented on Figure 3 (b) has relaxor character. Temperature of permittivity maximum is offset to right and maximum value is decreased as measuring field frequency rise.

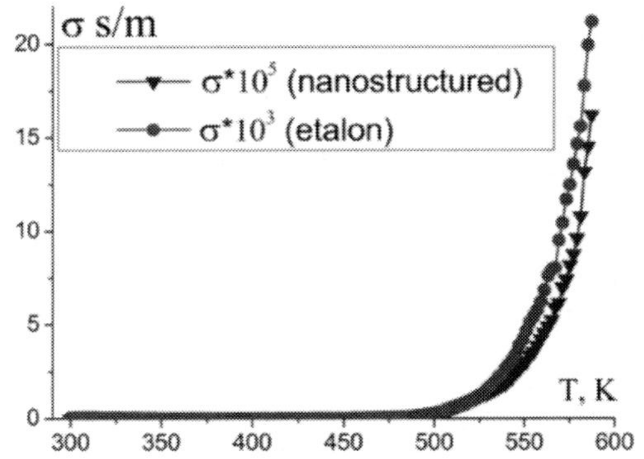

Figure 4. $SmFeO_3$ ceramic specific conductivity temperature dependence for conventional and mechanoactivated samples.

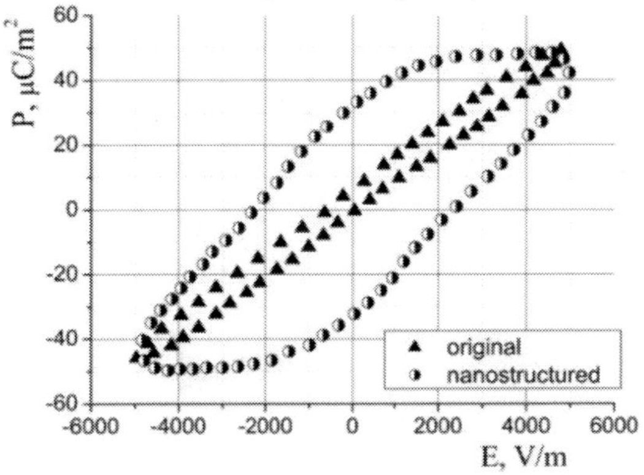

Figure 5. $SmFeO_3$ ceramic hysteresis loop for etalon and mechanoactivated samples.

The possibility of spontaneous polarization P switching by electric field or by mechanical pressure is the main ferroelectric property. Spontaneous polarization P hysteresis loops as a function of applied electric field E, are presented on Figure 5. For unactivated $SmFeO_3$ has a weak and distorted loop, characteristic for defective crystals-antiferroelectric, ferroelectric properties are mild. $SmFeO_3$ mechanoactivation lead to spontaneous polarization P rising and hysteresis loop saturating, loop becomes similar to loop of barium titanate, as a classical ferroelectric.

4. Conclusions

The authors have proposed technical solution for purposeful control physical properties on fabrication stage. Both conventional and mechanoactivated $SmFeO_3$ samples structure and dielectric properties are investigated, presented diagrams proves ferroelectric properties. Electrical conductivity is rather stable in wide positive temperature range. It have been demonstrate, that samarium orthoferrite mechanoactivated sample electro physical characteristic is sufficient higher, than etalon sample. Dosed pressure nanostructured mechanoactivation after synthesis is universal modifications method for any complex material without alloying element adding.

Acknowledgements
This work was partially supported by Southern Federal University (grant 213.01-2014/011-VG).

References

[1] Tasaki T, Takase S and Shimizu Y 2012 Fabrication of Sm-based perovskite-type oxide thin-films and gas sensing properties to acetylene *J. Sensor Technol.* **2** 75
[2] Niu X and Du W 2004 *Sensors and Actuators* B 99, p. 399
[3] Ramesh R, Spaldin N 2007 *Nature Materials* v. 6, p. 21
[4] Aono H, Sato M, Traversa E, Sakamoto M and Sadaoka Y 2001 *J. Am. Ceram. Soc.* 84, 341
[5] Nascimento M P, Weber I T, Barrozo P, Oliveira A A M and Aguiar J A 2009 Studies of structural, microstructural, electrical and magnetic properties of REFeO3 (RE = Gd, Eu, Sm) *11th International Conference on Advanced Materials (ICAM 2009) Rio de Janeiro, Brazil. 20-25 September*
[6] Kraus W and Nolze G 1999 Powder Cell for Windows (version 2.3). *Federal Institute for Materials Research and Testing (Berlin; Germany)*
[7] Abdulvakhidov K G, Vitchenko M A, Mardasova I V, Oshaeva E N and Abdulvakhidov B K 2007 *Technical Physics. Rus. J. of Applied Physics* **52**/11 1458
[8] Arinstein A E 1997 *Physical Chemistry* **355(**4-6) 237

The research of structure and mechanical properties of superhard electro-spark coatings for hardwearing mining tools

P A Bajin, A P Chijikov, D V Leybo, K O Chuprunov, A G Yudin, M A Alymov, D V Kuznetsov

National University of Science and Technology "MISiS"
Moscow, Leninsky prospekt 4, 119049, Russian Federation

Email: amstolin@ism.ac.ru

Abstract. The development of low cost and hardwearing mining tools is one of the most important areas in mining industry. It is especially important for technologies of rare and rare earth metals mining due to high hardness of related ores. Coatings for electrodes, produced by extrusion of self-propagating high temperature synthesis (SHS) products from hard-alloyed materials with nanosized structure, for further application in processes of electrospark alloying and deposition were studied in this work. The results of microstructure and properties of deposited layers, interaction of support with SHS produced electrodes, comparison of frictional properties of obtained materials as well as some industrial testing results are presented in this work.

1. Introduction

Electrospark alloying (ESA) methods are widely used for hardwearing mining tools protection and repairing after deterioration. Novel method of SHS extrusion for production of electrodes for ESA and electrospark deposition (ESD) can be used in order to improve existing technological process [1]. The method includes exothermic combustion of initial components mixture and shear deformation of still hot products of synthesis. Application of SHS extrusion method for electrodes production substantially simplifies the process since the synthesis of materials takes only several seconds (instead of hours) in one technological stage. Detailed studies of compositions, properties (density, hardness, electrical resistivity etc.), grain size of electrodes and ranges of their variations at fixed parameters were performed in previous works [1, 2]. The results of application of ESA electrodes produced by HSH extrusion method from hard-alloyed materials with nanosized structure are presented in this work. The microstructure of layers and study of interaction between support material and SHS electrodes as well as frictional properties were done.

2. Materials and methods

Nowadays more than 50 different compositions of electrode materials for different applications have been synthesized by SHS extrusion method [1]. Coating deposition by ESA method with electrodes synthesized via SHS extrusion was done on SE-5.01 device (Tomsk, Russia). The device allows to prepare coatings with up to 0.2 mm thickness depending on deposition parameters, electrode material and support. Carbon steel 45 was used like model support material. In order to compare frictional

Content from this work may be used under the terms of the Creative Commons Attribution 3.0 licence. Any further distribution of this work must maintain attribution to the author(s) and the title of the work, journal citation and DOI.
Published under licence by IOP Publishing Ltd

IOP Conf. Series: Materials Science and Engineering **112** (2016) 012021 doi:10.1088/1757-899X/112/1/012021

characteristics of coatings, obtained with SHS electrodes, commercially available tungsten titanium carbide electrodes were also used for coatings deposition.

Study of structure was done on scanning electron microscope (SEM) JSM 6610LV (Jeol, Japan) using 20kV accelerating voltage.

3. Results and discussions

Study of microstructure of coatings deposited by ESD using SHS electrodes showed that hardened layers had complex structure and consisted of at least two regions. Sizes of wear resistant structural components on the surface of coating corresponded to the ones of initial electrodes (figure 1a). Coating region located closer to the support consisted of smaller structural components with sizes of 20 – 100 nm (figure 1b). This can be explained by different heat removal rates. The support cools down more intensively and consequently cooling rate of structural components is higher on support/coating boundary. Thus there is not enough time for grain sizes to increase substantially. Constant action of sparks on the surface decreases heat removal rate, temperature of melt is higher and grain sizes reach values up to 1 – 2 μm. Also one can see grains broken off electrode surface with initial sizes. Wear-resistant grains uniformly distributed inside metallic support along whole region of support/coating interface which increases adhesion of electrode material with the support.

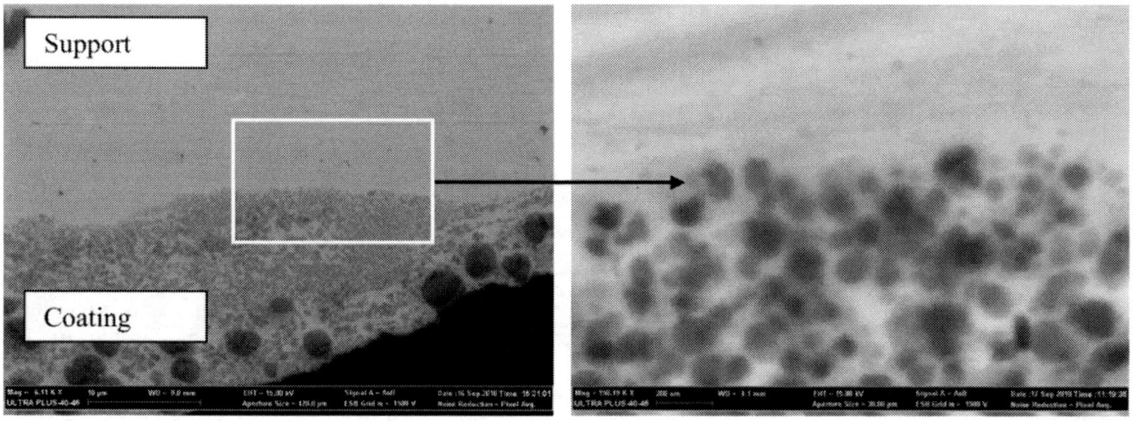

a) *b)*

Figure 1. Microstructure of SHS electrodes: a – STIM-2/30, b – STIM 3/10H, c – STIM-4

The presence of transition layer with inclusions of coating material indicates that ESD process proceeds throw the diffusion of cathode and anode materials at the interface. We observed processes of electrode material particles penetration inside steel support on the depth of 15 - 20 μm (figure 2 a, c, d) as well as outward diffusion of iron into the coating material (figure 2 b). Thus interdiffusion processes positively affect adhesion which in turn provides high sustainability of coated instrument against abrasion in extreme conditions.

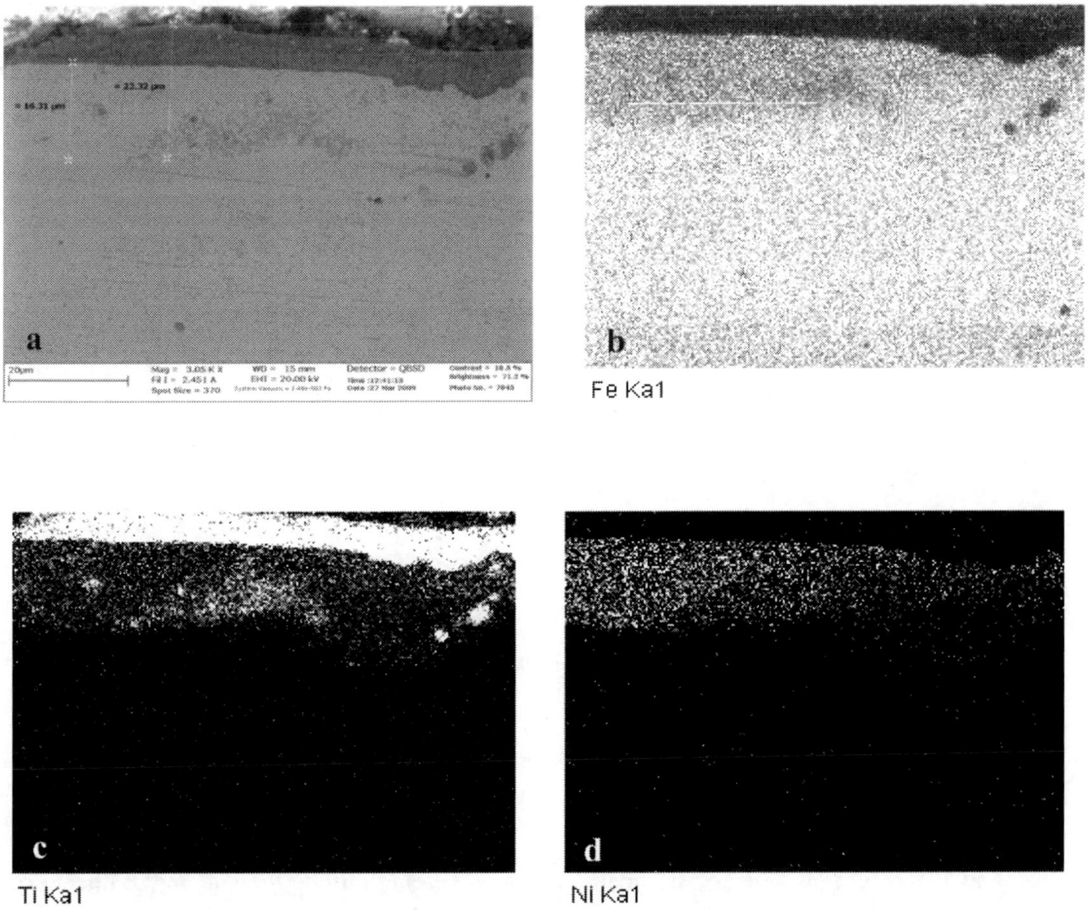

Figure 2. a – Microstructure of coating STIM – 2/30 on steel 45. Distribution of elements in the coating: b – Fe, c – Ti, d - Ni

Microhardness measurements were done for coated material from the surface down to the core of the support under the loading of 50 g. Sample with traces of indenter as well as plot of microhardness vs. depth are shown in figure 3.

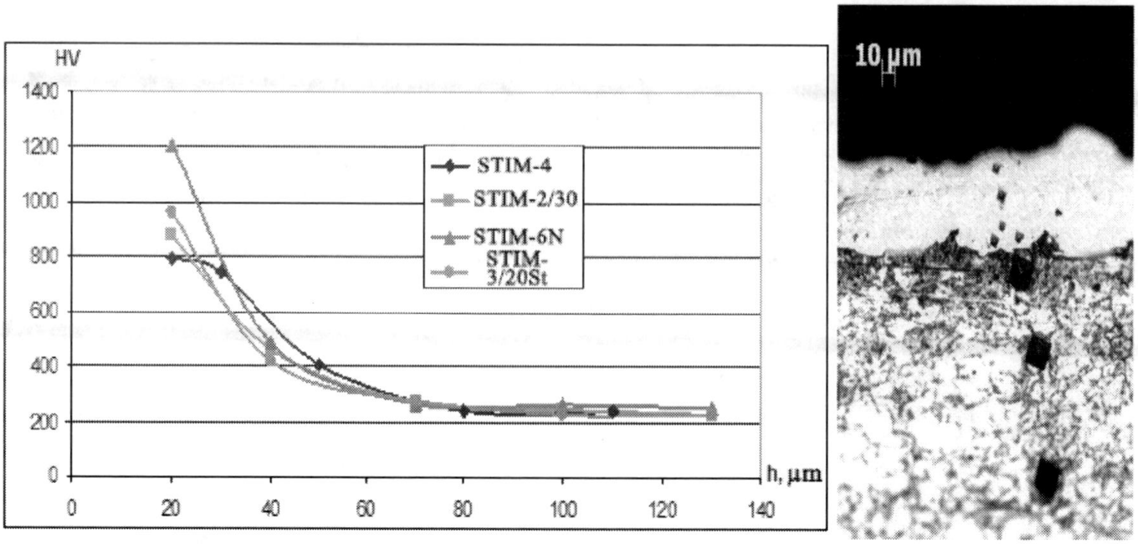

Figure 3. Sample with coating deposited using SHS electrodes and microhardness of coatings

The results of microhardness measurements showed that at the surface of the support layer with microhardness of 1200 HV forms. Microhardness decreases from the surface down to the core of the support which indicates presence of transition layer on the surface of the support with composition similar to electrode material. This ensures correlation of mechanical properties of electrode material and coating. As the result microhardness of metallic support surface increases up to 4 – 4.5 times.

In order to test industrial application of coatings obtained using SHS electrodes deposition on the surface of jaw crusher was done. Testing was performed at Nikopol Ferroalloy Plant. It was shown that deterioration rate of coated crusher surface was 1.7-2 times lower than in the case of uncoated surface made from steel grade 40X (Russian, GOST). During testing 5800 tons of manganese agglomerate was processed.

Therefore the application of electro-spark coating for longwearing mining tools is one of unexpensive and promising ways of increasing their durability. SHS extrusion method is known [3] as the effective method for production of electrodes with nanosized structure made of wear-resistant component and oxides (without metallic binder) initially in the form of micrometer-sized powders. Usually in order to obtain composite nanostructured ceramics by SHS ultrafine powders are used, or chemical and mechanical treatment is done after synthesis [4]. Another option is to add material with low melting point and make long mechanical treatment [5]. In case of SHS extrusion on the other hand nanosized structure is formed due to combination of exothermic combustion of initial components mixture with shear deformation.

4. Conclusions

High potential of SHS electrodes for coating deposition on longwearing mining tools was shown in this work. ESA and ESD treatments with SHS electrodes can be successfully applied at different mining plants where wear resistance of steel instruments is of great importance.

Acknowledgements

The project is supported by Ministry of Science and Education of Russian Federation (project RFMEFI57614X0014)

References

[1] Mosina T V 2014 *Refractories and industrial ceramics* **54** 384
[2] Kudryashov A E, Levashov E A, Aksenov L B and Petrov V M 2010 *Metallurgist* **54** 514

[3] Merjanov A G 1998 *Combustion processes and synthesis of materials* (Chernogolovka: ISMAN) [in Russian]

[4] Stolin A M and Bazhin P M 2014 *International Journal of Self-Propagating High-Temperature Synthesis* **23** 65

[5] Panteleenko F I, Sarantsev V V, Stolin A M, Bazhin P M and Azarenko E L 2011 *Surf. Eng. Appl. Electrochem.* **47** 328

IOP Conf. Series: Materials Science and Engineering **112** (2016) 012022 doi:10.1088/1757-899X/112/1/012022

Quantum chemical study of the structure and properties of isotopically pure lead chalcogenides

O Kh Poleshchuk[1,2], N B Egorov[1], D V Akimov[1], I I Zherin[1] and N A Zhuravlev[1]

[1]National Research Tomsk Polytechnic University, 30 Lenina ave., Tomsk, 634050, Russia
[2] Tomsk State Pedagogical University, 60 Kievskaya street, Tomsk, 634061, Russia,

E-mail: egorov@tpu.ru

Abstract. In the present work the theoretical methods B3LYP/SDD, GGA and BP86/TZ2P were used for quantum-chemical calculations of lead chalcogenides. It is shown that these levels of theory are applicable for assessment of their geometric parameters, Raman and IR spectra and thermodynamic characteristics. It is shown that there are correlations between the experimental and calculated characteristics of lead sulphide, selenide and telluride. The influence of different isotopes of lead, sulphur, selenium and tellurium on the thermodynamic parameters and the Raman spectra for the lead chalcogenides is shown.

1. Introduction

Lead chalcogenides compounds of PbS, PbSe, PbTe are narrow band-gap semiconductors used in infrared detectors, thermoelectric devices and photoresistors. In spite of the fact that the electronic and optical properties of lead chalcogenides were studied long ago and are rather well known, in these materials they continue to reveal interesting properties. For example, PbSe semiconductors display unusually high temperatures of continuous emission that are of great interest for engineering infra-red lasers that work at room temperatures [1]. Data from recent research indicates that elements of high-sensitivity receivers in the distant infra-red range (20–200 microns) based on lead alloy chalcogenides can successfully compete with germanium and silicon [2]. Unlike those elements, lead chalcogenides possess considerable radiation resistance caused by their high density, stabilizing Fermi's level. Moreover lead chalcogenides are interesting as objects of research, namely studying the influence of isotope composition of semi-conductor materials on their physical properties. Recently interesting results in this direction were obtained in papers [3, 4].

Several attempts have been made to develop theoretical settlement models using various approaches, in order to accurately reproduce the properties of lead chalcogenides. Previously, geometrical, spectral and thermodynamic characteristics of sulphur and lead were calculated by density functional method using various values of functionals and basic conditions, such as B3LYP (GAUSSIAN), BP86 (ADF), GGA (FLAPW) [5].

The purpose of this work is the analysis of applicability of the theory of a functional of density to an assessment of frequency shifts in Raman spectra and thermodynamic characteristics of chalcogenides with various isotope compositions of lead, selenium and tellurium, and also an elementary cell PbS, PbSe and PbTe.

Content from this work may be used under the terms of the Creative Commons Attribution 3.0 licence. Any further distribution of this work must maintain attribution to the author(s) and the title of the work, journal citation and DOI.
Published under licence by IOP Publishing Ltd

2. Experimental part

For carrying out quantum and chemical calculations we used the standard software package GAUSSIAN'03W [6]. Calculations were carried out using a hybrid method of a functional of density of B3LYP, with an exchange functional to Becke B3 and a correlation functional Lee, Wang and Parr (LYP) [7, 8]. This method is now generally accepted to be the best for describing thermodynamic properties and vibrational spectra [9]. As a basic condition we used the SDD pseudopotential for an atom of lead and the D95 double-dzeta as the basis for other atoms [10]. Calculated geometries of the molecules were fully optimized; the absence of imaginary vibration frequencies confirmed the stationary character of the structures. Energy calculations were corrected for zero-point vibrational energy and reduced to normal conditions (298.15K, 1 atm) using thermal corrections to enthalpy and free energy. We have also calculated the ADF'2004 (Amsterdam density functional) program [11]. The OPTX exchange functional was used in combination with the PBE correlation functional using an all electron triple-ζ + polarization basis set of Slater orbitals (BP86/TZ2P+) [12]. To estimate the enthalpy of formation the semi-empirical method PM3 was used [13].

All the above-stated calculations were carried out for compounds of PbS, PbSe and PbTe in the gaseous phase. The calculations of electronic structure and characteristics of PbSb, PbSe and PbTe, were in the crystalline state through the full-potential linear augmented plane wave method (code WIEN2k [14]) with the generalized gradient approximation of exchange and correlation potential [15], and full structural optimization. The density of electronic conditions was calculated by the method of tetrahedrons [16]. In this case cubic monosulphide, monoselenide and monotelluride of lead PbS, PbSe, PbTe (the B1 type, spatial group № 225, Fm-3m) with atomic coordinates: Pb (0; 0; 0) and S, Se, Te (1/2; 1/2; 1/2) were used for calculations.

3. Results and discussion

To assess the "quality" of the calculations it is necessary to rely on some physico-chemical properties of similar compounds, for which sufficiently accurate experimental values are known. As experimental values, we used the bond lengths of compounds of lead, the frequency of stretching and deformation vibrations in the IR spectra of two- and four-coordinated lead compounds, as well as the band gap of crystalline halides, and the entropy and enthalpy of formation and dissociation energies of compounds of lead in the gas phase (Table 1).

From the results of the calculation it follows that the valent angles of Hal-Pb-Hal optimized in the GAUSSIAN program are close to the experimental values [10] (about 98–100 °), and the bond lengths of Pb-Hal (Hal=F, Cl, Br, I, Me) Pb (II) and Pb (IV) compounds have a good correlation ratio (fig. 1).

Table 1. The calculated and experimental [12] enthalpy of formation (kJ/mol), entropy (J/mol·K), dissociation energy (kJ/mol) of lead halides

Compound	ΔH°, PM3	ΔH°, exp.	S°, calc.	S°, exp.	$-D_o$, calc.	$-D_o$, exp.
PbF_2	-375	-435.14	293	292.66	484	388
$PbCl_2$	-166	-174.05	318	317.19	388	304
$PbBr_2$	-119	-104.39	346	339.39	353	262
PbI_2	-18	-3.18	362	359.55	315	209
PbF_4	-617	-1133.45	345	333.52	341	327
$PbCl_4$	-259	-552.41	399	381.65	260	252
$PbBr_4$	-109	-456.36	445	426.20	229	199
PbI_4	56	-224.47	477	466.26	200	164

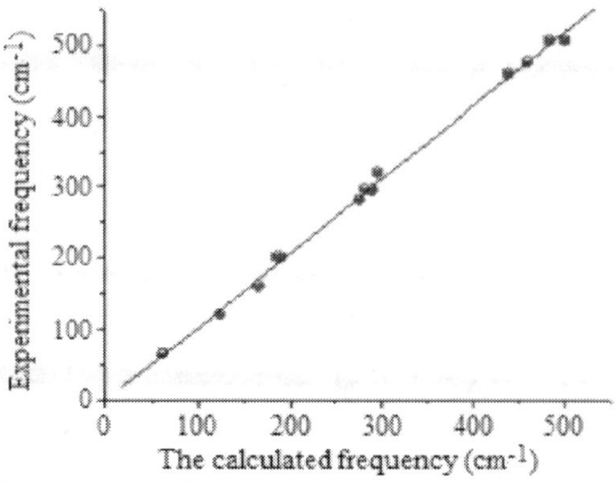

Figure 1. The relationship between experimental bond lengths of compounds of Pb (II) and Pb (IV) and those calculated by the B3LYP/SDD method.

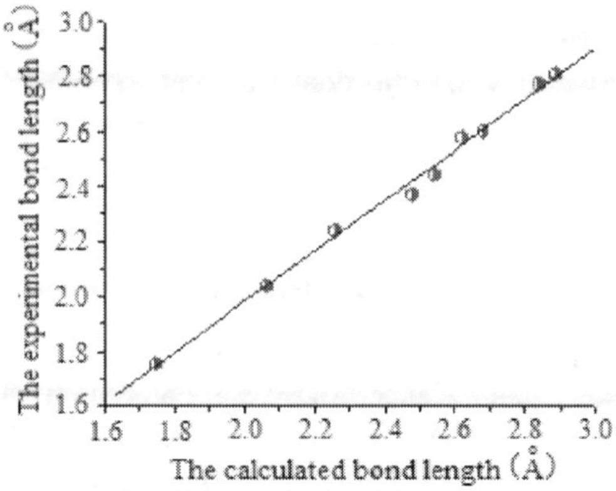

Figure 2. The relationship between experimental IR-frequencies of lead halides and those calculated by the B3LYP/SDD method.

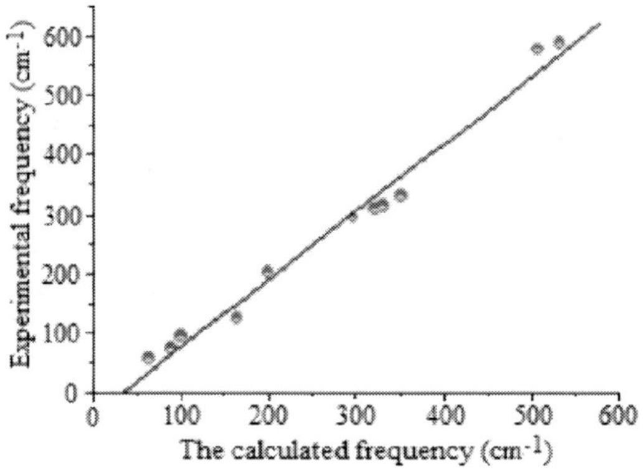

Figure 3. The relationship between experimental IR-frequencies of lead halides and those calculated by the ADF method.

In addition, the frequencies calculated by the two methods of IR and Raman spectra of lead halides are close to the experimental values (Fig. 2, 3) with a coefficient of correlation of 0.998 and a standard deviation of 9 %.

We observed quite reliable linear correlation (1–8), which showed that the calculations for lead halides by the PM3 and B3LYP/SDD methods correctly describe not only the geometry, but also the thermodynamic parameters such as enthalpies ($\Delta H°$) and entropies ($S°$) of formation, and the dissociation energies (D_0). The dissociation energy of the lead compounds was determined as the difference between the energy of the compound and the energies of their constituent atoms, as amended by the enthalpy of formation.

The calculated dissociation energy was related to the number of bonds, compared with the experimental average binding energy of lead-halogen [15]. It should be noted that if the entropy falls on the curve for compounds Pb (II) and Pb (IV), then the relationship between the calculated and experimental enthalpies of formation and dissociation energies are observed separately for the coordination of various lead compounds on the basis of the two methods of calculation GAUSSIAN and ADF programs:

GAUSSIAN: $S°\ (exp.) = -23 + 0.914S°\ (calc.)$ $r = 0.996$; $s = 5$; $n = 8$ (1)

ADF: $S°\ (exp.) = 10.6 + 0.97S°\ (calc.)$ $r = 0.998$; $s = 4$; $n = 7$ (2)

IOP Conf. Series: Materials Science and Engineering **112** (2016) 012022 doi:10.1088/1757-899X/112/1/012022

For Pb(II):

PM3: $\Delta H°$ (exp.) = 29 +1.23 $\Delta H°$ (calc.) r = 0.999; s = 12; n = 4 (3)

GAUSSIAN: D_o (exp.) = -108 +1.04 D_o (calc.) r = 0.993; s = 11; n = 4 (4)

ADF: D_o (exp.) = -44 +1.54 D_o (calc.) r = 0.990; s = 13; n = 4 (5)

For Pb(IV):

PM3: $\Delta H°$ (exp.) = -289 +1,35 $\Delta H°$ (calc.) r = 0.997; s = 35; n = 4 (6)

GAUSSIAN: D_o (exp.) = -62 +1.14 D_o (calc.) r = 0.992; s = 11; n = 4 (7)

ADF: D_o (exp.) = -28 +0.83 D_o (calc.) r = 0.986; s = 14; n = 4 (8)

Here and below, r - correlation coefficient, s - standard deviation, n - number of compounds.

The values of the energy distance between the occupied and unoccupied molecular orbitals ΔE(B3MO-HCMO) calculated by both methods agree well with the band gap in crystalline chloride, bromide and iodide of lead [17], with a coefficient of correlation 0.999 and a standard deviation of 0.02 eV.

By analogy with the sulphur compounds [5], for selenium and tellurium compounds were used as a "quality assessment" on the base on the bond lengths, frequencies in the vibrational spectra, and experimental values of the band gap of crystalline chalcogenides, as well as the dissociation energy of selenium and tellurium in the gaseous phase (Table 2).

Table 2. The experimental and calculated dissociation energy of selenium and tellurium compounds

Molecule	Experimental bond dissociation energy, kJ/mol	Bond dissociation energy calculated by ADF, kJ/mol	Bond dissociation energy calculated by B3LYP/SDD, kJ/mol
TeB	354	485	477
SeAl	338	468	438
TeAl	268	343	372
SeIn	247	359	414
TeIn	218	326	364
SeP	364	505	574
SeAs	352	490	564
TeAs	312	445	519
SeS	371	547	490
TeS	339	502	469
SeTe	292	468	446
Molecule	Experimental bond dissociation energy, kJ/mol	Bond dissociation energy calculated by ADF, kJ/mol	Bond dissociation energy calculated by B3LYP/SDD, kJ/mol
SeO	465	694	623
SeC	590	869	802
Se_2	333	510	468
SeH	311	493	439
SeSi	534	731	694
SeGe	485	694	673
SeSn	401	635	631
Te_2	260	390	422
TeSi	448	631	623
TeO	376	550	602
TeCu	229	320	351
TeGe	430	610	614

On the other hand, the results of the calculations imply that bond lengths optimized by the B3LYP/SDD method for the compounds of selenium and tellurium are proportional to the experimental values [17], and there is a good correlation between them:

For selenium compounds

$$R(\text{exp.}) = 0.957\ R(\text{calc.})\ r = 0.995;\ s = 0.04;\ n = 18 \quad (9)$$

For tellurium compounds

$$R(\text{exp.}) = 0.969\ R(\text{calc.})\ r = 0.993;\ s = 0.04;\ n = 10 \quad (10)$$

Table 3. The energy of the unit cell of lead chalcogenides (relative to 5619 u.a.m.u. for PbS, 176 u.a.m.u. for PbSe and 158 u.a.m.u. for PbTe) and isotopic composition

M_{Pb}	M_S	M_{Se}	M_{Te}	E_0, u.a.m.u.
204	32	76	122	0.636289 0.217974 0.611721
206	32	76	122	0.636300 0.217984 0.611731
207	32	76	122	0.636305 0.217989 0.611735
208	32	76	122	0.636311 0.217993 0.611740
204	33	78	124	0.636365 0.218014 0.611746
206	33	78	-	0.636376 0.218024
207	33	78	124	0.636381 0.218029 0.611760
208	33	78	124	0.636387 0.218033 0.611765
M_{Pb}	M_S	M_{Se}	M_{Te}	E_0, u.a.m.u.
204	34	80	126	0.636438 0.218053 0.61177
206	34	80	126	0.636449 0.218062 0.611779
207	34	80	-	0.636454 0.218067
208	34	80	-	0.636460 0.218072
204	36	82	128	0.636575 0.218090 0.611793
206	36	82	128	0.636586 0.218099

				0.611803
207 (204 for PbTe)	36	82	130	0.636591 0.218104 0.611816
208	36	82	-	0.636597 0.218109

In addition, the calculated frequency of the IR and Raman spectra compounds of selenium and tellurium are close to the experimental values [18] (Fig. 4, 5):

For selenium compounds

$$\omega(\text{exp.}) = 59 + 0.996\ \omega(\text{calc.})\ r = 0.993;\ s = 57;\ n = 91 \quad (11)$$

For tellurium compounds

$$\omega(\text{exp.}) = 21 + 1.07\ \omega(\text{calc.})\ r = 0.993;\ s = 27;\ n = 67 \quad (12)$$

It should be noted that correlation coefficients close to one of the calculated parameters in (9–12) indicate the high reliability of the use of the SDD basis set.

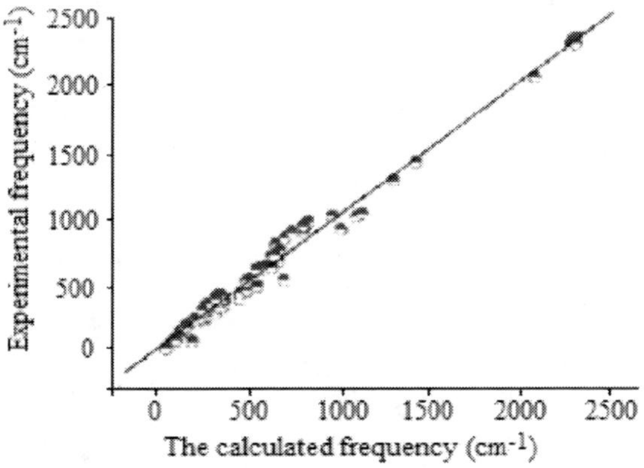

Figure 4. Relationship between experimental IR and Raman frequencies and those calculated by the B3LYP/SDD method for selenium compounds.

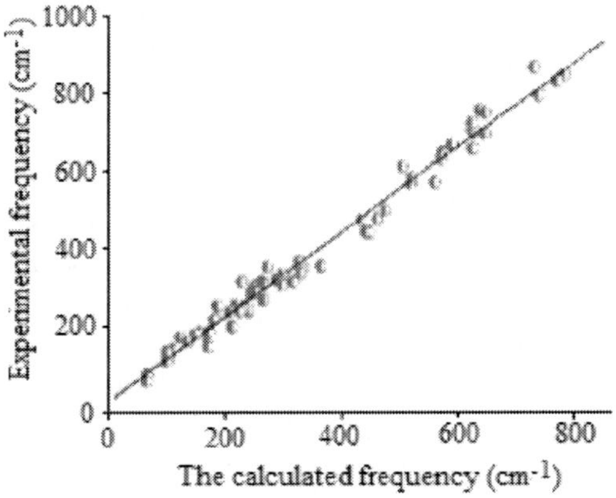

Figure 5. Relationship between experimental IR and Raman frequencies and those calculated by the B3LYP/SDD method tellurium compounds.

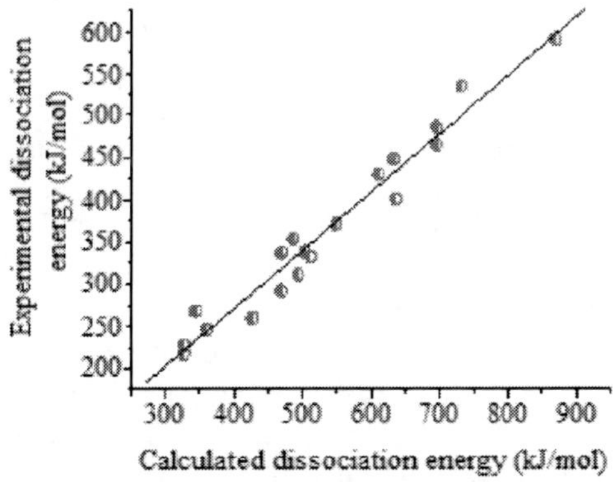

Figure 6. Relationship between experimental dissociation energies and those calculated by the B3LYP/SDD method of selenium and tellurium compounds.

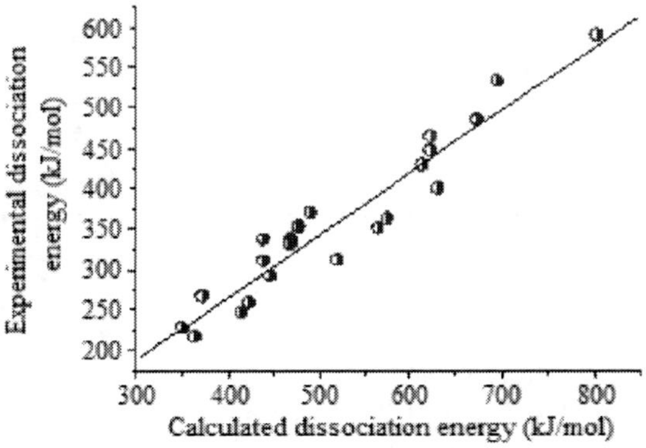

Figure 7. Relationship between experimental dissociation energies and those calculated by the ADF method of selenium and tellurium compounds

We have also obtained quite reliable linear correlation (13 and 14), which suggests that the calculations for compounds of selenium and tellurium by the B3LYP/SDD and BP86/TZ2P+ methods correctly describe not only the geometry, but also the thermodynamic parameters, in particular the dissociation energy (D_0). The energy of dissociation of selenium and tellurium compounds was determined as the difference between the energy of the compound and the energies of their constituent atoms, as amended by the enthalpy of formation. The calculated dissociation energy, relative to the number of bonds, was compared with the experimental average binding energy of selenium- selenium and tellurium-tellurium [19] (Table 3).

As a result we obtained a correlation ratio based on the two methods of calculation using the GAUSSIAN and ADF programs (Fig. 6, 7):

GAUSSIAN:

$$D_o\ (exp.) = -44 + 0.77\ D_o\ (calc.)\ r = 0.952;\ s = 30;\ n = 23 \quad (13)$$

ADF:

$$D_o\ (exp.) = 6 + 0.68\ D_o\ (calc.)\ r = 0.982;\ s = 18;\ n = 23 \quad (14)$$

On the basis of optimized structures for PbS, PbSe and PbTe in the gas phase, we have assessed the effect of different isotopes of sulphur (^{32}S-^{34}S), selenium (^{76}Se-^{82}Se), tellurium (^{122}Te-^{130}Te) and lead (^{204}PbS-^{208}Pb) on the spectral and thermodynamic parameters of lead chalcogenides. For the most intense bands of lattice oscillations in the Raman spectra calculated by the B3LYP/SDD method: (~ 120 cm^{-1}) for the lead sulphide; (~ 68 cm^{-1}) for lead selenide; and (~ 81 cm^{-1}) for lead telluride; we obtained multiple correlation ratios between the calculated Raman frequencies (ω) and the masses of the isotopes of lead, sulphur, selenium, and tellurium (M_{Pb}, M_S, M_{Se}, M_{Te}):

$$\omega = 167 - 0{,}26M_S - 0{,}74M_{Pb};\ r = 0.980;\ s = 0.2;\ n = 11 \quad (15)$$

$$\omega = 100 - 0{,}335M_{Se} - 0{,}027M_{Pb};\ r = 0.99996;\ s = 0.017;\ n = 16 \quad (16)$$

$$\omega = 121 - 0{,}25M_{Te} - 0{,}044M_{Pb};\ r = 0.999;\ s = 0.02;\ n = 12 \quad (17)$$

For all chalcogenides the error of calculation does not exceed 0.02 %, indicating that the calculation is correct. For lead chalcogenides we also obtained multiple correlations between the energy of the unit cell PbS (Se) (Te) with a correction to the energy (E_0) and the masses of the isotopes of lead, sulphur, selenium and tellurium:

$$E_0 = 0.63 + 7.13\cdot10^{-5}M_S + 5.46\cdot10^{-6}\ M_{Pb};$$
$$r = 0.999;\ s = 2.96\cdot10^{-6};\ n = 16 \quad (18)$$

$$E_0 = 0.216 + 1.925\cdot10^{-5}M_{Se} + 4.77\cdot10^{-6}\ M_{Pb};$$
$$r = 0.9997;\ s = 8.84\cdot10^{-7};\ n = 16 \quad (19)$$

$$E_0 = 0.6 + 1.19\cdot10^{-5}M_{Te} + 4.77\cdot10^{-6}\ M_{Pb};$$
$$r = 0.9996;\ s = 6.46\cdot10^{-7};\ n = 12 \quad (20)$$

A similar dependence is also observed for a lattice of four elemental isotopes of lead:

$$E_0 = -48.2 - 1.5\cdot10^{-5}\ M_{Pb};\ r = 0.995;\ s = 2.9\cdot10^{-6};\ n = 4 \quad (21)$$

To further examine the effect of the isotopes on the physico-chemical properties, we performed optimization of the unit cell of PbS, PbSe and PbTe using the PM3 method. The calculated bond lengths for Pb-S, Pb-Se and Pb-Te totalled 3.5 Å, 3.57 Å and 3,45 Å, close to the experimental values of 2.97 Å, 3.06 Å and 3.23 Å. It should be noted that our calculation using the FLAPW program for crystalline PbS, PbSe and PbTe gave the best value of the lattice constants at 6.01 Å, 6.23 Å and 6.57 Å, close to the experimental values of 5.94 Å, 7.1 Å and 6.75 Å, respectively.

One of the most important characteristics of semiconductors is the value of the energy gap. The calculation in the solid phase for PbS, PbSe and PbTe gave values of 0.5, 0.44 and 0.70 eV and in the gas phase 0.38, 0.36 and 0.28 eV using the GAUSSIAN program. The experimental width of a single-crystal PbS is 0.41 eV [17]. Thus, we can assume that the calculations for the gas phase and crystal lead chalcogenides gave values fairly close to the experimental values for the energy characteristics. Deviation of the calculated energy gap in gas phase from the FLAPW calculation for PbTe apparently is due to the inaccuracy of the parameters for Te by the PM3 method.

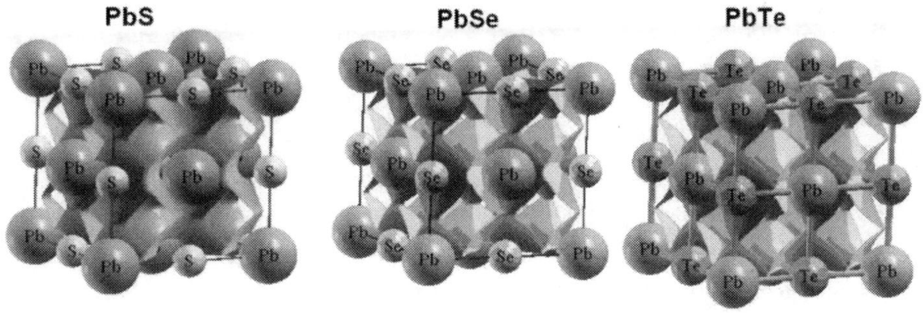

Figure 8. Isoelectronic surfaces (ρ = 0.15 e/Å3) cubic (type B1) PbS, PbSe and PbTe.

Based on the calculations in the solid phase (FLAPW) and in the gaseous phase by ADF we can estimate a bond in terms of ionic (electrostatic) and covalent (orbital) components. FLAPW data indicate a mixed ionic-covalent type (Fig. 8). The ionic component is created by the charge polarization in the direction of Pb → S, Pb → Se and Pb → Te in an ideal ion model of the charge states of $Pb^{2+}Hal^{2-}$. In reality, the effective charges of the atoms are smaller due to the covalent contribution. The covalent constituent is formed by the partially overlapping Halnp - Pb6s, 6p states. The covalent constituent can be illustrated as in Fig. 8, which shows the overlapping of isoelectronic surfaces in the direction Pb-(S, Se, Te). In the ADF program the energy of the molecule is split into several contributions. Energy members can be identified by the three main components of the chemical bond: repulsive Pauli, electrostatic attraction and covalent interactions. Frenking and co-workers [20] suggested that $\Delta E_{elstat.}$ could be used to assess the strength of the electrostatic binding and $\Delta E_{orb.}$ for covalent binding. These results for the calculation of lead chalcogenides in the gaseous phase also indicate that the chemical bond is composed of ionic and covalent components with a predominance of the latter (65% for S, 79% for Se and 56% in the case of Te).

4.Conclusions

Thus, the results suggest that the use of the density functional method in various versions with different functionals and basic conditions, such as B3LYP (GAUSSIAN), BP86 (ADF), GGA (FLAPW) presents an opportunity to evaluate not only the spectral isotope shifts, but also the thermodynamic parameters of the structures of lead chalcogenides containing different isotopes of lead, selenium and tellurium.

Acknowledgements

This work is financially supported by Tomsk Polytechnic University within the Program of Improving the Competitiveness.

References

[1] Dashevsky Z, Kasiyan V, Radovsky G, Shufer E and Auslender M 2008 *Proceedings of SPIE* **7142** 71420

[2] Tsidil'kovskii I M 1972 *Electrons and holes in semiconductors* (Moscow, Nauka)

[3] Sherwin R, Clark R J H, Lauck R and Cardona M 2005 *Solid State Communications* **134** 565

[4] Etchegoin P G, Cardona M, Lauck R, Clark J H, Serrano J and Romero A H 2008 *Physica Status Solidi B* **245** 1125

[5] Andrienko O S, Gusev A L, Egorov N B, Zherin I I and Indyk D V 2011 *Perspektivnye Materialy* **10** 130

[6] Frisch M J, Trucks G W, Schlegel H B, Scuseria G E, Robb M A, Cheeseman J R, Montgomery J J A, Vreven T, Kudin K N, Burant J C et al 2003 *Gaussian 03 Revision B03* (Pittsburg: Gaussian Inc.)

IOP Conf. Series: Materials Science and Engineering **112** (2016) 012022 doi:10.1088/1757-899X/112/1/012022

[7] Becke A D 1993 *Journal Chemical Physics* **98** 5648
[8] Lee C, Wang Y and Parr R G 1988 *Physical Review* **37** 785
[9] Curtiss L A, Raghavachari K, Redfern P C and Pople J 1997 *Journal Chemical Physics* **106** 1063
[10] Cao X Y and Dolg M 2002 *Journal of Molecular Structure* **581** 139
[11] *ADF2004.01. SCM. Theoretical Chemistry* (Amsterdam: Vrije Universities)
[12] Velde G, Bickelhaupt F M, Baerends E J, Guerra C, Van Gisbergen S J A, Snijders J G and Ziegler T 2001 *J. Comput.Chem.* **22** 931
[13] Stewart J J P 1989 *J. Comput. Chem.* **10** 209
[14] Blaha P, Schwarz K, Madsen G K H, Kvasnicka D and Luitz J 2001 *WIEN2k, An Augmented Plane Wave Plus Local Orbitals Program for Calculating Crystal Properties* (Vienna: Vienna University of Technology)
[15] Perdew J P, Burke K and Ernzerhof M 1996 *Phys. Rev. Let.* **77** 3865
[16] Blochl P E, Jepsen O and Anderson O K 1994 *Phys. Rev. B* **49** 16223
[17] Batsanov S S 2000 *Structural Chemistry. Facts and Dependences* (Moscow: Dialog-MGU)
[18] Nakamoto K 1997 *Infrared and raman spectra of inorganic and coordination compounds* part 1 (New York: Wiley)
[19] Linstrom P J and Mallard W G 2010 *NIST Chemistry WebBook, NIST Standard Reference Database Number 69* (Gaithersburg: National Institute of Standards and Technology)
[20] Frenking G, Wichmann K, Fronlich K, Loschen C, Lein M, Frunzke J and Rayon V M 2003 *Coordination Chem. Rev.* **55** 23

IOP Conf. Series: Materials Science and Engineering **112** (2016) 012023 doi:10.1088/1757-899X/112/1/012023

Nucleation and crystallization behavior of RE – doped tellurite glasses

V Goncharuk[1,2] *, A Mamaev[2], V Silant'ev[2], P Starodubtsev[3] and I Maslennikova[1,2]

[1] Far-Eastern Federal University, 8, Sukhanova Str., Vladivostok, 690950, Russia
[2] Far Eastern Branch of Russian Academy of Sciences, 159 Pr. 100-letiya Vladivostoka, Vladivostok, 690022, Russia
[3] Pacific S O Makarov Higher Naval School, 6 Per. Kamskyi, Vladivostok, 690062, Russia

*E-mail: gon@ich.dvo.ru

Abstract. The microstructure and crystallization of the glasses with composition (100-x-y)TeO_2-xPbO·P_2O_5-yPbF_2:zMF_3 (M= Er, Eu, Nd; x=42.5-30, y=5-30, z=0.5-3.0) were investigated by transmission electron microscopy (TEM) and luminescence methods. It was found that the doping with the rare-earth (III) fluorides promoted nucleation in the bulk glasses. The sizes of generated particles are about 2–5 nanometers and their shapes are close to spherical. The growth rate of crystallites depended on the lead fluoride content and glass forming rate. The heat treatment of the samples promotes the glass ceramic formation, where the crystalline phase is $Pb_2P_2O_7$.

1. Introduction

Tellurium dioxide based glasses are of increasing interest due to a possibility of their application as components of the optical data storage devices and telecommunication systems [1, 2]. It is possible to use these glasses in production of lasers because of the high solubility of rare-earth ions and relatively low phonon energy among oxide glass-formers [3]. Doped glass ceramics are the most perspective materials for these applications. Rare-earth fluorides often act as nucleating centers [4].

Preparation of oxyfluoride glasses in the ternary system TeO_2-PbF_2-CdF_2 was reported previously [5]. Recently the $Pb(PO_3)_2$-TeO_2 glass system was studied in the whole range of glass composition [6]. Furthermore, a glass formation of the TeO_2-PbO·P_2O_5-PbF_3: MF_3 system was investigated [7].

We analyzed in [7] the thermal properties and the ionic mobility in the system TeO_2-PbO·P_2O_5-PbF_3: MF_3 (M = Er, Eu, Nd) by the DTA and NMR methods. This paper is a continuation of the work mentioned above. Our purpose was to consider an effect of the rare-earth (III) fluorides on crystallization of the TeO_2-PbO·P_2O_5-PbF_3: MF_3 system.

2. Experimental section

Glass samples compositions of (100-x-y)TeO_2-xPbO·P_2O_5-yPbF_2 : zMF_3 (M=Er, Eu, Nd, x=42.5-30 mol. %, y=5-30 mol. %, z=0.5-3.0 mol. %) were examined. Influence of rare-earth fluorides concentration on glass nucleation was investigated. Samples contained 5 and 20 mol. % of PbF_2 for

Content from this work may be used under the terms of the Creative Commons Attribution 3.0 licence. Any further distribution of this work must maintain attribution to the author(s) and the title of the work, journal citation and DOI.
Published under licence by IOP Publishing Ltd

$52.5TeO_2$-$42.5PbO \cdot P_2O_5$-$5PbF_2$ and $45TeO_2$-$35PbO \cdot P_2O_5$-$20PbF_2$, respectively, were doped with 0.5÷3.0 mol. % of ErF_3 (in addition to 100 % of the basic glass). Reagents preparation, glass synthesis, data of differential thermal analysis (DTA) and X-ray diffraction (XRD) discussed in more details in [**Ошибка! Закладка не определена.**].

2.1. Samples preparation
Thin plates of glasses were prepared by two techniques: 1) slow cooling to the room temperature of the glass melt that was poured on a bulky nickel plate heated to the temperature below then Tg on 10÷15 °C (to 220-300°C against to constitution); 2) fast cooling between two cold nickel plates.

2.2. Luminescence spectra
Luminescence spectra were recorded in the IR region on spectrometer SDL-1 (Russia). Xenon lamp DKsSH-150 with combined filter SZS23+SZS25 was used for illumination of the samples. Luminescence intensity was measured on 1054 and 1550 nm for Nd(III) and Er(III), respectively.
Visible spectra of the glasses were recorded on RF-5301 (Shimadsu, Japan) at 300 K.

2.3. TEM
TEM images were obtained using Libra-200 (Carl Zeiss, Germany). Samples preliminarily were dispersed in acetone through the Bandelin Sonopuls HD-2070 (Germany) with wedge-shaped tip MS-73. Ultrasonic dispersing mode was 30 min and 5 cycles. Droplet of this dispersion was placed onto a cooper grid covered with an amorphous polymer formvar and dried at 40 °C.

3. Results
Compositions and thermal properties of the samples $(100-x-y)TeO_2$-$xPbO \cdot P_2O_5$-$yPbF_2$: zMF_3 (M= Er, Eu, Nd, x=42.5-30 mol. %, y=5-30 mol. %, z=0.5-3.0 mol. %) were considered in [**Ошибка! Закладка не определена.**]. Their x-ray amorphous structures confirmed the existence of glassy phases. All glass samples with 5 mol. % of rare-earth (III) fluorides were transparent and slightly yellow. The color became slightly pink after erbium fluoride doping, light violet with the addition of neodymium fluoride and slightly green in the case of europium fluoride.

The $52.5TeO_2$-$42.5PbO \cdot P_2O_5$-$5PbF_2$ glass remained transparent with addition up to 2.0 mol. % of ErF_3. Thickness of these samples was 3 mm. The opaque appeared after introduction of 3.0 mol. % of ErF_3. Compositions with 20 mol. % of PbF_2 showed opalescence after addition of 1 mol. % of erbium fluoride (thickness of the glasses was ~ 4 mm). Further ErF_3 doping up to 2.0 mol. % led to non-transparency of the samples.

The effect of non-transparency indicates indirectly on the formation of crystallites in the bulk of glass. It is interesting to note that it was impossible to recognize crystallization visually. XRD data confirmed amorphous structures of the samples.

According to luminescence data increasing the PbF_2 (figure 1) and rare-earth (III) fluorides (figure 2) concentrations led to rise of both neodymium(III) and erbium(III) luminescence intensity in the IR range. Similar dependence observed for europium(III) luminescence in the visible range (figure 3). Luminescence of erbium(III) at typical wavelength of 550 nm was very weak. But another intensive emission band appeared at 450 nm. The ErF_3 doped glasses didn't show visual luminescence under UV exposure.

Increasing of luminescence intensity evidences that crystallite structures appear in samples and rare-earth ions segregate into crystallite particles. These particles must be nanosized due to the absence of crystal phases on XRD spectrum.

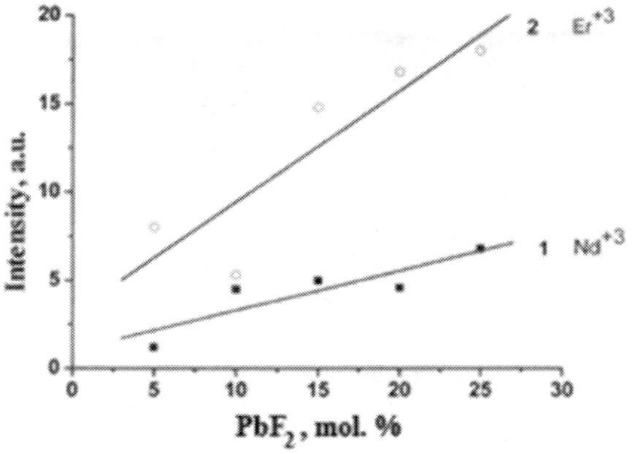

Figure 1. Dependences of the luminescence intensities at wavelengths 1054 (Nd^{+3}) and 1550 (Er^{+3}) nm on PbF_2 concentration in the samples (100-x-y)TeO_2-xPbO·P_2O_5-yPbF_2:0.5MF_3 (M=Er, Nd, x=42.5-30, y=5-30 mol. %)

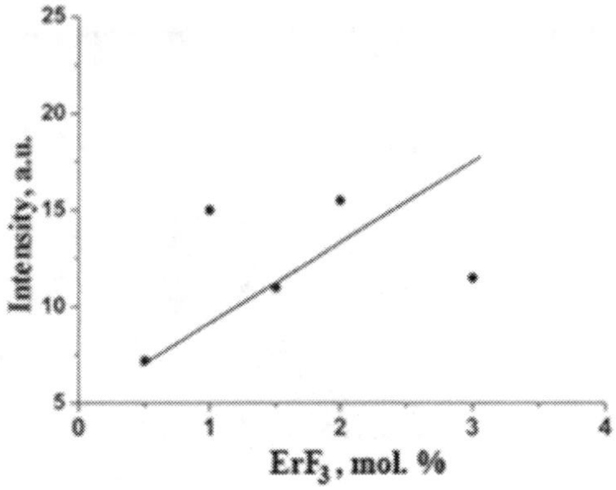

Figure 2. Dependence of the luminescence intensities at wavelength 1550 nm (Er^{+3}) on ErF_3 concentration in the samples 52.5TeO_2-42.5PbO P_2O_5-5PbF_2: xErF_3 (0.5<x<3.0).

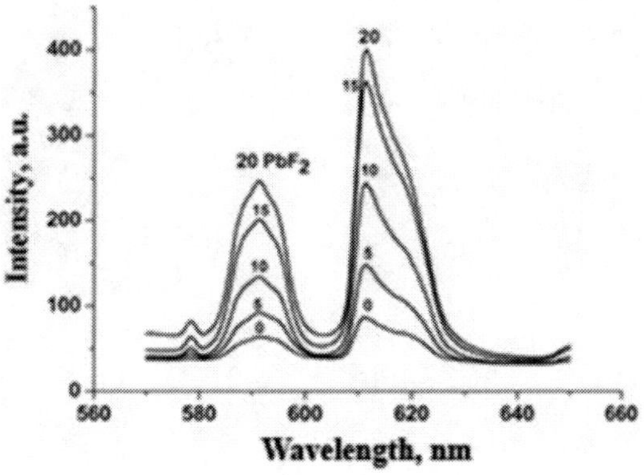

Figure 3. Dependences of the luminescence intensities of Eu^{+3} on PbF_2 concentration in the samples (100-x-y)TeO_2-xPbO·P_2O_5-yPbF_2 :0.5EuF_3 (x=42.5-30, y=5-30 mol. %).

TEM images confirm the existence of the structures with approximate dimensions 2–5 nm. It is seen from the photos (figures 4–7) that shape of the particles is closed to spherical. Moreover, their number and dimensions depend on concentration of the rare-earth (III) fluorides, lead fluoride and the rate of glass formation. Increasing of the ErF_3 doping results in growing amount of the nucleation centers. There are a lot of particles with small dimensions (figure 5). Increasing of the PbF_2 concentration in glass compositions with small ErF_3 doping results in growing of particle dimensions (figure 6). Fast cooling on cold plates decreases dramatically the number of particles and the rate of their growing (figures 8, 9). On the other hand, slow cooling results in aggregation of crystallite nanoparticles and in their growing rate.

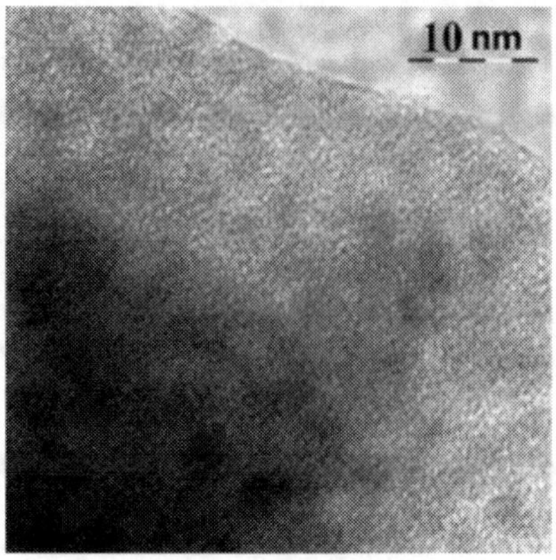

Figure 4. TEM image of the $52.5TeO_2$-42.5PbO P_2O_5-$5PbF_2$:$0.5ErF_3$ sample, hot plate.

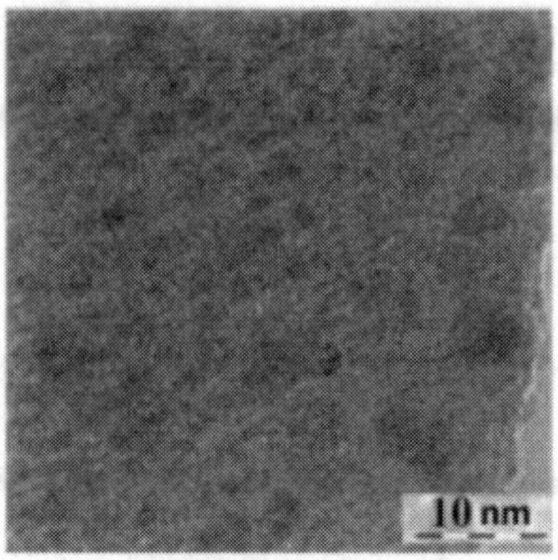

Figure 5. TEM image of the $52.5TeO_2$-42.5PbO P_2O_5-$5PbF_2$:$3.0ErF_3$ sample, hot plate.

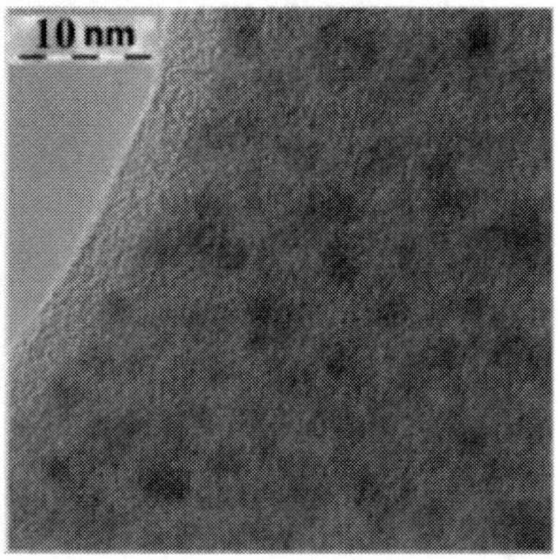

Figure 6. TEM image of the $40TeO_2$-30PbO P_2O_5-$30PbF_2$:$0.5ErF_3$ sample, hot plate.

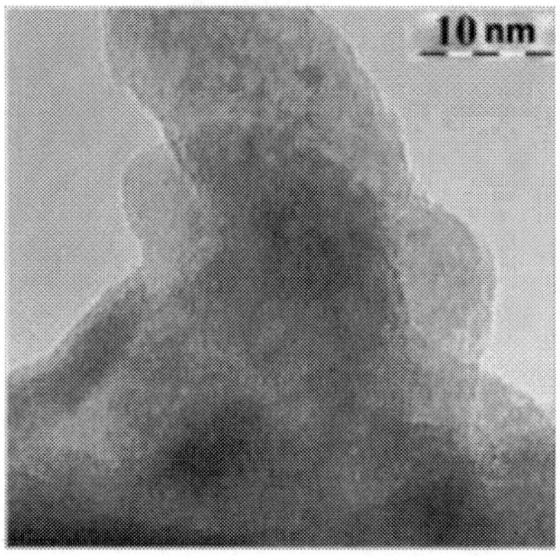

Figure 7. TEM image of the $40TeO_2$-30PbO P_2O_5-$30PbF_2$:$0ErF_3$ sample, hot plate.

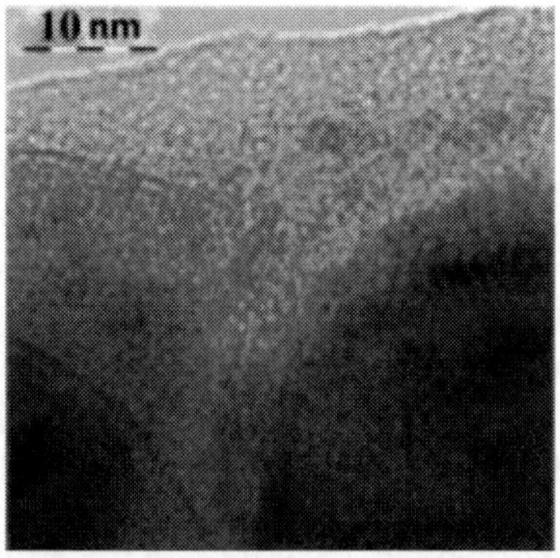

Figure 8. TEM image of the $40TeO_2$-30PbO P_2O_5-$30PbF_2$:$0.5ErF_3$ sample, cold plate.

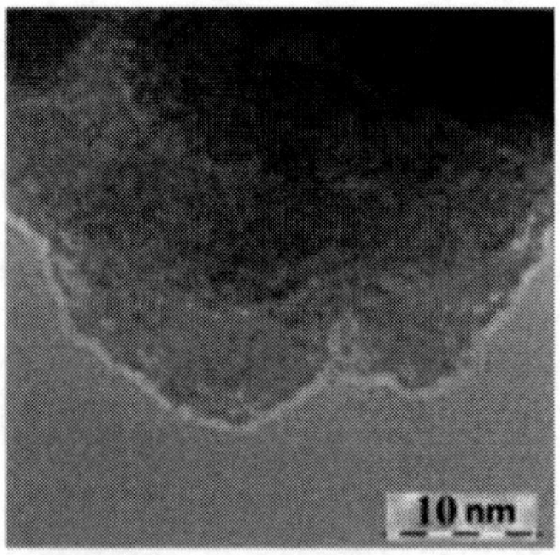

Figure 9. TEM image of the $52.5TeO_2$-42.5PbO P_2O_5-$5PbF_2$: $3.0ErF_3$ sample, cold plate.

Thus, erbium fluoride plays the role of nucleation agent. Previously, this fact was noted for other glass systems [8, 9, 10].

Heat treatment of the glasses consisted of 5–10 mol. % of PbF_2 and doped with neodymium or europium fluorides during 1–1.5 hours at 490 °C (the range Tg-Tx) promoted glass ceramics formation. Composition of the crystal phase corresponded to $Pb_2P_2O_7$ that was confirmed by the XRD data. Probably, actual composition of the crystal nanoparticles corresponded to $Pb_2P_2O_7$:REF_3.

We kept the 52.5 TeO_2–42.5 $PbO{\cdot}P_2O_5$-5 PbF_2 : 0.5 NdF_3 glass at 490 °C during 0.5; 1, 2 and 3 hours to determine the influence of annealing time on glasses stability. After 1 hour of heat treatment, weak reflexes of $Pb_2P_2O_7$ appeared on XRD spectrum. Their intensity rises with the increase of annealing time. Samples remained transparent up to 2 hours of heat treatment. After 3 hours they became opaque.

The crystal phase formation of $Pb_2P_2O_7$ occurred in the case of the $(PbO)_{0.50}(P_2O_5)_{0.50-x}(TeO_2)_x$ system [11]. This phase was produced during the annealing of the glasses with molar ratio $(PbO)/(P_2O_5)>1$.

4. Conclusions

Nucleation and crystallization phenomena in the (100-x-y)TeO_2-x$PbO{\cdot}P_2O_5$-yPbF_2 : zMF_3 (M= Er, Eu, Nd, x=42.5-30 mol. %, y=5-30 mol. %, z=0.5-3.0 mol. %) glasses were investigated. It was found that the rare-earth (III) fluorides doping results in the formation of spherical nanosized crystallites in the bulk glass and segregation of rare-earth ions from glassy phase to crystallites. ErF_3 plays the role of nucleation agent in glasses. Heat treatment of the investigated glasses at suitable conditions enables to produce transparent glass ceramics.

Acknowledgments

The study was supported by the Ministry of Education and Sciences of the Russian Federation, Agreement # 14.578.21.0024, RFMEFI57814X0024.

References

[1] Ribero C, Richardson K, Stegeman R, Cardinal T, Fargin E, Couzi M and Rodriguez V 2004 *J. Non-Cryst. Solids*. **345–346** 396
[2] Zhao S, Wang X, Fang D, Xu S and Nu L 2006 *J. Alloys Compd.* **427** 243
[3] Xudong Z, Tiefeng X, Qiuhua N, Shixun D, Xiang S and Xianghua Z 2006 *J. Rare Earth*. **24** 771
[4] Yu H, Zhou K, Chen R, Song J, Hou Ch and Zhao L 2008 *J. Non-Cryst. Solids.* **354** 3649
[5] Silva M A P, Messaddeq Y, Briois V, Poulain M, Villain F and Ribero S J L 2002 *J. Phys. Chem. Solids*. **63** 605
[6] Apakova I E, Goncharuk V K, Maslennikova I G, Polyakova I G, Anan'ev A V, Maksimov LV, Markova T S and Yanush O V 2010 *Glass Phys. Chem.* **36** 637
[7] Alexeiko L N, Kavun V Ya, Kharchenko V I, Slobodyuk A B, Maslennikova I G and Goncharuk V K 2014 *J. Non-Cryst. Solids*. **401** 119
[8] Mortier M and Patriarche G 2000 *J. Mater. Sci.* **35** 4849
[9] Dantelle G, Mortier M, Patriarche G and Vivien D 2006 *J. Solid State Chem.* **179** 1995
[10] Kukkonen L L, Reaney I M, Furniss D, Pellat M G and Seddon A B 2001 *J. Non-Cryst. Solids.* **290** 25
[11] Sudarsan V, Mishra R and Kulshreshtha S K 2004 *J. Non-Cryst. Sol.* **342** 160

Hydrofluoride decomposition of natural materials including zirconium–containing minerals

N Laptash[1] **and I Maslennikova**[1,2]
[1] Institute of Chemistry, Far Eastern Branch of Russian Academy of Sciences, 159 Pr. 100-letiya Vladivostoka, 690022 Vladivostok, Russia
[2] Far-Eastern Federal University, 8, Sukhanova Str., 690950 Vladivostok, Russia

E-mail: laptash@ich.dvo.ru

Abstract. Recently, interest in ammonium hydrogen difluoride (NH_4HF_2) as a versatile fluorinating agent for the decomposition of natural materials resumed. It is considered to be a new and more efficient than hydrofluoric acid (HF) reagent in analytical chemistry. Thermodynamically possible fluorination reactions with NH_4HF_2 are exothermic and proceed even at room temperature with the entropy reserve. The fluorination products are of high symmetry phases (tetragonal or cubic) with partial substitution of fluoride ion for oxide (or hydroxide). The fluorination of refractory silicate zircon ($ZrSiO_4$) is kinetically hindered, and its complete decomposition requires the use of a Teflon autoclave at 200°C. The fluorination products are cubic $(NH_4)_3Zr(OH)_xF_{7-x}$ ($x \leq 0.3$) and tetragonal double salt $(NH_4)_3SiF_7$, which can be separated due to incongruent sublimation of $(NH_4)_2SiF_6$. The mechanism of the latter process is proposed.

1. Introduction

It is known that geological samples are not easy to dissolve, due to the presence of resistant minerals, such as garnets, spinel, and zircon. Usually, geological sample dissolution of refractory silicates requires hydrofluoric acid and a mixture of oxidizing acids, such as nitric and perchloric, in the case of open acid digestion. Then, the choice of decomposition method depends on the mineralogical composition of the rock [1]. In recent years, a simple, effective, and comparatively safe dissolution method with the use of ammonium fluorides (NH_4HF_2 or NH_4F) shows great potential for the digestion of geological samples. NH_4HF_2 was used in innovative method for preparation of mullite-based refractories from aluminosilicate quartz-topaz ores [2] or for the synthesis of mullite from kaolin [3]. A new technique of decomposing beryl ore [4] or a treatment of gold-containing graphite-bearing ores with NH_4HF_2 to concentrate gold [5] was recently suggested. Tantalite dissolution using NH_4HF_2 as alternative to HF [6] or processing of tantalite ore from Mozambique with molten NH_4HF_2 [7] were also described. For decomposition of refractory zircon, an autoclave breakdown with ammonium fluorides [8] or microwave digestion with NH_4HF_2 were studied [9, 10].

Today, the NH_4HF_2 acid digestion is recognized as a new method in analytical chemistry which is practical and effective and has lower toxicity and cost, compared with traditional hydrofluoric acid digestion [11–13]. However, we suggested this method earlier for analysis of silicon-containing materials [14, 15].

Content from this work may be used under the terms of the Creative Commons Attribution 3.0 licence. Any further distribution of this work must maintain attribution to the author(s) and the title of the work, journal citation and DOI.
Published under licence by IOP Publishing Ltd

2. Fluorination reactions with NH_4HF_2

Ammonium hydrogen difluoride is a solid, so it is more convenient in handling than HF. It melts at 126°C and sublimes incongruently at 239°C. During its decomposition, adducts of HF with NH_3 are formed [16]:

$$NH_4HF_2(s) \rightarrow H_3N\text{–}HF(g) + HF(g);\ 2NH_4HF_2(s) \rightarrow H_3N\text{–}HF(g) + HF(g) + H_3N\text{–}HF\text{–}HF(g) \quad (1)$$

Table 1 represents fluorination reactions of silicon-containing minerals of different compositions and structure with NH_4HF_2. All thermodynamically possible reactions with the formation of silicon double salt $(NH_4)_3SiF_7 = (NH_4)_2SiF_7{\cdot}NH_4F$ are exothermic.

Table 1. Thermodynamic function changes of fluorination reactions of silicon-bearing minerals with NH_4HF_2

Mineral	Fluorination reaction	$-\Delta H^o_{298}$ kJ	$-\Delta G^o_{298}$ kJ
quartz	$SiO_2 + 3.5NH_4HF_2 = (NH_4)_3SiF_7 + 0.5NH_3 + 2H_2O$	21.3	65.0
wollasto-nite	$CaSiO_3 + 4.5NH_4HF_2 = CaF_2 + (NH_4)_3SiF_7 + 1.5NH_3 + 3H_2O$	44.9	140.8
larnite	$Ca_2SiO_4 + 5.5NH_4HF_2 = 2CaF_2 + (NH_4)_3SiF_7 + 2.5NH_3 + 4H_2O$	112.9	61.5
zircon	$ZrSiO_4 + 7NH_4HF_2 = (NH_4)_3ZrF_7 + (NH_4)_3SiF_7 + NH_3 + 4H_2O$	53.0	134.5
kaolinite	$Al_2Si_2O_7{\cdot}2H_2O + 13NH_4HF_2 = 2(NH_4)_3AlF_6 + 2(NH_4)_3SiF_7 + NH_3 + 9H_2O$	488.0	605.8
topaz	$Al_2SiO_4(OH)_2 + 9.5NH_4HF_2 = 2(NH_4)_3AlF_6 + (NH_4)_3SiF_7 + 0.5NH_3 + 6H_2O$	467.0	583.3
chlorite	$Mg_6Si_4O_{10}(OH)_8 + 20NH_4HF_2 = 6MgF_2 + 4(NH_4)_3SiF_7 + 8NH_3 + 18H_2O$	39.0	580.3
diopside	$CaMgSi_2O_6 + 9NH_4HF_2 = CaF_2 + MgF_2 + 2(NH_4)_3SiF_7 + 3NH_3 + 6H_2O$	56.9	249.9
forsterite	$Mg_2SiO_4 + 5.5NH_4HF_2 = 2MgF_2 + (NH_4)_3SiF_7 + 2.5NH_3 + 4H_2O$	58.1	204.1
sphene	$CaTiSiO_5 + 8NH_4HF_2 = CaF_2 + (NH_4)_3TiF_7 + (NH_4)_3SiF_7 + 2NH_3 + 5H_2O$	28.7	164.9
anortite	$CaAl_2Si_2O_8 + 14NH_4HF_2 = CaF_2 + 2(NH_4)_3AlF_6 + 2(NH_4)_3SiF_7 + 2NH_3 + 8H_2O$	539.9	680.3

Indeed, a well pronounced exoeffect can be observed on DTA curves of a framework-structure quartz (figure 1), ribbon amphibole (figure 2), chain chrome diopside (figure3) with NH_4HF_2. The layered minerals such as chlorite (figure 4) or vermiculite (figure 5) are fluorinated easier. In the latter case, the exoeffect does not appear because of complete fluorination of vermiculite during the grinding of initial components at room temperature. The same concerns the natural mineral scheelite. Elemental silicon just inflames under grinding with NH_4HF_2.
The corresponding reactions can be expressed as follows:

$$SiO_2 + 3.5NH_4HF_2 = (NH_4)_3SiF_7 + 0.5NH_3 + 2H_2O \quad (2)$$

$$Ca_2Mg_3Fe^{2+}{}_2Fe^{3+}{}_{0.7}Al_2Si_6O_{22}(OH)_2 + 37.1NH_4HF_2 = 2CaF_2 + 3MgF_2 + 2NH_4FeF_3$$

$$+ 0.7(NH_4)_3FeF_6 + 2(NH_4)_3AlF_6 + 6(NH_4)_3SiF_7 + 9NH_3 + 24H_2O \quad (3)$$

$$CaMgSi_2O_6 + 9NH_4HF_2 = CaF_2 + MgF_2 + 2(NH_4)_3SiF_7 + 3NH_3 + 6H_2O \quad (4)$$

$$Mg_2Al_4Si_3O_{10}(OH)_8 + 24.5NH_4HF_2 = 2MgF_2 + 4(NH_4)_3AlF_6 + 3(NH_4)_3SiF_7 + 3.5NH_3 + 18H_2O \quad (5)$$

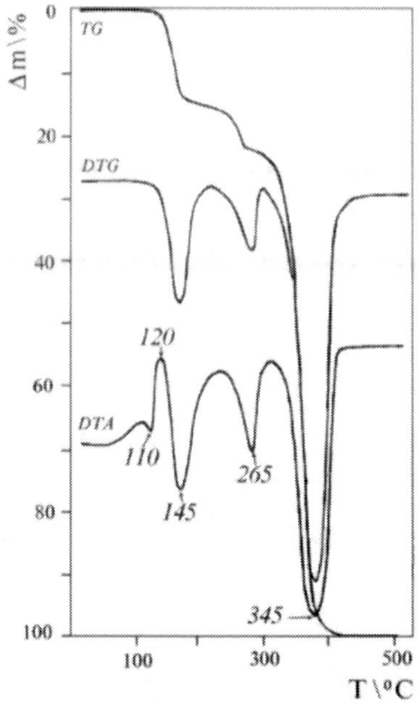

Figure 1. Thermal curves of the mixture of quartz with NH_4HF_2 in molar ratio of 1 : 3.5.

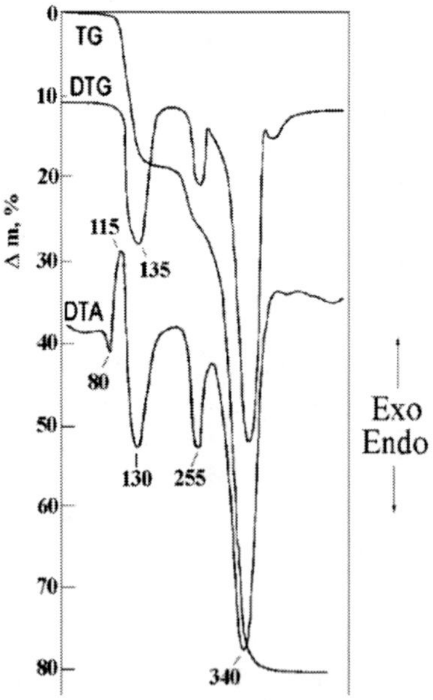

Figure 2. Thermal curves of amphibole with NH_4HF_2.

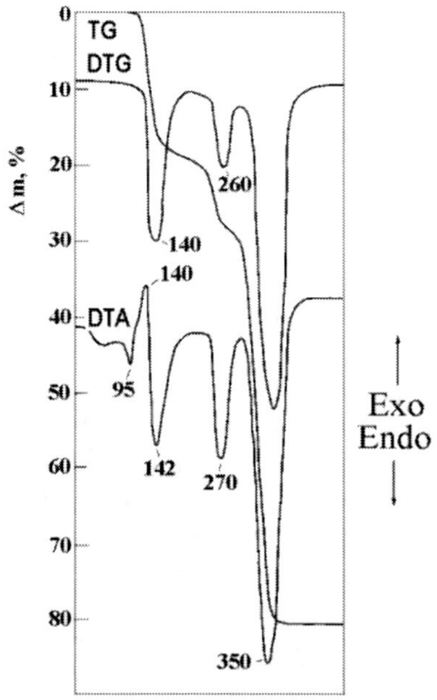

Figure 3. Thermal curves of chrome diopside with NH_4HF_2.

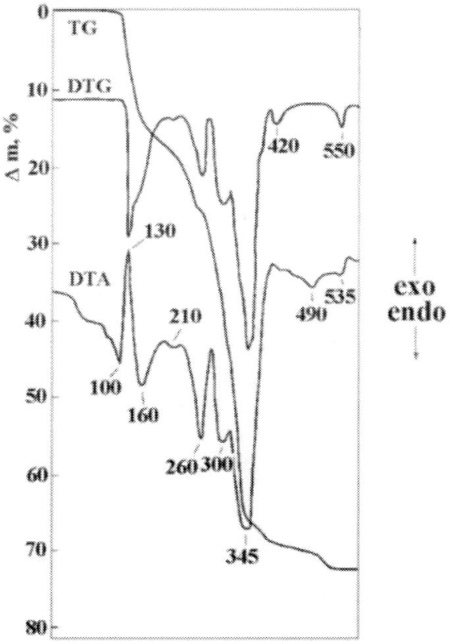

Figure 4. Thermal curves of chlorite with NH_4HF_2.

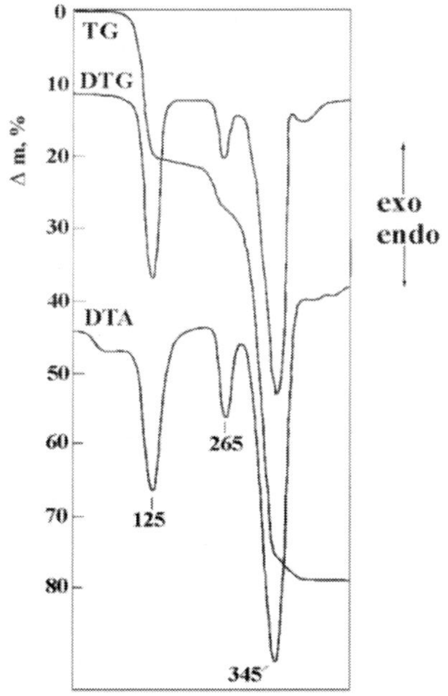

Figure 5. Thermal curves of vermiculite with NH_4HF_2.

$$CaMg_5FeAl_3Si_5O_{20}(OH)_4{\cdot}8H_2O + 35.5NH_4HF_2 = CaF_2 + 5MgF_2 + (NH_4)_3FeF_6 + 3(NH_4)_3AlF_6 + 5(NH_4)_3SiF_7 + 8.5NH_3 + 32H_2O \quad (6)$$

$$2CaWO_4 + 7NH_4HF_2 = 2CaF_2 + (NH_4)_2WO_3F_2 + (NH_4)_3WO_3F_3 + 2NH_4HF_2 + HF + 2H_20 \rightarrow 2CaF_2 + 2(NH_4)_3WO_2F_5 + NH_3 + 4H_2O \text{ [17]} \quad (7)$$

$$Si + 3.5NH_4HF_2 = (NH_4)_3SiF_7 + 2H_2 + 0.5NH_3. \quad (8)$$

We presented here stoichiometric formulae of the fluorination reaction products. In fact, they were nonstoichiometric due to the water molecules isolated during the reagent interactions, as we clearly demonstrated in the case of ilmenite fluorination with NH_4HF_2 [18, 19]:

$$0.8FeTiO_3{\cdot}0.1Fe_2O_3 + 4.4NH_4HF_2 = 0.8(NH_4)_3Ti(OH)_{0.4}F_{6.6} + 0.2(NH_4)_{2.8}Fe(OH)_{0.2}F_{5.6} + 0.8NH_4H_{0.4}FeO_{0.2}F_3 + 0.64NH_3 + 2.18H_2O \quad (9)$$

Ammonium fluoro- or oxofluorometallates (the fluorination products) are the phases of high symmetry: tetragonal double salts of silicon, germanium, tin, titanium and cubic fluoroperovskites or elpasolites characterized by strong dynamic orientational disorder [20, 21]. On cooling, these compounds undergo phase transitions (PT) of order-disorder type with a rather large value of entropy change. This means that fluorination reactions with NH_4HF_2 occur spontaneously with a reserve of entropy.

3. Fluorination of zircon-containing natural materials

It should be noted that natural mineral zircon ($ZrSiO_4$) with island structure is the most difficult to complete recovery with NH_4HF_2. Melnichenko et al. [22] mentioned that it was not decomposed even during 48 h at 200°C. Since its interaction with NH_4HF_2 is thermodynamically possible (see table 1), it is evident that the reaction is kinetically hindered. We managed to overcome this barrier using Teflon autoclave at 200°C (Teflon inner vessel and an outer stainless steel pressure jacket were used). For two hours, the samples were decomposed completely. Zircon was presented by the Brazil sample of Pocos de Caldas deposit. On the other hand, fine dispersive and amorphous sample of gel zircon of Algoma deposit (Khabarovsk region, Russia) interacts with NH_4HF_2 with a strong heating (exoeffect) under simple mixing of reagents. The main fluorination products are stoichiometric double salt $(NH_4)_3SiF_7$ and nonstoichiometric seven-coordinated $(NH_4)_3Zr(OH)_xF_{7-x}$ ($x \leq 0.3$). We refined the elpasolite-like crystal structure of the latter [23] to avoid abnormally short F–F distances in pentagonal bipyramid previously determined by Hurst and Taylor [24]. The structure is highly disordered (figure 6) with six independent orientations of $[ZrF_7]^{3-}$ and eight orientations of one ammonium group.

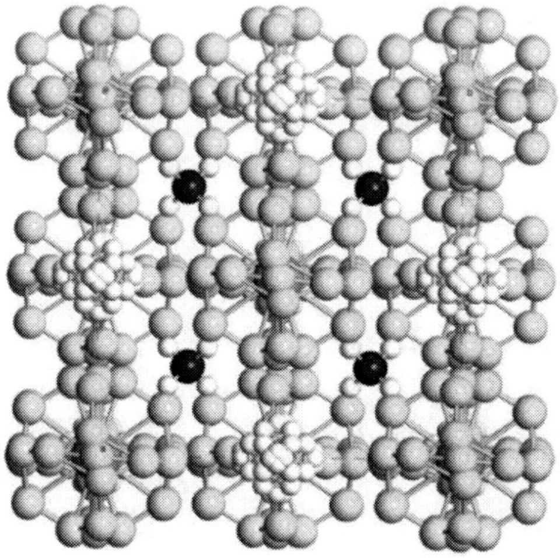

Figure 6. Dynamically disordered crystal structure of $(NH_4)_3ZrF_7$: sp. gr. *F*23, a = 9.4185(3) Å, Z = 4.

The former complex $(NH_4)_3SiF_7$ decomposes above 200°C with the formation of $(NH_4)_2SiF_6$ which is a rather unique compound. It does not hydrolyze and pyrohydrolyze, corresponds fully to its stoichiometry. The solubility in water is 18.75 %. It can be obtained in a highly pure form by incongruent sublimation and/or recrystaliization. The history of compound numbers more than 200 years but its thermal properties are still under discussion.

4. Sublimation of $(NH_4)_2SiF_6$

Davy was the first who described thermal properties of $(NH_4)_2SiF_6$ [25]. He mentioned that the compound "just perceptibly reddens litmus paper and appears to sublime unaltered when heated". Marignac described cubic (cubooctahedra) and hexagonal (thin plate or large flat crystal) [26] forms of $(NH_4)_2SiF_6$. "Heated up to 100°C, the latter became white and volatilized without melting, while cubooctahedra cracked and volatilized without albication". Mel'nichenko *et al.* suppose that $(NH_4)_2SiF_6$ sublimes mainly as a hole molecule and decomposes only partly through NH_4SiF_5, which is moisture-sensitive and hydrolyzes with the formation of volatile ammonium oxofluorosilicates [27]. However, the more accepted statement is that it decomposes with the formation of five gas molecules [28, 29]:

$$(NH_4)_2SiF_{6(s)} = SiF_{4(g)} + 2NH_{3(g)} + 2HF_{(g)} \tag{10}$$

Our tensimetric investigation of thermal behavior of cubic $(NH_4)_2SiF_6$ (cryptohalite) showed that, in fact, it decomposes with the formation of four gas molecules. Since the $(NH_4)_2SiF_6$ vapours have an acid reaction, it is logically to suppose that HF isolates and adduct of silicon fluoride with ammonia is formed. We tried to scavenge HF by elemental silicon, and in the mass-spectrum of the reaction product detected the fragment with the mass of 223 that corresponded to dinuclear silicon fluoride complex $Si_2F_7NH_2NH_3$, so we suppose that the process of incongruent sublimation of $(NH_4)_2SiF_6$ can be expressed as follows:

$$(NH_4)_2SiF_6 = Si(NH_3)_2F_4 + 2HF \quad (11)$$

$$Si(NH_3)_2F_4 = SiF_4 + 2NH_3 \quad (12)$$

$$Si(NH_3)_2F_4 + SiF_4 = Si_2F_7NH_2NH_3 + HF \quad (13)$$

The overal reaction is following:

$$2(NH_4)_2SiF_6 = Si_2F_7NH_3NH_2 + 5HF + 2NH_3 \quad (14)$$

Our preliminary quantum chemical calculations of dinuclear complex give the only viable structural model consisting of silicon octahedron and tetrahedron joined by the bridging NH_2 group (figure 7).

Figure 7. Structure of dinuclear silicon complex $Si_2F_7NH_2NH_3$ with bridging NH_2 group.

Sublimation properties of $(NH_4)_2SiF_6$ were used for separation of zirconium and silicon from Tugan zircon deposit of Tomsk region (Russia) [8]. We performed the trace elements analysis of precious opals of deposit Raduzhnoe (Primorsky region, Russia) with respect to the problem of their coloring using sublimation of $(NH_4)_2SiF_6$ [15]. Using NH_4HF_2 we could obtain the balanced results during the analysis of zircon raw material of Pocos de Caldas deposit (Brazil), when classical decomposition methods were very tedious and did not lead to balanced results.

5. Conclusions

The ease of use NH_4HF_2 digestion method suggests that it will become the method of choice in the future. This simple, effective, and comparatively safe dissolution method shows great potential for the decomposition of natural raw materials. Complete dissolution of Indian coal sample has been just achieved for the quantitative determination of major to ultra-trace elements using NH_4HF_2 instead of highly toxic HF [30].

Acknowledgments

The study was supported by the Ministry of Education and Sciences of the Russian Federation, Agreement # 14.578.21.0024, RFMEFI57814X0024.

References

[1] Pinto F G, Escalfoni R Junior and Saint'Pierre T D 2012 *Anal. Lett.* **45** 1537–56

[2] D'yachenko A N, Andreev A A, Buinovskii A S and Kraidenko R I 2006 *Refractories and Industrial Ceramics*. **47** 146–8
[3] Pogrebenkova V V, Vakalova T V, Gorbatenko V V and Grekhova M V 2010 *Refractories and Industrial Ceramics*. **51** 197–201
[4] Thorat D D, Tripathi B M and Sathiyamoorthy D 2011 *Hydrometallurgy*. **109** 18–22
[5] Medkov M, Khanchuk A, Molchanov V, Epov D, Krysenko G and Plyusnina L 2011 *Dokl. Chem*. **436** 15–8
[6] Nete M, Purcell W and Nel J T 2014 *J. Fluorine Chem*. **165** 20–6
[7] Kabangu M J and Crouse P L 2012 *Hydrometallurgy*. **129–130** 151–5
[8] Guzeev V V and D'yachenko A N 2006 *Russ. J. Appl. Chem*. **79** 1757–60
[9] Nel J T, du Plessis W, Nhlabathi T N, Pretorius C J, Jansen A A and Crouse P L 2011 *J. Fluorine Chem*. **132** 258–62
[10] Nhlabathi T N, Nel J T, Puts G J and Crouse P L 2012 *Int. J. Miner. Processing*. **114–117** 35–9
[11] Zhang W, Hu Z C, Liu Y S, Chen H H, Gao S, Richard M and Gaschnig R M 2012 *Anal. Chem*. **84** 10686–93
[12] Meng Z C and Zhang N 2012 *Anal. Letters* **45** 2086–95
[13] Zhang N, Gao X H, Shen Y, Wang C and Wang J T 2013 *Anal. Sci*. **29** 441–6
[14] Laptash N M and Kurilenko L N 2000 *Method of Determination of Silicon* RU Patent 2157523
[15] Laptash N M and Kurilenko L N 2007 *Method of Determination of Impurities in Silicon-Containing Materials* RU Patent 2306546
[16] Hunt S W, Higgins K J, Craddock M B, Brauer C S and Leopold K R 2003 *J. Am. Chem. Soc*. **125** 13850–60
[17] Laptash N M, Melnichenko E I, Polyshchuk S A and Kaidalova T A 1992 *J. Therm. Anal*. **38** 2335–2341
[18] Laptash N M and Maslennikova I G 2012 *Adv. Mater. Phys.Chem*. **2** 21–24
[19] Laptash N M, Maslennikova I G, Kurilenko L N and Mishchenko N M 2001 *Russ. J. Inorg. Chem*. **46** 28–34
[20] Udovenko A A, Laptash N M and Maslennikova I G 2003 *J. Fluorine Chem*. **124** 5–15
[21] Udovenko A A and Laptash N M 2008 *Acta Cryst*. **B64** 305–11
[22] Melnichenko E I, Krysenko G F, Epov D G and Ovsyannikova A A 1994 *Russ. J. Appl. Chem*. **67** 737–41
[23] Udovenko A A and Laptash N M 2008 *Russ. J. Struct. Chem*. **49** 500–6
[24] Hurst H J and Taylor J C 1970 *Acta Crystallogr*. **B26** 417–21
[25] Davy J 1812 *Phil. Trans*. **102** 352–69
[26] Marignac M 1860 *Ann. Chim. Phys*. **60** (3) 257–307
[27] Mel'nichenko E I, Krysenko G F, Epov D G and Marusova E Y 2004 *Russ. J. Inorg. Chem*. **49** 1803–6
[28] Abdel Rehim A M 1992 *J. Them. Anal*. **38** 475–86
[29] Abdel Rehim A M 1997 *J. Them. Anal*. **48** 177–202
[30] Balarama Krishna M V, Chandrasekaran K, Chakravarthy S and Karunasagar D 2015 *Fuel* **158** 770–78

Lead ferroniobat ceramic and films structures manufacture features and dielectric characteristic

S H Alikhadjiev[1], M A Kazaryan[2], A V Pavlenko[1,3] and P S Plyaka[3,4]
[1] Southern federal university, 105/42 B Sadovaya str., Rostov-on-Don, 344006, Russia
[2] P.N. Lebedev Physical Institute of Russian Academy of Sciences, 53 Lenin ave., Moscow, 119991, Russia
[3] Southern Scientific Center of Russian Academy of Sciences, 41 Chekhova str., Rostov-on-Don, 344006, Russia

[4] E-mail: pavelsp08@gmail.com

Abstract. Lead ferroniobat lithium doped ceramic samples have been prepared and investigated. Ceramic high quality has confirmed by obtained structural and electro physical properties. The attempt to thin films synthesis by deposition in a capacitive radiofrequency discharge from ceramic target is described; MgO substrate and pure oxygen ambient have been used. Synthesized films structure properties have been investigated, X-ray pattern with lead ferroniobat weak peak and cross-section SEM-image are presented.

1. Introduction

Lead ferroniobat (PFN) is a representative of the class of materials called "multiferroics" [1]. Its rather high values of the Neel and Curie temperatures make PFN a very interesting object from an applied point of view [2]. PFN have great perspectives for UHF and sensor technique [3], especially in thin film form [2, 4]. A series of papers is devoted to growth and characterization of lead ferroniobat thin films, but achieved structural properties are not high yet [5, 6], even when one uses specially prepared substrate [7]. Almost all PFN thin films have been prepared by pulsed laser depositions [4 - 6], film parameters are critical to substrate temperature and ambient oxygen pressure [5].

Thin films deposition technology in oxygen capacitive radiofrequency discharge is well known [7, 8], one stage film synthesis is realized in high energies electron beam influence discharge region [9]. Electron beams transmit energy to deposited film and increase atoms mobility, facilitating to dislocations reduction. Barium-strontium titanate [8], barium-strontium niobate [10], bismuth ferrite [7] and other thin films have just been successfully synthesized on MgO simple substrate and demonstrated high quality structure and electro physical parameters.

Attempt to fabricate PFN thin films by discharge sputtering deposition from specially prepared ceramic target is the aim of presented work. Target technology and properties have been also investigated.

2. Experimental part

The ceramic targets of 38 mm in diameter have been synthesized by the method of solid-state reactions from the high-purity oxides PbO, Fe_2O_3 and Nb_2O_5 using two-stage annealing with intermediate milling at temperatures $T1=T2=1123$ K for 4 hours ($\tau1=\tau2$). At each temperature the

Content from this work may be used under the terms of the Creative Commons Attribution 3.0 licence. Any further distribution of this work must maintain attribution to the author(s) and the title of the work, journal citation and DOI.
Published under licence by IOP Publishing Ltd

synthesized products were sintered at *Tsint*= 1373 K for 2 hours. 1 mass % lithium carbonate has been added at synthesis stage for lithium doping.

X-ray powder diffraction patterns were recorded on DRON-3 diffractometer in the Bragg–Bretano geometry with filtered CuKα radiation. The dielectric spectra were investigated using Aglient E4980A LCR meter in the temperature and frequency ranges of 300–500 K and 20 Hz–2.0 MHz, respectively. Piezoelectric constants have been measured by YE2730A d33 meter.

Fore thin film depositions we have chosen technology based on ceramic target sputtering in capacitive radiofrequency discharge without using magnetron [8]. Film rising and crystallizations occurred in pure oxygen, RF power, initial substrate temperature, substrate-target distance and oxygen pressure varied and for the best films sample were equals to 240 W, 460 °C, 12 mm and 0.38 Torr respectively.

3. Results and discussion

The analysis of experimental results evidences the obtaining of high-density, impurity-free samples of the PFNL ceramics (the relative density, ρ_{rel}>95%), which enables one to consider them as reliable and trustworthy. The microstructure of the surface and ceramic cleavages, presented in Figure 1, were studied in detail by microanalysis.

Figure 1. PFN+Li_2CO_3 target microstructure fragments. The markers in fragments are 10 μm

The deep etching of the surface allowed assessing the grain size (~10–20 μm) and their shape (polyhedrons). Fractures in the cleavage occured mostly along grains and, to a lesser extent, along their boundaries. The character of the cleavage and the features of the etching likely indicate the closeness of the strengths of intergrain boundaries and the grains themselves.

Investigated PFN ceramic permittivity has demonstrated variance absence until 400 K temperature. Measured piezoelectric constants had the next values: $|d_{31}|$= 82 pQ·N^{-1}, d_{33} = 230 pQ·N^{-1}, and Q_M = 262. Obtained results confirm enough high material quality. Ceramic dielectric parameters plots are presented in Figure 2. The well defined, frequency independent $\varepsilon'(T)$ maximum at T=367 K is indicative of the fact that it is just the temperature which is T_C.

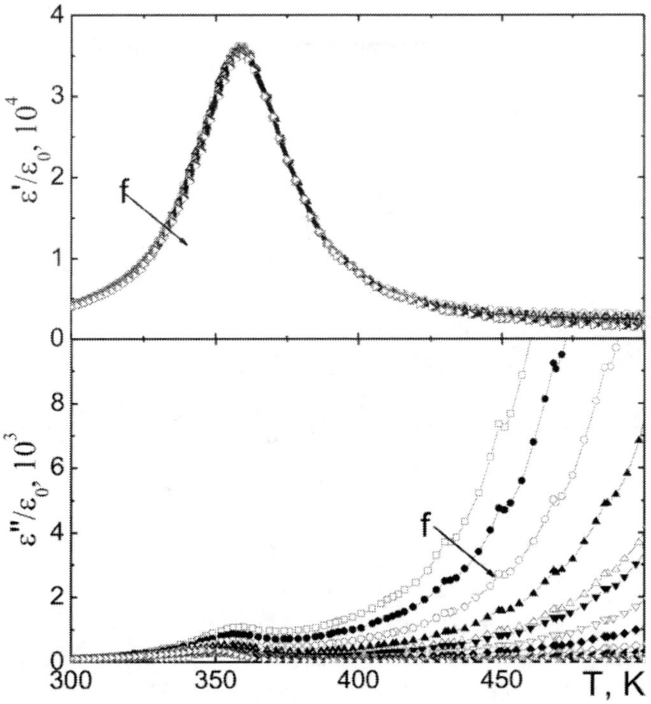

Figure 2. PFNL ceramic temperature dependences of relative permittivity ε' (top) and dielectric loses $tg\delta$ (bottom) for lot of frequencies in the range of $10^2 \div 10^6$ Hz

Obtained thin film samples were not convincing enough. X-ray pattern example is presented in Figure 3. As can be seen lead ferroniobat thin film reflex is smaller than MgO substrate peak. The major well-known difficulty in lead containing material film sputtering is lead volatility. Lead evaporation from target and deposited film surfaces results in a thin films violation. A lot of high intensity atomic lead spectral lines have been observed in discharge plasma during PFNL thin films deposition. Grown films composition investigation confirmed the lead deficiency. To prevent lead evaporation losses we plan to change ceramic target technology.

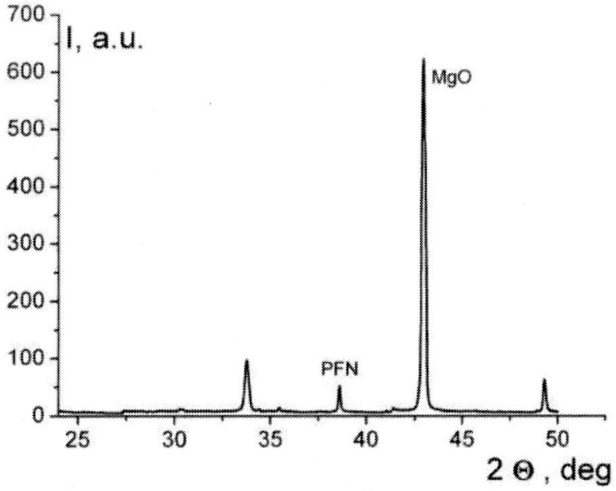

Figure 3. Synthesized thin films X-ray pattern

Electron microscope investigations by FE-SEM Zeiss SUPRA 25 of deposited on substrate film chipping demonstrated monolithic structure layer on substrate surface. Results are presented in the figure 4 and confirmed suboptimal technology regimes. As can be seen monolithic layer has thickness equal to 127 nm, but it is covered by loose layer with 17 nm thickness. Thin films growth rate was about 4 nm/sec, thus only films layer, formed during last 4 minutes, is non monolithic. Evaporation of lead and oxygen can result in top layer structure defects. Works on PFN thin film radiofrequency discharge sputter depositions optimal regimes will be continued, the main efforts will be focused on film deposition final stage including pressure rising and RF power changing.

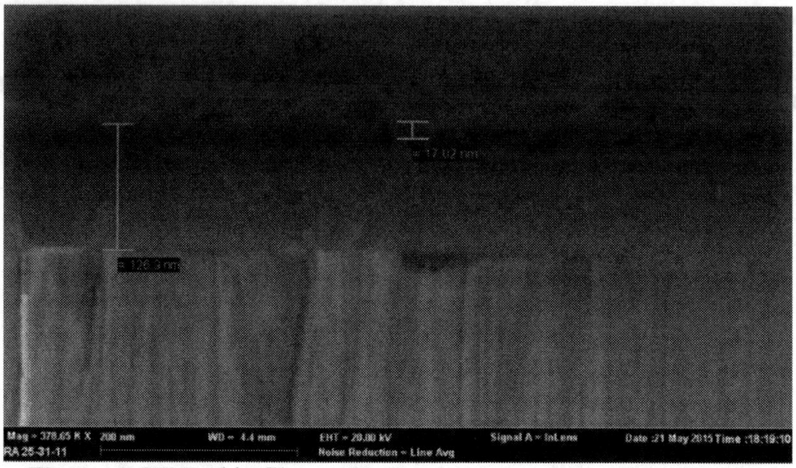

Figure 4. PFNL thin film on MgO substrate cross-section SEM-image

4. Conclusions

Obtained lead ferroniobat ceramic samples have demonstrated high ferroelectric characteristics. The main measured electro physical parameters are presented. An encouraging result has been reached in lead ferroniobat thin film on MgO substrate deposition. Film samples have X-ray reflex corresponding to perovskite crystal structure and basically monolithic layer.

Acknowledgments

This study was supported by the Ministry of Education and Science of the Russian Federation (the base and project parts of state order: no. 3.1246.2014/K) and the Russian Foundation for Basic Research (project no. 150805711 A)

References

[1] Lente M H, Guerra J D S, De Souza G K S, Fraygola B M, Raigoza C F V, Garcia D and Eiras J A 2008 *Phys. Rev. B.* **78** 5421-5425

[2] Gao X S, Chen X Y, Yin J, Wu J, Liu Z G and Wang M 2000 *Journal of Materials Science.* **35** 054109

[3] Zvezdin A K, Loginov A S, Meshkov G A and Pyatakov A P 2007 *Proc. Rus. Academy of Sci. Phys.* [Izvestiya Rossiyskoy Akademii Nauk. Fizika – in Russian] **71/1** 1604

[4] Peng W, Lemée N, Karkut M, Dkhil B, Shvartsman V V, Borisov P, Kleemann W, Holc J, Kosec M and Blinc R 2009 *Appl. Phys. Lett.* 94 012509

[5] Wei Z, Shu Ya W and XiangMing C 2013 *Chinese Science Bulletin* **58** 27 3398-3402

[6] Yan L, Lia J and Viehland D 2011 *Journal of Materials Research* **23** 3 663-670

[7] Mukhortov V M, Golovko Y I and Yuzyuk Y I 2009 *Advances in Phys. [Uspekhi Fiziki – in Russian]* **52/8** 856

[8] Mukhortov V M, Golovko Yu I, Tolmachev G N and Mashchenko A I 1999 *Rus. J. App. Phys.* **44/12** 1477

[9] Schulze J, Heil1 B G, Luggenhölscher D, Mussenbrock T, Brinkmann R P and Czarnetzki U 2008 *J. Phys. D: Appl. Phys.* 41/4 042003
[10] Tolmachev G N, Kovtun A P, Zakharchenko I N, Aliev I M, Pavlenko A V, Reznichenko L A and Verbenko I A 2015 *Phys. Solid State* **57**/10 2106

REES-2015 IOP Publishing
IOP Conf. Series: Materials Science and Engineering **112** (2016) 012026 doi:10.1088/1757-899X/112/1/012026

Experimental determination of energy transfer in Eu(III) complexes, based on pyrazole substituted 1,3-diketones

E A Varaksina[1, 2, *], S A Ambrozevich[1], N P Datskevich[1], I V Taydakov[1] and A G Vitukhnovsky[1, 2, 3]

[1]S.I. Vavilov Department of Luminescence, P.N. Lebedev Physical Institute of the Russian Academy of Sciences, 53 Leninsky Prospekt, Moscow, 119991, Russia
[2]Department of Problems of Physics and Energetics, Moscow Institute of Physics and Technology, 9 Institutsky per., Dolgoprudny, Moscow Region, 141700, Russia
[3]National Research Nuclear University (Moscow Engineering Physics Institute), 31 Kashirskoe shosse, Moscow, 115409, Russia

[*]E-mail: janiy92@yandex.ru

Abstract. We succeeded in synthesis of the novel Eu-based luminophore with almost 100% efficiency of energy transfer from ligands to the lanthanide ion. The luminophores with β-diketonate ligands were characterized by means of optical spectroscopy. Impact of different types of β-diketonates on the quantum yield of luminescence was revealed. A β-diketonate complex exhibiting the highest luminescence quantum yield and prospective for OLED applications was determined. The reasons for decreasing quantum efficiencies for other types of investigated ligands are discussed.

1. Introduction

Lanthanides have characteristic luminescence spectral lines corresponding to transitions inside the partially filled 4f-shell. Screening by the filled $5s^2 5p^6$ shell leads to narrow luminescence bands of the lanthanides. Moreover, these materials can be an excellent substitute for iridium in the production of organic light-emitting diodes (OLED) as the lanthanides technology is by far less expensive. This fact inspires intensive research to find new types of lanthanide luminophores with high quantum efficiencies.

Effective luminescence of lanthanides is limited by weak light absorption due to forbidden f-f transitions. Combining lanthanide ion with organic ligands having broad absorption bands strongly increases a probability to absorb photons with the transition to the first excited singlet state and subsequent intersystem crossing of the ligands to the triplet state. To make an energy transfer from the ligand to the ion possible, the triplet level of the ligand should be just above the excited level of the ion. The correct matching of the ligands allows us to observe effective luminescence of lanthanides. The process of transfer of absorbed energy from ligands to an emission center is called "antenna effect". Furthermore, ligands also influence its luminescence spectrum by means of developing the electric field which leads to a redistribution of the intensity of different transitions bands and their splitting [1].

We synthesized europium complexes based on a new set of β-diketonates ligands containing a pyrazole groups (figure 1). The production of these luminophores is relatively straightforward and it

Content from this work may be used under the terms of the Creative Commons Attribution 3.0 licence. Any further distribution of this work must maintain attribution to the author(s) and the title of the work, journal citation and DOI.
Published under licence by IOP Publishing Ltd

can be manufactured on a commercial scale. We measured energy of the triplet states of the β-diketonates ligands and researched their influence on the quantum yield of luminescence of the europium complexes. We found out a complex exhibiting a high quantum yield due to efficient energy transfer of electronic excitation from the ligands to the europium ion.

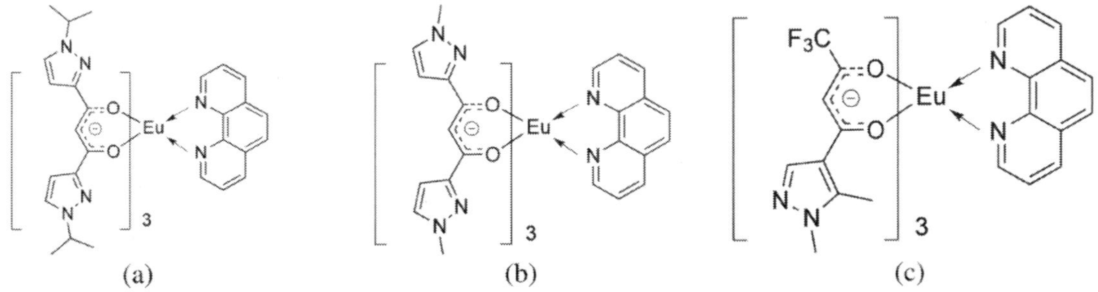

Figure 1. The chemical structure of complexes (a) $Eu(L^1)_3Phen$; (b) $Eu(L^2)_3Phen$ and (c) $Eu(L^3)_3Phen$.

2. Quantum yield

The luminescence quantum yield is the ratio of the number of emitted photons per absorbed photons. To describe the luminescence of lanthanide complexes it is essential to introduce two types of quantum yields: an overall Q_{Ln}^{L} and an intrinsic Q_{Ln}^{Ln} quantum yields. The overall quantum yield is related to the intrinsic one by the following equation:

$$Q_{Ln}^{L} = \eta_{sens} Q_{Ln}^{Ln},$$

where coefficient η_{sens} is the sensitization efficiency [2].

The intrinsic quantum efficiency can be calculated with the use of radiative A_{rad} and nonradiative A_{nrad} rate constants:

$$Q = \frac{A_{rad}}{A_{rad} + A_{nrad}}. \quad (1)$$

This calculation requires evaluating the rate constant of a transition between an initial state Ψ_J and a final state $\Psi_{J'}$. Taking into account only the dipole magnetic and electric transitions, this rate can be expressed as

$$A(\Psi_J, \Psi_{J'}) = \frac{64\pi^4 \nu^3}{3h(2J+1)} \left[n^3 D_{MD} + \frac{n(n^2+2)^2}{9} D_{ED} \right], \quad (2)$$

where ν is energy of the transition in cm^{-1}, h is Planck's constant, n is the refractive index, $2J+1$ is the degeneracy of the initial state, D_{ED} and D_{MD} are the electric and magnetic dipole strengths in $esu^2\ cm^2$, respectively [3]. Forbidden electric dipole f-f transitions become partially allowed due to influence of the ligand field which leads to the mixing of electronic states of opposite parity. These transitions called induced electric dipole transitions [4] are much weaker than the conventional electric dipole transitions.

To find out the intrinsic quantum yield using (2) it is required to determine electric and magnetic dipole strengths which are quite difficult to calculate. The europium spectra, in contrast to other rare earth complexes have unique features giving a way to avoid these complicated calculations. As the transition $^5D_0 \rightarrow ^7F_1$ has only magnetic dipole nature, its intensity does not depend on the ion's field environment. Thus radiative rate A_{01} for this transition behave as constant equal to 14.65 s^{-1}. This value is used as an intensity gauge for the calculation of the intrinsic quantum yield. The overall radiative rate constant can be calculated by

$$A_{rad} = A_{01} n^3 \frac{I_{tot}}{I_{01}}, \tag{3}$$

where I_{tot} is the integrated emission from the $^5D_0$ to the ^7F_J (J = 0–6), I_{01} is the integrated intensity of the magnetic dipole transition.

The nonradiative rate constant A_{nrad} can be determined from the luminescence decay measurements. The observed lifetime is related to the radiative and nonradiative rate constant by the following equation:

$$\tau = (A_{rad} + A_{nrad})^{-1}.$$

Thus, the intrinsic quantum efficiency is given by

$$Q_{in} = A_{rad}\tau,$$

where A_{rad} is determined from (3) [4].

3. Experimental part

We investigated polycrystalline samples of the different β-diketonate europium complexes. To determine the intrinsic quantum yield, we carried out two experiments. In the first experiment, we observed luminescence spectra when the sample was excited at the wavelength corresponding to absorption of the europium ion. It gave us an opportunity to determine correctly the intrinsic quantum efficiency of the ion, taking into account the influence of the ligand environment, which did not participate directly in light absorption. The radiative rate constants were evaluated from this spectrum by (3). In the second experiment, we evaluated time constants from luminescence intensity decay measurements. All the luminescent experiments were carried out with the use of SDL-1 double monochromator.

The experimental setup for measurement of luminescence spectrum is shown in the figure 2.

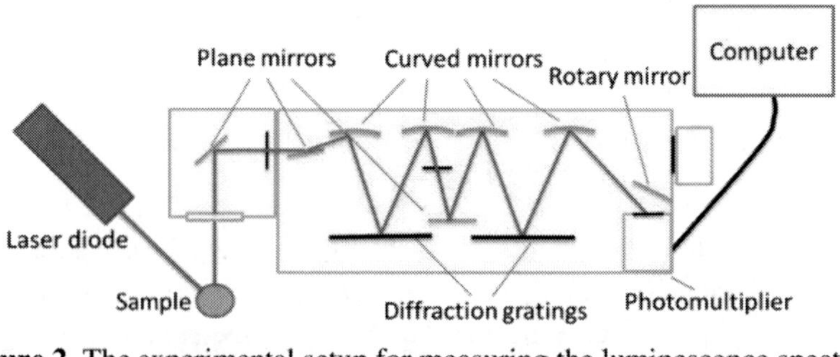

Figure 2. The experimental setup for measuring the luminescence spectra.

Selective excitation of the europium ions of the sample placed in the cuvette was performed with laser diode emission at a wavelength of 470 nm. Two gratings with 600 grooves per mm were installed in the monochromator. The luminescence was detected with a Hamamatsu H8259-01 photomultiplier. The measured spectra were corrected taking into account detector sensitivity and monochromator transmission functions and saved in a computer.

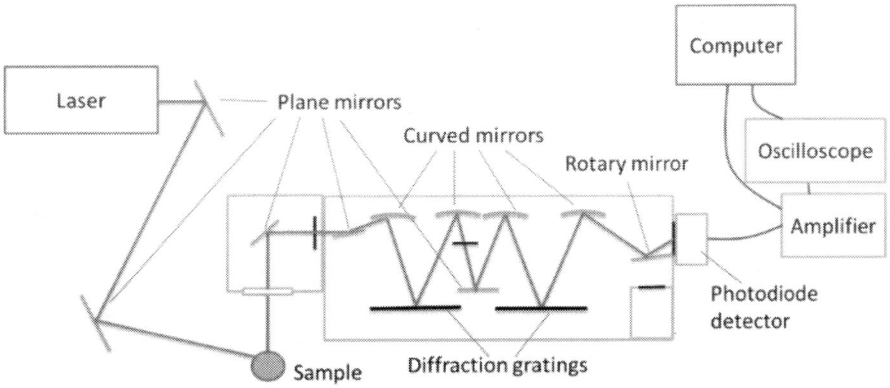

Figure 3. The experimental setup for measuring the luminescence decay kinetics.

When measuring decays, the excitation of the sample was performed with 355 nm emission from a YAG:Nd^{3+} pulse laser with 12 Hz repetition frequency and 20 mJ average energy. As the pulse duration (6 ns) is much smaller than the luminescence decay time it does not affect the measurement results. The monochromator (figure 3) was tuned to the wavelength 615 nm corresponding to the maximum intensity of the europium luminescence spectrum. The luminescence signal was detected with photodiode and then amplified with Stanford Research SR570 current amplifier. The amplified signal was directed to an oscilloscope which averaged over 1000 decay curves for different laser shots. This averaging technique significantly increased signal-to-noise ratio and provided a way to obtain more accurate results for decay times.

Measurements of the overall quantum yield were carried out using an integrating sphere [5].

4. Results and discussions

For β-diketonate ligands using in the study the energy of triplet states was measured [6]. The results represent in the table 1. The excited level of europium $^5D_0$ is about 17300 cm^{-1}.

Table 1. The energy of triplets states of the β-diketonate ligands.

Ligand	L^1	L^2	L^3
Energy of triplet level, cm^{-1}	21276	19638	19764

The measured luminescence spectra of the complexes are shown in the figure 4.

Lifetime, radiative and nonradiative rate constants, overall and intrinsic quantum yields and sensitization efficiency calculated from the experimental for the investigated europium complexes are given in the table 2. The error of measurement of the quantum yields did not exceed 10%.

The highest luminescence quantum yield was achieved for $Eu(L^3)_3$Phen. Moreover, this complex demonstrated the highest sensitization efficiency - almost all the energy absorbed by ligand was transferred to the europium ion.

Complexes with ligands L^1 and L^2 exhibited low quantum yields. For complex $Eu(L^1)_3$Phen decrease of the quantum yield possibly emerges from the large differences in energy between the triplet levels of the ligands and the resonance level of the lanthanide ion (table 1). Triplet levels of ligands L^2 and L^3 have nearly the same energy but the quantum yield of complex $Eu(L^2)_3$Phen is by an order of magnitude less than the quantum yield of complex $Eu(L^3)_3$Phen. This difference can be caused by multiphonon relaxation of energy on the C-H vibrations in the complex $Eu(L^2)_3$Phen. The replacement of the C-H vibrations by C-F vibrations (figure 1) reduces nonradiative deactivation of the exited state of the europium ion. In addition, the fluorine in the inner coordination sphere screens the ion from the Van-der-Waals forces leading to the quenching of luminescence.

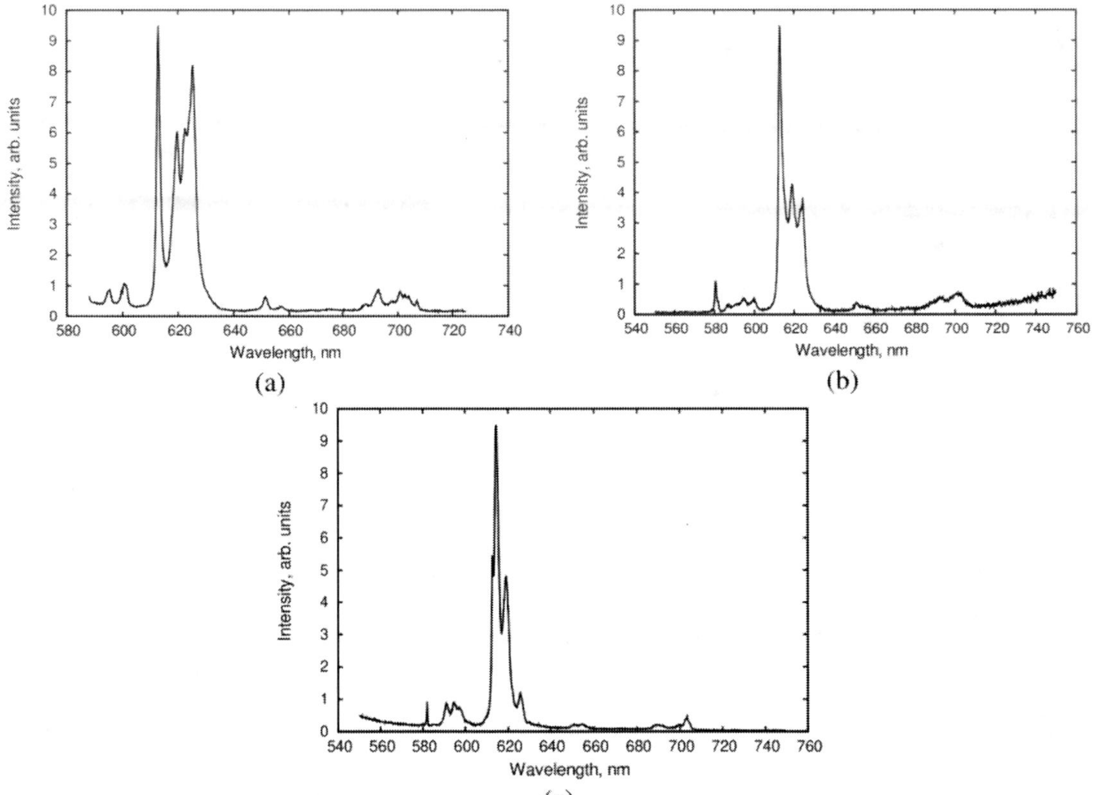

Figure 4. The luminescence spectra of complexes (a) $Eu(L^1)_3Phen$; (b) $Eu(L^2)_3Phen$ and (c) $Eu(L^3)_3Phen$.

Table 2. The experimental value of the lifetime (τ), radiative (A_{rad}) and nonradiative (A_{nrad}) rate constants, overall (Q^L_{Ln}) and intrinsic (Q^{Ln}_{Ln}) quantum yields and sensitization efficiency (η_{sens}) for β-diketonate europium complexes.

	τ , mks	A_{rad} , s^{-1}	A_{nrad} , s^{-1}	Q^{Ln}_{Ln} , %	Q^L_{Ln} , %	η_{sens} , %
$Eu(L^1)_3Phen$	279	1011	2573	28	6	21
$Eu(L^2)_3Phen$	243	873	3243	21	4	19
$Eu(L^3)_3Phen$	754	558	769	42	40	96

5. Conclusion

For a number of β-diketonates europium complexes investigated in this work we found how efficiencies of energy transfer from ligands to the europium ion depends on the ion environment. Complex $Eu(L^3)_3Phen$ was proved to exhibit the highest quantum yield due to the almost 100% sensitization efficiency. It gives a possibility to use it as an efficient luminophore in organic light emitting diodes.

IOP Conf. Series: Materials Science and Engineering **112** (2016) 012026 doi:10.1088/1757-899X/112/1/012026

Acknowledgement

This work has been carried out in the framework of Russian Fund for Basic Research (projects 14-02-90452 Ukr and 14-02-31269 mol_a); additional support was provided by grant of President of Russian Federation (project MK-7514.2015.2).

References

[1] Binnemans K 2005 *Handbook on the physics and chemistry of rare earths* vol 35 ed K A Gschneidner, J-C G Bünzli and et al (Celestijnenlaan: Elsevier Science B. V.) p 111

[2] Bünzli J-C G and Eliseeva S E 2011 Basics of Lanthanide Photophysics *Lanthanide Luminescence* ed P Hänninen and H Härmä (Berlin: Springer) p 15-20

[3] Walsh B M 2006 *Advances in spectroscopy for lasers and sensing* ed B Di Bartolo and O Forte (Brooklyn: Springer) p 403-28

[4] Werts M H V, Jukes R T F and Verhoev J W 2002 *Phys. Chem. Chem. Phys.* **4** 1542–48

[5] Timofeev Y P, Kokin S M, Martinov V N, Datskevich N P Andreev A I and Krotov AV 2004 *Materials Electronic Technology* **3** 52-58

[6] Taidakov I V 2013 Pyrazole substituted 1.3-diketones: synthesis, complexing properties and application in the rare earth coordination chemistry *Rare Earth: New Research* ed Z Liu (Nanjing: Nova Science Publishers) p 95-115

Problem of the lithium peroxide thermal stability

R A Nefedov[1,*], Yu A Ferapontov[1] and N P Kozlova[2]
[1]Tambov State Technical University, 106, Sovetskaya Str., Tambov, Russia
[2]JSC Corporation "Roskhimzashchita", 19, Morshanskoe Highway, Tambov, Russia

[*]E-mail: kliorik@mail.ru

Abstract. The behavior of lithium peroxide and lithium peroxide monohydrate samples under heating in atmospheric air was studied by the method of thermogravimetric analysis (TGA) and differential thermal analysis (DTA). It was found that in the temperature range of 32^0C to 82^0C the interaction of lithium peroxides and steam with the formation of lithium peroxide monohydrate occurs, which was confirmed chemically and by X-ray Single-qualitative analysis. It was experimentally found that lithium peroxide starts to decompose into the lithium oxide and oxygen in the temperature range of 340 ÷ 348^0C. It was established that the resulting thermal decomposition of lithium oxide, lithium peroxide at the temperature of 422 ^{0}C melts with lithium carbonate eutecticly. The manifestation of polymorphism was not marked(seen or noticed) under the heating of studied samples of lithium peroxide and lithium peroxide monohydrate in the temperature range of 25 ^{0}C ÷ 34 ^{0}C.

1. Introduction

Lithium peroxide (Li_2O_2) is a compound that contains 34.87% of active oxygen which is released in contact with water or steam. According to this criterion, Li_2O_2 concedes only to hydrogen peroxide and sodium superoxide from all of the peroxide compounds which are stable under normal conditions, and its stoichiometric capacity for carbon dioxide (488 l / kg) is the maximum total number used for chemisorbents in this purpose. Therefore, it is widely used as a component of chemical products applied for the regeneration of the air in the life-support systems (LSS), which is forced to work in the conditions isolated from the environment [1 - 4]. Not only detailed knowledge of the mechanisms of interaction of chemisorbents with water vapor and carbon dioxide, the main kinetic parameters of these processes, but also the knowledge of factors that may affect these parameters is required to make a right choice of chemical composition of chemisorbents, construction systems of LSS etc. Furthermore the goal of that choice is to increase the efficiency of LSS systems and create more comfortable conditions for the user. Previously mentioned factors include phase transitions, polymorphic transformations and others phenomena that occur in chemisorbents during the thermal exposure. However, there is not enough information about the problem of lithium peroxide. Moreover, any information on it is quite contradictory. Authors [5 - 6] report about the presence of irreversible exothermic effect on the DTA curve at about 225 ^{0}C, which they interpret as a polymorphic transformation of α-Li_2O_2 to β-Li_2O_2 that was accompanied by an increase in the reactivity of the sample, confirming the results of X-ray diffraction. At the same time, authors [7 - 8] did not confirm this information in later works. Therefore, it was significant scientific and practical interest to study the behavior of various samples of lithium peroxide at the thermal exposure in air, at the conditions, the occurrence of which is possible in the life - support systems.

Content from this work may be used under the terms of the Creative Commons Attribution 3.0 licence. Any further distribution of this work must maintain attribution to the author(s) and the title of the work, journal citation and DOI.
Published under licence by IOP Publishing Ltd

2. Experimental part

Different samples of lithium peroxide were taken for this study. They were obtained at the temperature of 33 ^{0}C by reaction of solid lithium hydroxide and 50% hydrogen peroxide aqueous solution of "*puriss. spec.*" marks at a molar ratio $LiOH/H_2O_2$ = 1.85, compartment of intermediate adduct of lithium peroxide monohydrate ($Li_2O_2 \cdot H_2O$) by filtration and its subsequent dehydration by heating in a muffle furnace at the temperature of 80 ^{0}C for 4 hours (sample #1), extended exposure (about three days) at a room temperature over a layer of zeolite (sample #2) [9] and dehydration in the ultra high frequency (UHF) (sample #3) [10 - 16]. Time of dehydration in the UHF was between 7 and 12 seconds, depending on the sample's weight. In cases of dehydration of $Li_2O_2 \cdot H_2O$ in the UHF and thermal treatment in the muffle furnace processes were carried out in the current of dry decarbonized air in order to intensify dehydration (accompanied by an increase in the content of the target product)and prevent the formation of lithium carbonate. After exposure at the previously mentioned temperature, resulting from the interaction of the system for 2 hours required to achieve the equilibrium between the phases, solid phase ($Li_2O_2 \cdot H_2O$) was separated by filtration, washed with distilled water and dried at a room temperature over a layer of adsorbent until the full removal of chemically unbound water [14, 16 - 17]. The temperature during the experiments was maintained with a water bath up to ±0.2 ^{0}C.

In order to determine the chemical composition, samples were analyzed for the content of active oxygen, lithium hydroxide and lithium carbonate. Determining the content of active oxygen, studied samples were preliminary stabilized with boric acid, followed by permanganometric titration [18]. Analyzing of the LiOH content, samples were boiled with water for 15 - 20 minutes to break down peroxide compounds, and then titrated 0.1N H_2SO_4 by phenolphthalein. Determination of lithium carbonate was performed by the method of Fresenius [19].

Compositions of the samples that were determined by chemical analysis and confirmed by data of X-ray qualitative diffraction are presented in Table 1. X-ray diffraction analysis was carried out on a DRON-6 with the filtered Cu K_α- radiation (λ = 1.54051 Å). Scan step 0.05^0, scanning range of $20^0 \leq 2\Theta \leq 120^0$, 3 seconds exposure at each point. In addition, Table 1 shows the dispersion characteristics of the studied samples, evaluated by a specially developed method of determining particle's size of peroxide products based on their inertial separation [20].

Table 1. The chemical composition of the samples

Sample number	Method of production	The median particle diameter, d_m, μm	Chemical composition, wt%			
			Li_2O_2	LiOH	Li_2CO_3	H_2O
1	thermal dehydration $Li_2O_2 \cdot H_2O$	23.4	97.7	0.4	1.4	0.5
2	dehydration $Li_2O_2 \cdot H_2O$ over a dehumidifier	23.2	98.0	0.3	1.3	0.4
3	dehydration $Li_2O_2 \cdot H_2O$ in the UHF	18.7	97.8	0.3	1.5	0.4
4	initial $Li_2O_2 \cdot H_2O$	21.6	70.3	0.4	1.3	28.0

Thermogravimetric and differential thermal analysis of the samples was carried out under non-isothermal conditions at a research complex TAG - 24 of "Setaram" company (France). Investigated samples were of 70 ÷ 80 mg. Temperature was recorded with an accuracy of 0.01^{0}C changes in the sample's weight were recorded with an accuracy of 0.01 mg. Studied samples were placed in a corundum crucible. A platinum - platinum-rhodium thermocouple, placed in the volume of the tested sample and calibrated on common reference points, were used to measure the temperature. Study was carried out in air at atmospheric pressure with a heating rate of 5 ÷ 10 degrees per minute. Mentioned heating rate provides the identity of temperatures throughout the whole volume of the tested sample during the experiment.

When the heating rate of the sample is less than 5 °C / min and above 12 °C / min, there is a significant deviation of the curve, changes in the temperature T from a straight line, leading to a significant error in the calculations of kinetic parameters of the studied process. This experimental fact is explained by a violation of temperatures' equality of external and internal layers of weighted quantity of tested samples, caused by insufficient thermal conductivity of lithium peroxide and lithium peroxide monohydrate [21, 22].

3. Results and discussion

During the experiments, it was found that the nature of the TGA and DTA curves of lithium peroxide samples obtained by various methods is similar. Therefore, not to make this text complicated with the same results, we decided to report only the results of the thermal tests of sample # 3 from Table 1 (figure 1).

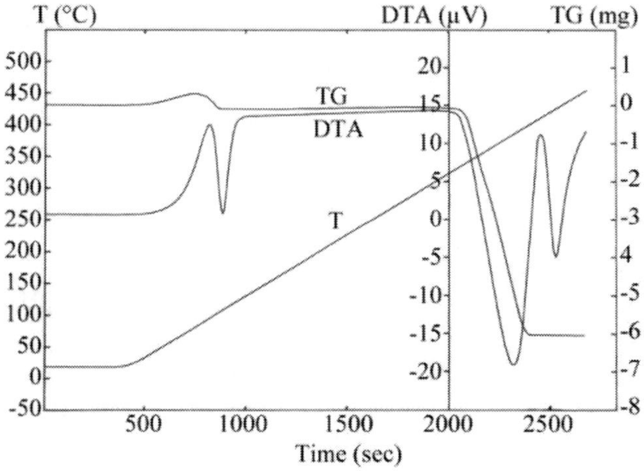

Figure 1. Complex thermal analysis of Li_2O_2

All the samples of lithium peroxide obtained by various methods showed exothermal effect on the DTA curves at the temperature range from of 32 ^{0}C to 82 ^{0}C accompanied by an increase in weight of the samples and related to the interaction of lithium peroxide with water vapor with the formation of lithium peroxide monohydrate. That fact was mentioned by authors [23 - 24], who studied the interaction of lithium peroxide and water vapor. To validate the assumption that was made about the nature of the exothermic effect, parent drug Li_2O_2 was heated in air at a temperature of 450 ^{0}C and a relative humidity $\varphi = 96$ % for 1 hour and then exposed to qualitative XRF analysis. The appearance of new reflexes on the diffractogram (at an angle of slip $2\Theta = 35.09$ - reflex corresponding to the interplanar spacing d = 2.543 Å, at an angle of slip $2\Theta = 30.48$ - reflex corresponding to the interplanar spacing d = 2.90 Å, at an angle of slip $2\Theta = 40.75$ - reflex corresponding to the interplanar spacing d = 2.20 Å) clearly indicates the appearance of a new phase composition $Li_2O_2{\cdot}H_2O$ in the tested sample [16, 25]. It should be noted that during this exothermic effect weight of samples #1 and #2 increases by approximately 1.7 %, while there is an increase in the weight of the tested sample №3 for 2.4% on the TGA curve. Peak on the DTA curve for sample №3 has more expressed nature than the peaks in the DTA curves for samples #1 and 2. This difference in the behavior of lithium peroxide samples, obtained in different ways, is due to differences in the surface area of the tested samples in other words because of the kinetics of the interaction between lithium peroxide and water vapor. Then, all Li_2O_2 samples have endothermic effect that has a maximum at a temperature of 106 ^{0}C, leading to a decrease in the mass of the tested samples for about 2.9 %.Following large endothermic effect on the DTA curves, starts at a temperature 348 ^{0}C, 344 ^{0}C and 341 ^{0}C with peaks at temperatures of 369 ^{0}C, 373 ^{0}C and 374 ^{0}C for samples № 1 - № 3, respectively. In addition, this effect is accompanied by a decrease in the weight of all tested samples by approximately 34%, corresponding to the decomposition of the lithium peroxide to lithium oxide and oxygen (stoichiometric content of active oxygen in the tested samples is about 34.6%). This assumption was confirmed by chemical and qualitative XRF analyzes that showed that original samples of lithium peroxide warmed at 390 ^{0}C (at a temperature slightly above this effect) for 3 hours contain only lithium oxide and lithium carbonate, but not the original Li_2O_2. Obtained results are in better agreement with the results of authors [7, 26], indicating the decomposition temperature lithium peroxide at 350 ^{0}C than with the results of authors [5, 6, 8], observing the beginning of decomposition of lithium at 315 ^{0}C. Finally, all tested samples clearly expressed endothermic effect on the DTA curves that has a maximum at 424 ^{0}C altogether with the corresponding eutectic melting of system Li_2O - Li_2CO_3, visually observable, that also was noted in [7 - 8].

Figure 2 shows the results of lithium peroxide monohydrate thermal analysis (sample #4).

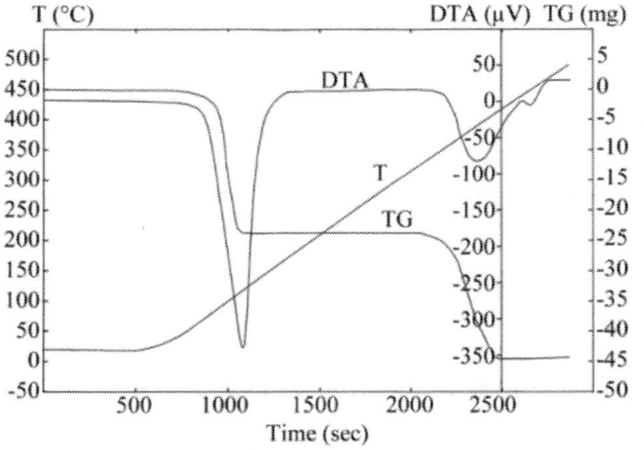

Figure 2. Complex thermal analysis of $Li_2O_2 \cdot H_2O$

During the process of lithium peroxide decomposition the mass of the tested sample is reduced by 24.1% due to the evolved oxygen (theoretical content of active oxygen in the tested sample was 24.5%). Data obtained on lithium peroxide monohydrate dehydration temperature and decomposition temperatures of lithium peroxide produced, as a result of this process, is similar to the results obtained T. A. Dobrynina [25]. However, as for samples of lithium peroxide, there are no effects on the DTA curve at the temperature range of 200 ^{0}C – 240 ^{0}C that allow argument about existence of polymorphic transformations, which contradicts with the data of the author [25]. This author registered a weak exothermic effect on the DTA curve at the temperature of 225 ^{0}C, which was interpreted as a manifestation of polymorphism - the transition α-Li_2O_2 in β-Li_2O_2. Last endothermic effect starting at 424 ^{0}C with a maximum at 436 ^{0}C corresponds to the eutectic melting of Li_2O - Li_2CO_3.

Carrying out thermal analysis in atmospheric air at the temperature range from of 25 ^{0}C to 340 ^{0}C lithium peroxide samples obtained by different ways with different chemical and dispersion composition as well as lithium peroxide monohydrate, we didn't find data indicating the manifestation of polymorphism in the tested samples, not only due to a change in the crystal lattice, but also to the transition from a nonequilibrium state to an equilibrium. Obtained experimental results do not confirm authors [5, 6, 25], who told about the existence of two polymorphic modifications of lithium peroxide.

4. Conclusion

The behavior of lithium peroxide samples, obtained by different methods, under the heating of atmospheric air was studied by the method of thermogravimetric and differential thermal analysis. The interaction of lithium peroxide with water vapor, forming lithium peroxide monohydrate at the temperature range from 32 ^{0}C to 82 ^{0}C was confirmed by the methods of thermal, chemical and qualitative XRF analysis. It was determined that temperatures of decomposition of lithium peroxide into lithium oxide and oxygen, obtained by different ways, are at the range from 340 ^{0}C to 348 ^{0}C.

It was found that the resulting thermal decomposition of lithium oxide, lithium peroxide melts with lithium carbonate eutecticly at 422 ^{0}C. There were not any signs of polymorphism in all tested samples of lithium peroxide during the experiments.

Acknowledgement
This article was supported by the Russian Science Foundation (Agreement No. 15-19-10028).

References

[1] Presti J, Wallman H and Petroctlli A 1967 *Undersea Technol.* **8** 20
[2] Voronin G I and Polivoda A I 1967 *Sustenance space crews* [Zhizneobespechenie ekipazhey kosmicheskikh korabley – in Russian](Moscow: Mashinostroenie Publ.)
[3] Petroctlli A and Capolesto A 1964 *Aerospace Med.* **31** 442
[4] Gladyshev N F, Gladysheva T V and Ferapontov Yu A 2007 *Regenerative Products of New Generation: Technology and Hardware Design* [Regenerativnyye produkty novogo pokoleniya: tekhnologiya i apparaturnoye oformleniye – in Russian] (Moscow: Mashinostroenie Publ.)
[5] Rode T V, Dobrynina T A and Gol'der G A 1955 *Iznestiya AN SSSR OKhN [Russian Academy of Sciences Library]* **4** 611.
[6] Dobrynina T A 1964 *Lithium Peroxide* [Perekis' litiya – in Russian] (Moscow: Nauka Publ.)
[7] Notz K and Bach R 1963 *Chimia* **17** 158
[8] Popova T I 1967 *Reactivity of Lithium Peroxide in the Solid State [*Reaktsionnaya sposobnost' perekisi litiya v tverdom sostoyanii – in Russian] PhD Thesis, Minsk
[9] Ferapontov Yu A, Kozadaev L E, Putin B V and Putin S B 2012 *RF Patent 2465195*
[10] Ferapontov Yu A, Simanenkov S I, Gladyshev N F and Putin B V 2002 *RF Patent 2193522*
[11] Ferapontov Yu A, Simanenkov S I and Gladyshev N F *Chemical Technology* [Khimicheskaya Tekhnologiya – in Russian] **4** 2
[12] Ferapontov Yu A, Ul'yanov M A and Sazhneva T V 2007 *Chemical Technology* [Khimicheskaya Tekhnologiya – in Russian] **5** 193
[13] Ferapontov Yu A, Ul'yanova M A, Sazhneva T V and Dvoretskiy S I 2007 *Abstracts of the International conference on chemical engineering "Chemical engineering 2007"* [in Russian] **1** 259
[14] Ferapontov Yu A, Ul'yanova M A, Sazhneva T V and Kozlova N P 2008 *ZHPKH [Russian Library of Science].* **81/1** 28
[15] Ferapontov Yu A, Ul'yanova M A and Sazhneva T V 2008 *RF Patent 2322387*
[16] Ferapontov Yu A, Ul'yanova M A and Sazhneva T V 2008 *ZHNKH [Russian Library of Science]* **53/10** 1749
[17] Ferapontov Yu A, Ulyanova M A and Sazhneva T V 2008 *Rus. J. Inor. Chem.* **53/10** 1635
[18] Seyb E and Kleinberg J 1951 *J. Am. Chem. Soc.* **73** 2308
[19] Alekseyevskiy G V, Gol'ts R K and Musakin A A 1955 *Quantitative Analysis* [Kolichestvennyy analiz – in Russian] (Moscow: Goskhimizdat Publ.)
[20] Ferapontov Yu A, Ul'yanova M A and Zhdanov D V 2008 *Chemical Technology* [Khimicheskaya Tekhnologiya – in Russian] **9**/1 1
[21] Ferapontov Yu A, Ferapontova L L and Putin S B 2010 *ZHOKH [Russian Library of Science]* **80/**4 539
[22] Ferapontov Yu A, Putin S B, Ferapontova L L and Putin P Yu 2009 *Vestnik TGTU* **15/**4 826
[23] Markowitz M, Baryta D and Stewart H 1964 *J. Inorg. Nucl.Chem.* **26** 2028
[24] Selezneva K I 1966 *ZHNKH* **5**/8 1688

[25] Dobrynina T A 1960 *Izv. AN SSSR OKhN [Russian Academy of Sciences Library]* **6** 961
[26] Meyer M and Markowitz M 1964 *Ind. Engng. Chem.* **3** 321

IOP Conf. Series: Materials Science and Engineering **112** (2016) 012028 doi:10.1088/1757-899X/112/1/012028

Estimation of the physico-chemical parameters of materials based on rare earth elements with the application of computational model

K Mamaev[1], **A Obkhodsky**[1,3] **and A Popov**[2]
[1] National Research Tomsk Polytechnic University, 30, Lenina ave., Tomsk, 634050, Russia
[2] LLC "Depos", 7a, Botanicheskij side Str, Tomsk, 634033, Russia

[3] E-mail: art707@tpu.ru

Abstract. Computational model, technique and the basic principles of operation program complex for quantum-chemical calculations of material's physico-chemical parameters with rare earth elements are discussed. The calculating system is scalable and includes CPU and GPU computational resources. Control and operation of computational jobs and also Globus Toolkit 5 software provides the possibility to join computer users in a unified system of data processing with peer-to-peer architecture. CUDA software is used to integrate graphic processors into calculation system.

1. Introduction

Universal and specialized program packages of simulation are applied while carrying out the research of new functional material creation. Significant computational resources are needed to model atomic structures properties of sold materials. The requirements to the volume of computational resources are being raised multifold with the development of mathematical apparatus techniques and the increase of model description precision.

Characteristics and properties of atomic structures are studied in the program complexes GIMM_FPEIP, MATLAB and also in program complexes GAMESS, Firefly, Gaussian, HyperChem, MOPAC, ORCA, ADF, PRIRODA, SPARTAN and others. All of them have individual peculiarities. In some cases, the broad functionality of these program packages or narrow specification on the contrary does not allow you to apply these programs to model characteristics of alloys with rare-earth metals additions in case of variable external actions in a wide range of values.

The automation of research process is especially attended while modeling the properties of atomic structures. A large volume of data is accumulated in the course of modeling and the data should be stored in the full amount. Specialized supercomputers and clusters are used to carry out quantum-chemical calculations what makes it possible to carry out calculations in acceptable time. In this connection, the way of data processing, storage, aggregation and analysis have to be automated as much as possible in order to solve problems of researching new material successfully [1-5].

The aim of the work was to create a program complex for modeling atomic structure characteristics of alloys with rare-earth metals additions. The program complex would provide users with relatively cheap computational and informational resources for automated data processing and for storage and analysis of modeling results.

Content from this work may be used under the terms of the Creative Commons Attribution 3.0 licence. Any further distribution of this work must maintain attribution to the author(s) and the title of the work, journal citation and DOI.
Published under licence by IOP Publishing Ltd

IOP Conf. Series: Materials Science and Engineering **112** (2016) 012028 doi:10.1088/1757-899X/112/1/012028

2. The data processing system structure

A lot of attention was paid to the unification of data format for interaction of Antares components in the course of its development. In turn, this made it possible to provide needed functionality and relatively easiness of its implementation.

The program complex has functions which allow carrying out model calculations on the basis of heterogeneous computing resources of GRID network [6]. Calculations can be carried out on GPU graphics accelerators which surpass streaming CPU joined as clusters in speed of carrying out mathematical operations.

In program complex Antares CUDA hardware-software architecture is used to parallel computational jobs [7]. Antares general structure is presented in Figure 1.

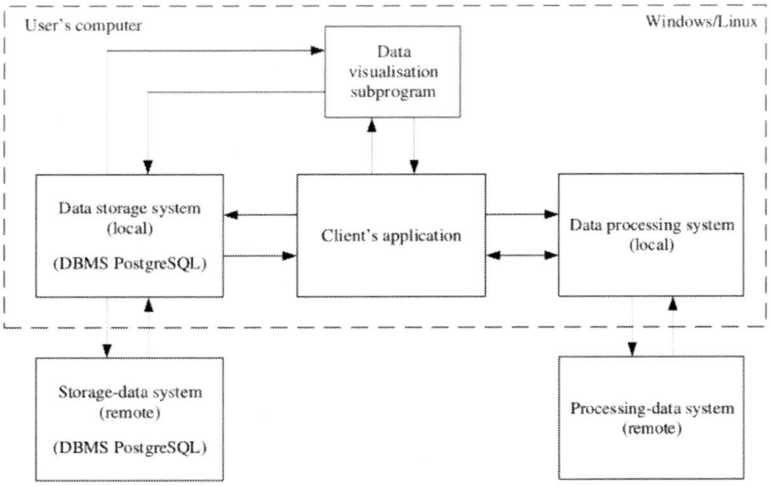

Figure 1. General structure of program complex Antares

The set of experimental data formed in the course of material research is stored in local and remote systems.

Local and remote data storage systems comprises DBMS server PostgreSQL and FTP file server.

The database structures in a user's computer and in remote DBMS server have an identical structure and set of tables.

Remote processing system is a remote Data Processing Center (DPC) which provides a user without sufficient computational resources with a possibility to operate data on GPU basis.

Internal data processing system in its structure and functions similar to the external one and allows data processing by means of a user's computer with the help of CPU or GPU, if available.

A user's computer is both a client and a server, which can give its informational and computational resources or it can use distributed network resources to solve the problems of modeling. In order to store experimental data, a user can engage own computer memory resources or the user can engage resources of dedicated remote data storage server that is included in DSS. The client application allows initiating data communication between the local and remote DSS servers, and also to initiate the visualization of experimental data in order to carry out analysis and it allows you to run computing jobs.

The results of the conducted experiment are stored as a set of files in the working directory of the client's application. Files contain the calculated functional of a model, calculation results, executable code of parallel calculations, and other metadata. There is the header XML-file in each directory of a certain experiment, this file contains the experiment identification number, the basic characteristics of the modeled material and pointers (links) to all files relating to a certain experiments.

The header XML-file is used to identify files in a separate experiment. Data from the header XML-file is copied into database while preserving the experimental results and that fact allows the user to search for required materials according to a set of supplied parameters. The possibility of synchronizing data between local and remote servers is introduced into DSS functionality of Antares program complex. The exchange of data between servers is carried out by means of PostgreSQL DBMS.

Data synchronization allows backing up user's data to ensure its safety and it gives the possibility to analyze the data by other users.

Synchronization request is sent to the DBMS of a remote server at the moment when user's computer connects to DSS network. The remote DBMS connects to the DBMS of a local DSS server and searches for new records in a user's database. In the case of finding new records, the remote DBMS copies from the local file server header XML-files which contain key information about the location of new data in the memory of a local file server A stored procedure (subprogram) executes the process of copying files with the results of new experiments; this subprogram is run by the DBMS on a remote server. At this time new records are created in the database of a remote DSS server. These records have information about main material parameters and links to experiment files.

The formation and the start of computational jobs in a local or remote data processing system (DPS) is done by means of a client application. The functions of user's interaction with the ODS include the following: the volume assessment of free computational resources, the sending of a job into DPS and the controlling of an assignment in current status. The interaction scheme of a client's application of program package Antares with DPS is presented in Figure 2.

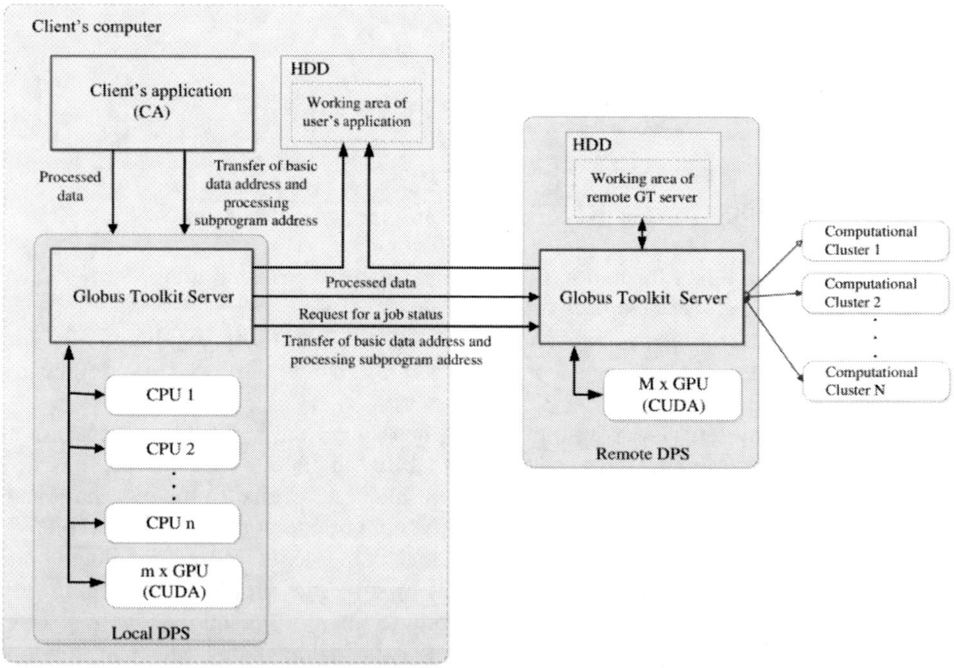

Figure 2. The scheme of interaction between a client's application and DPS

Globus Toolkit software is the basis of the local and remote DPS server. The local DPS server distributes and starts computational job on the basis of information about available computational resources. This is performed on available CPU and GPU (by means of CUDA) if there is a graphics accelerator. Computational jobs are performed remotely on DPS servers if a user has an access to the network [8].

Toolkit functions are based on four protocols [9, 10]:

- Link layer – security (GSI, Globus Security Infrastructure);
- Resource layer – resources control (GRAM, Grid Resource Access and Management), information service (GRIP, Grid Resource Information Protocol), data transfer (Grid FTP).

Globus Toolkit provides an access to computing processes by means of GRAM5 component (Grid Resource Allocation and Management). Computational jobs in GRID-network are controlled, added and canceled with the help of this component [11]. Gatekeeper process is run on a random computational node to initiate the computing program on the same node. A client application sends requests to GRAM, the requests are written in Resource Specification language (RSL) [12].

Basic data is transferred into GT with the help of Grid FTP in order to carry out calculations [13].

The Globus Toolkit software includes the following components which are necessary for the operation of program complex for material modeling: the file server "globus-gridftp-server", client "globus-url-copy" and the set of libraries for developers [14, 15].

Interaction between the client application and DPS is carried out according to the algorithm which is shown in Figure 3.

A transport file of UTF experiment is formed in the working memory area of a client's application before the job sending function into DPS is run. Further, the resources initialization of local and remote DPS is performed by means of local GT server. Local and remote DPS are chosen on the basis of information about available resources.

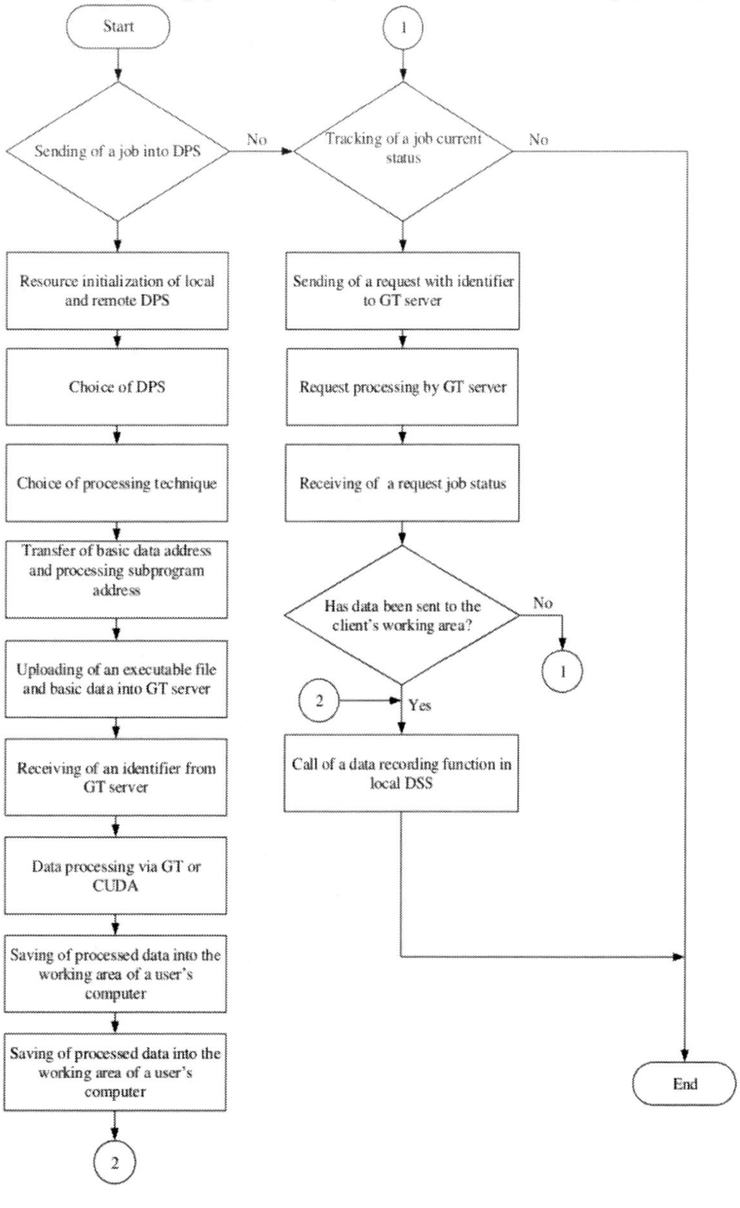

Figure 3. The flowchart of calculating the material properties in DPS

Two modes of data are available to a user: static and dynamic. After selecting a mode, the client application passes a pointer to the basic data files and processing program into the local GT server. If you select the local DPS, an executable program file and files with the basic data are loaded into the local GT server. Computations in the remote DPS are performed in the same way.

After files uploading GT server forms the executive job and assigns it an identification number. After calculations the local GT server collects files with processed data and transfers them into the working area of a user's application.

User functions of client's applications interaction with DPS implemented by means of commands which are transmitted to Globus Toolkit server via a terminal. The whole list of interaction functions with DPS which is available to the user in the client's application of Antares is shown in Table 1.

Table 1. The list of user's functions which interact with DPS

User's functions	GT program components for implementation of functions	Description of GT component
Function of sending a job to DPS	globusrun	It serves to control jobs which are performed on a local or remote host with use of GRAM.
	globus-job-run	It forms a job on the basis of parameters of a using command and it transfers the job to GRAM service
	globus-job-manager	It allows you to start and control GRAM service which work is performed by the system that controls local resources directly on a computational node. It is run with globus-gatekeeper.
	globus-gatekeeper	Metaserver which runs different services after authentication of TCP connection. At the same time it correlates client's accounting data of a local account
	globus-gatekeeper-admin	It controls data which is used by globus-gatekeeper.
	globus-job-submit	It is similar to globus-job-run, but it is used to add the whole package of jobs.
	globus-job-cancel	It cancel jobs that have been already created
	globus-url-copy	It transfers data (data of a file or the whole file directory) between two addresses. It supports the following URL-prefixes :file:// (only local machine) ftp://, gsiftp://, http://, https://.
Tracking of a job current status	globus-job-status	It checks the status of jobs performance in GRAM service sending status requests to a job manager. In the case of successful job, the information about job status displayed in a terminal window.

IOP Conf. Series: Materials Science and Engineering **112** (2016) 012028 doi:10.1088/1757-899X/112/1/012028

3. Experiment

The properties calculation of Al_3Ti system atomic structure was carried out in order test the functions of the program complex. Mathematical model that describes the properties of Al_3Ti atomic structure was specified with the help of specialized file entering into the UTF composition.

Mathematical calculations were taken from famous LCAO-CO method (crystal orbital as a linear combination of atomic orbitals) that is an analogue of the Hartree-Fock-Roothaan method for crystals. On the basis of method results it was possible to connect atom properties which form a crystal with the crystal itself.

In LCAO-CO one-electron functions for the crystal are constructed as linear combinations of atomic orbitals sums in blocks:

$$\varphi_{nk}(r) = \sum_p C_{np}(k) b_{pk}(r).$$

Coefficients $C_{np}(k)$ are found with the help of variational principle from the Hartree-Fock-Roothaan equations:

$$\sum_p F_{pq}(k) C_{np}(k) = E_n(k) \sum_p S_{pq}(k) C_{np}(k),$$

and besides it comes out as N systems for crystals instead of such a system for a molecule.

Matrix $F_{pq}(k)$ is found as a sum of matrixes $H_{pq}(k)$ and $G_{pq}(k)$, the elements of which are determined in the following way:

$$H_{pq}(k) = \int b^*_{pk}(r) \left[-\frac{1}{2}\Delta - \sum_A \frac{Z_A}{|r - R_A|} \right] b_{qk}(r),$$

$$G_{pq}(k) = \sum_{s,t} \sum_{k'} P_{st}(k') \left[(b_{kp} b_{kq} \mid b_{k's} b_{k't}) - \frac{1}{2} (b_{kp} b_{k's} \mid b_{k't} b_{kq}) \right].$$

Density matrix for crystal is calculated with the formula:

$$P_{st}(k') = 2 \sum_m C_{ms}(k') C^*_{mt}(k')$$

Lattice sums of matrix elements calculated with functions from the zero (central cell):

$$F_{pq}(k) = \sum_n F_{p0,qn} \exp(ika_n),$$

$$S_{pq}(k) = \sum_n S_{p0,qn} \exp(ika_n).$$

While solving the Hartree-Fock-Roothaan equations for crystals it is necessary to know $C_{np}(k)$ of the whole reduced Brillouin zone for fixed value because the specified form of matrix elements as follows:

$$F_{p0,qn} = H_{p0,qn} + G_{p0,qn},$$

$H_{pq}(k)$ are Hamiltonian elements for the crystal which contains energy of electron interaction with cores *V(r)*, and also it contains kinetic energy. Two-electron component $G_{pq}(k)$ contains summation according to filled state:

IOP Conf. Series: Materials Science and Engineering **112** (2016) 012028 doi:10.1088/1757-899X/112/1/012028

$$G_{p0,qn} = \frac{1}{N}\sum_{rs}\sum_{k'} P_{rs}(k')\sum_{n'n''}\exp\{ik(a_{n'} - a_{n''})\}[(p0,qn\,|\,rn',sn'') - \frac{1}{2}(p0,rn\,|\,qn',sn'')]$$

As a result, for self-consistent calculation of the electron crystal structure it is necessary to summarize a large number of occupied electronic states on every stage of integration process.

The given method was implemented in the client's application.

In the course of the experiment, special attention was paid to the problems of elaboration of a client's application interaction with data processing system.

The following software was installed and adjusted on the clients computer with OS Linux:

- S/W DBMS PostgreSQL 9.4;
- S/W FTP-server File Zilla Server 0.9.39;
- S/W GlobusToolkit 6.0.

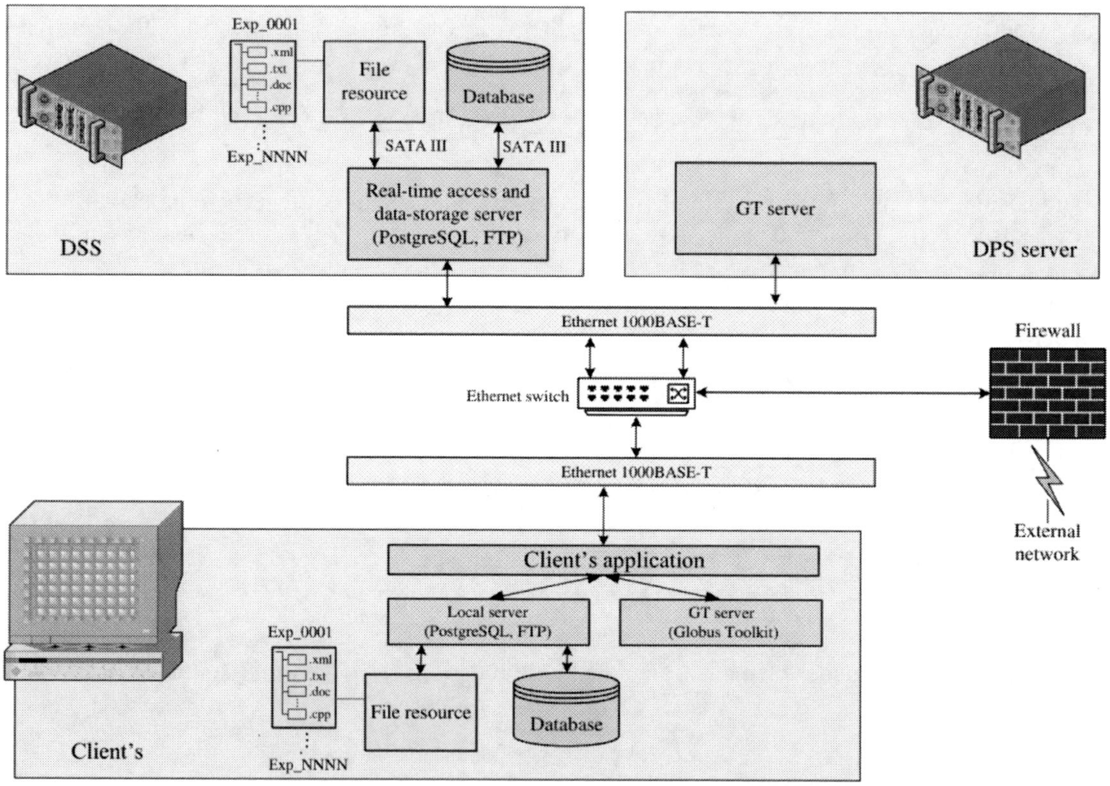

Figure 4. The TestStand scheme

The client's computer and two remote servers interact in local network. The scheme of the TestStand is presented in Figure 4. DPS and DSS servers function under OS Linux.

Data which represents five experiments was written in local and remote DSS. The further work was to check how the interaction protocol of PC components functions on the level of program features.

Connection to local database:

conn = PQconnectdb("user=postgres password=12345 dbname=materials hostaddr=127.0.0.1 port=5432");

Verification of status connection with the following identiifier:

```
if (PQstatus(conn) != CONNECTION_OK)
{
fprintf("Connection to database failed"),
CloseConn(conn);
}
```

The search for the fourth experiment in the local database:

```
res = PQexec(conn, "SELECT * FROM experiment1 WHERE id=4 ");
if (PQresultStatus(res) != PGRES_TUPLES_OK)
  {
fprintf("select failed");
PQclear(res);
CloseConn(conn);   }
```

A framework for command execution is a set of functions which includes SQL-request.

4. Conclusion

The model described by the given equations was applied in program package Antares for atomic system Al_3Ti. It was carried out the calculation of geometrical parameter a of the crystal grid; in the result we got the value which is equal to 3.997 Å, what is 0.161 Å more than the 3.836 Å experimental value [16]. According to the result we can make a conclusion that the Hartree-Fock method for crystals which is used jointly with STO-3G basis calculates geometrical parameters of the chosen atomic system Al_3Ti with relative error 4% because of error accumulation during the system self-consistency procedure.

Acknowledgements

The work is performed with financial support of the Ministry of Education and Science of the Russian Federation. Grant agreement No. RFMEFI57814X0095 as of November 28, 2014.

References

[1] Aleksandrov E I, Amirhanov I V, Zemljanaja E V, Zrelov P V, Zuev M I, Ivanov V V, Podgainyi D V, Sarcar N R, Sarhadov I S, Streltsova O I, Tukhliev Z K and Sharipov Z A 2014 Principles of software for the simulation the physical processes in the hybrid computing systems (for example, the complex GIMM_FPEIP) *Vestnik RUDN* **2** 197 (in Russian)

[2] Bondarenko A A, Kononov Je M, Kosolapov O A, Poljakov S V and Jakobovskij M V 2013 Software package GIMM_NANO *Scientific service on the Internet: all facets of parallelism [Nauchnyj servis v seti internet: Vse grani parallelizma]* 333 (in Russian)

[3] Jel'kin P M, Stepanovich E Ju, Pulin V F, Mozhaeva N A and Bulatov M F 2010 The use of computer technology "Gaussian" to simulate the vibrational states of polymers *Vestnik SGTU* **1** 149-155 (in Russian)

[4] Artyshhenko S V, Butyrskij A M, Kuchin E L and Chernov R A 2014 Features Gaussian to calculate various properties of the molecules that determine their interaction with the laser field. Some aspects of the calculation of the scattering tensor of molecules *Vestnik VGTU* **10** 81-85 (in Russian)

[5] Grishaeva T N and Maslij A N 2012 Comparison of quantum chemical software package GAUSSIAN 09, ORCA 2.8 and PRIRODA 11 the example of calculation of structures of complexes of nickel(II) and copper(II) *Vestnik Kazanskogo tehnologicheskogo universiteta* **15** 7-10 (in Russian)

[6] Koren'kov V V 2010 Grid technology: current status and prospects *"Herald of the International Academy of Science. Russian Section"* **1** 41-44

[7] Astaf'ev A C, Afanas'ev A P and Lazarev I V 2009 Scientific service-oriented environment based on Web technology and distributed computing *Proceedings of the All-Russian Scientific Conference (Novorossiysk)* 463-467 (in Russian)
[8] Boreskov A V 2010 The basics of working with CUDA technology [Osnovy raboty s tehnologiej CUDA] (Moscow: DMK Press) (in Russian)
[9] Doroshenko A E Ruhlis K A and Mohnica A S 2008 Distributed resource management platform for heterogeneous cluster *Problemy programmirovanija* **2-3** 150-156 (in Russian)
[10] Brunett S, Czajkowski K, Fitzgerald S, Foster I, Johnson A, Kesselman C and Leigh J 1998 Application Experiences with the Globus Toolkit Proc. *7th IEEE Symp. on High Performance Distributed Computing* 81-89
[11] Savin G I, Korneev V V, Shabanov B M, Telegin P N, Semenov D V, Kiselev A V, Kuznecov A V, Vdovikin O I, Aladyshev O S and Ovsjannikov A P 2008 Creation of a distributed infrastructure for Supercomputing Applications *Programmnye produkty i sistemy* **2** 2-7 (in Russian)
[12] Koren'kov V V, Kutovskij N A and Semenov R N 2012 Applied experience in adapting software packages for use in grid environments *Computer Research and Modeling* **4** 339-344 (in Russian)
[13] Bresnahan J, Link M, Kettimuthu R, Fraser D and Foster I 2007 Gridftp pipelining *Proceedings of TeraGrid Conference*, Madison, WI, 2007.
[14] Allcock W, Bresnahan J, Kettimuthu R, Link M, Dumitrescu C, Raicu I and Foster I 2005 The globus striped GridFTP framework and server *Proceedings of the 2005 ACM/IEEE Conference on Supercomputing, SC'05* 54-64
[15] Kettimuthu R, Liu W, Link J and Bresnahan J 2008 A GridFTP transport driver for Globus XIO *International Conference on Parallel and Distributed Processing Techniques and Applications*, 843-849
[16] Nikol'skij B P 1966 Directory chemist [Spravochnik himika] (Moscow: Himiya) (in Russian)

Recent advances and future perspectives of nanosized zero-valent iron for extraction of heavy elements from metallurgical sludges

I Yu Mikhailov, V V Levina, E A Kolesnikov, K O Chuprunov, A A Gusev, A Yu Godymchuk, D V Kuznetsov

National University of Science and Technology "MISiS", 4 Leninsky prospect, Moscow, 119049, Russia

E-mail: ivan.mikhailov@misis.ru

Abstract. Advanced oxidation processes with nanosized zero-valent iron have presented great potential in wastewater treatment technology and now experience both increasing popularity and reliable technical improvements. Besides wastewater treatment, there is another promising application for an emerging technology of iron nanoparticles – as Fenton-like catalyst for extraction of valuable elements from poor and secondary raw materials such as metallurgical sludges. In present research, we carried out a set of experiments with emphasis on the physicochemical mechanisms and their relationship to the performance. In particular, we examined complex acidic – hydrogen peroxide leaching of zinc from blast furnace sludge with nanosized zero-valent iron as Fenton-like catalyst. Results of the experiments showed promising potential for subsequent application in extraction of heavy and rare-earth elements.

1. Introduction

Ongoing development of state-of-art technologies and production is the increasing trend of modern industry. Essential role of heavy and rare-earth elements in magnets, batteries, catalysts, metallurgy, ceramics and other applications leads to rapidly increasing consumption. That makes search of new and reliable sources of rare-earth elements one of most important strategic issues [1-2]. Industrial wastes are promising secondary source of hard and rare-earth elements. For example, valuable information on the presence of rare-earth oxides in flux powders used in the continuous steel casting as well as in slags and sludges resulting from the smelting of rare-earth-containing iron ores was presented by various researchers [3-10].

On the other hand technologies and methods for effective recovery are still an issue due to presence of goal elements in hard soluble phases, requiring high oxidation-reduction potential besides of high concentrations of acids. To promote the oxidation conditions, so-called Fenton-like process is widely used [11]. Nanosized zero-valent iron (nZVI) has shown significant results as highly reactive Fenton-like catalyst for production of hydroxyl radicals [12]. High values of specific surface area makes it more reliable compared to other sources of ferrous ions for Fenton reaction [13].

A significant example of large scale industries generating valuable heavy and rare-earth elements is ferrous metallurgy. During conventional steel-making process, a various types of dusts and sludges are formed. Traditionally one of the most problematic metallurgical wastes is blast furnace sludge, which is basically produced during wet scrubbing of top furnace gas. In Europe alone, annual production

Content from this work may be used under the terms of the Creative Commons Attribution 3.0 licence. Any further distribution of this work must maintain attribution to the author(s) and the title of the work, journal citation and DOI.
Published under licence by IOP Publishing Ltd

of solid blast furnace sludge exceeds as high as 500 000 tons. Blast furnace sludge is mostly not suitable for direct recycling in blast furnace process due to zinc content, higher than 0.5 % by mass: during the metallurgical process, zinc partially condensates at the upper levels of blast furnace, which creates huge risks of damage to refractory coatings of the furnace. On the other hand, continuous tightening of environmental laws make such waste unacceptable for landfilling due to moderately high content of hard metals and alkalis, that may cause environment and public health risks [14]. Various types of sludges generated during steel production are shown in table 1.

Table 1. Types of sludge waste generated by steel plants [15]

Waste	Source	Production (kg/t of metal)
Coke sludge	Coke oven	-
Blast furnace sludge	Blast furnace	25-30
Sintering plant sludge	Sintering plant	-
LD sludge	Steel melting	15-17
ACP/GCP sludge	Steel melting	-
Carbide sludge	Acetylene plant	-
Mill sludge	Rolling mills	10-12
Water treatment sludge	Water treatment plant	-

As is shown above (table 1) blast furnace sludge is also generated in the highest quantities compared to other sludges from steel plants. Besides zinc, blast furnace sludge and slug could also contain lanthanum (La), cerium (Ce), praseodymium (Pr), yttrium (Y), europium (Er), dysprosium (Dy), neodymium (Nd) and indium (In) [16-18].

In this study we focused on the effect of nZVI additions on complex leaching of zinc and indium from blast furnace sludge in presence of 0.4 M sulfuric acid and hydrogen peroxide at various concentrations.

2. Materials and methods

In present study we used a sample of air-stable nZVI, synthesized by chemical precipitation and metal recovery in continuous hydrogen flow with subsequent passivation in technical grade nitrogen, as described in [19]. nZVI sample was characterized by BET adsorption (Quantachrome Nova 1200e instrument) and TEM (JEOL JEM-2000 EX with accelerating voltage of 100 kV).

In order to prepare nZVI sample for TEM analysis, small quantity of nZVI (below 10^{-8} kg) was added to 50 ml of 98% ethanol and thoroughly mixed using ultrasonic irradiation (Hielscher UP400S) for approximately 1 min. Then, a few drops of nZVI suspension were added on a carbon grid. Finally, prepared grid was dried under N_2 (nitrogen) atmosphere at room temperature for 15 min.

Particle size was calculated (assuming that all particles have spherical morphology) as follows:

$$d_{SSA} = \frac{\rho}{6 \cdot S} \tag{1}$$

where ρ – density, kg/m^3 and S – specific surface area, m^2/kg.

As mentioned above, as an example of possible source of valuable elements, we used a sample of blast-furnace sludge, provided by Cherepovetsky metallurgical plant of JSC “Severstal” (Vologda region, Russia). Before experiments, sample of blast furnace sludge was dried at room temperature and carefully homogenized using agate mortar.

All leaching experiments were carried with the solid-liquid (S/L) ratio of 1:10, in opened 300 ml quartz flask with cooling jacket. Constant temperature 20 °Cduring all experiments was maintained using Lauda RE420 thermostat. To achieve different initial pH and ORP levels, sulfuric acid and hydrogen peroxide solutions with concentrations 0.4 M and 0.1, 0.05, 0.01, 0.005 M were used, respectively. Amount of nZVI addition was 1 g/dm^3 for all experiments. Mixing was carried out using Heidolph RZR-1 mechanical stirrer at continuous rate 500 RPM.

After treatment, leachates were separated from nZVI using 0.22 μm nylon membrane (Millipore), heated for 15 min at 60 °C in 1 M nitric acid to prevent possible precipitation and analyzed by ICP-AES (Thermo Scientific iCAP 6300 Radial View).

3. Results and discussions

Results for specific surface area measurement of nZVI sample are shown in table 2. TEM image of synthesized sample is presented in figure 1.

Table 2. Specific surface area and calculated particle size of nZVI sample

Sample	Material	Specific surface area, m^2/g	Calculated particle size, nm
1	nZVI	13.9±1	55

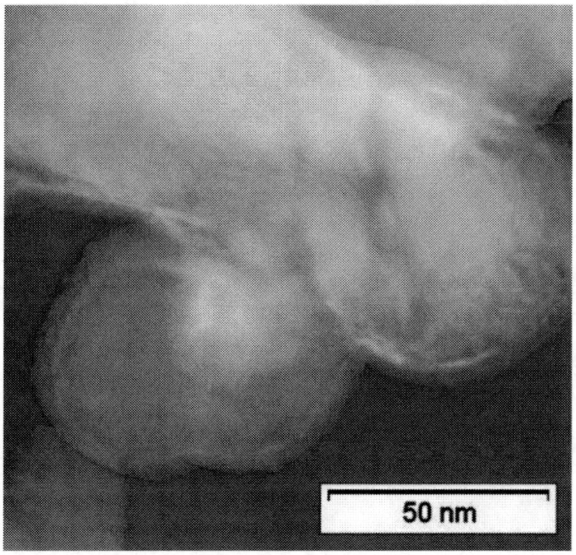

Figure 1. TEM image of the air-stable nZVI

Specific surface area for nZVI sample (table 1) correlates with diameter estimated from TEM image (fig. 1). It should be noted that iron nanoparticles have thin protective oxide shell that prevents air and water oxidation without significant loss of reactive properties.

Typical and sample element compositions of blast furnace sludge obtained by ICP-AES are shown in table 3. As it could be seen, sample contains significant amounts of Fe, Zn, Al, Pb, Si, alkali metals and about 0.008 % of In. Such amount of In makes blast furnace sludge a promising secondary source upon condition of its effective recovery into solution.

Table 3. Typical [14, 18] and sample element compositions of blast furnace sludge

Component	Typical (%)	Sample (%)
Fe	21-32	36.7
Zn	1.0-3.2	3.38
In	0-0.017	0.008
Pb	0.3-1.2	0.22
Cr	$\leq$ 0.024	0.02
Al	-	0.39
Cu	-	<0.01
Ni	-	<0.01
Mg	<2	1.23
Ca	<13	3.13
Si	-	2.13
Mn	-	0.11
Loss on ignition	-	36.4

Possible Fenton-like reactions, taking place during the leaching of blast furnace sludge are presented in table 4. Conventional Fenton process includes reactions #1-9. In presence of nZVI as catalyst and source of ferrous ions, Fenton-like process includes also reactions #10-12. High reactivity of nZVI ensures fast dissolution and formation of extremely oxidative hydroxyl radicals.

Table 4. Possible chemical reactions during Fenton-like process

#	Reaction	Source
1	$H_2O_2 \rightarrow 2OH^*$	
2	$Fe^{2+} + H_2O_2 \rightarrow Fe^{3+} + OH^* + OH^-$	
3	$Fe^{3+} + H_2O_2 \rightarrow Fe^{2+} + HO_2^* + H^+$	
4	$OH^* + H_2O_2 \rightarrow HO_2^* + H_2O$	
5	$OH^* + H_2O_2 \rightarrow HO_2^* + H_2O$	[20]
6	$OH^* + Fe^{2+} \rightarrow Fe^{3+} + OH^-$	
7	$Fe^{3+} + HO_2^* \rightarrow Fe^{2+} + O_2H^+$	
8	$Fe^{2+} + HO_2^* + H^+ \rightarrow Fe^{3+} + H_2O_2$	
9	$2HO_2^* \rightarrow H_2O_2 + O_2$	
10	$Fe^0 + H_2O_2 + H_2SO_4 \rightarrow Fe^{2+} + SO_4^{2-} + 2H_2O$	
11	$Fe^{2+} + H_2O_2 \rightarrow Fe^{3+} + OH^* + OH^-$	[21]
12	$Fe^{2+} + H_2O_2 \rightarrow Fe\ (IV) + H_2O$	

Results of the leaching experiments in presence of nZVI additions are shown in figure 2. As could be seen, addition of nZVI provides a significant catalytic effect on the process, which particularly noticeable at a concentration of H_2O_2 equal to 0.05 and 0.1 M. Thus, more than 70 % of Zn and 65 % of In are dissolved after treatment with 0.1 M of H_2O_2. It should be noted that further investigations are required for kinetics and mechanisms of Zn and In leaching processes.

Taking into account possible reactions (table 4) it is clear that decreasing of H_2O_2 concentration should lead to appropriate decrease of metals extraction due to decrease of oxidation-reduction potential. But for most of concentrations observed slow fall in In and especially Zn dissolution. We assume that effect takes place due to Fenton process: continuous production of water radicals allows to maintain necessary oxidative conditions for extraction.

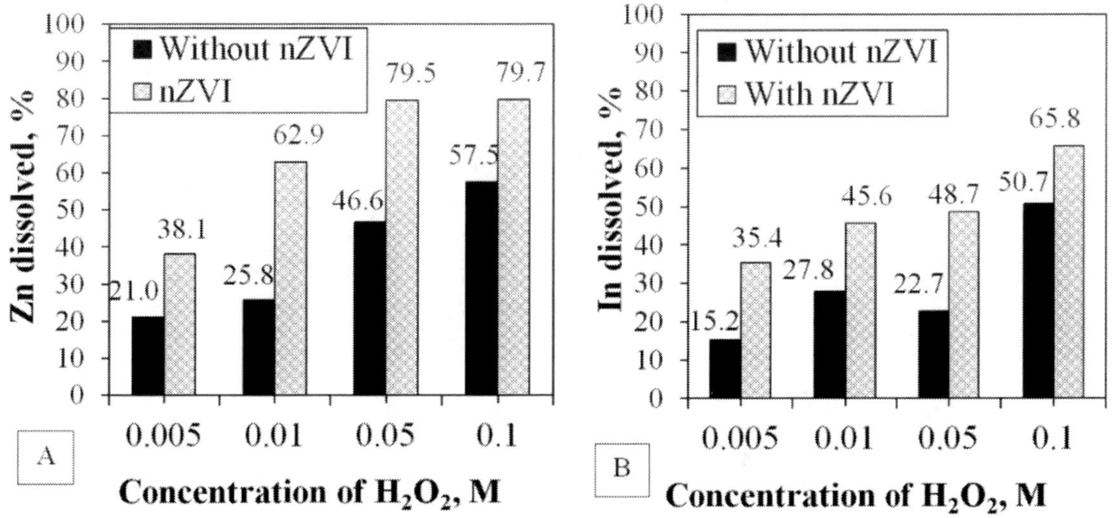

Figure 2. A: Dissolution degrees of zinc (Zn), B: dissolution degrees of indium (In)

4. Conclusions

The presented research focused on effect of nanosized zero-valent iron showed that Fenton-like process has a great potential in extraction of heavy and rare-earth elements from secondary sources such as blast furnace sludge.

Additions of nZVI lead to promotion of oxidative conditions and therefore more effective extraction of metals into solution. Achieved dissolution degrees observed during experiments makes Fenton-assisted leaching with nZVI as Fenton catalyst a possible solution for other applications in the field of extraction valuable rare-earth elements form poor ores and sludge wastes.

Acknowledgements

Authors gratefully acknowledge the Federal Program for Research and Development of Russian Federation (Ministry of Science and Education of Russian Federation, project №RFMEFI57514X0102) for financial support.

References

[1] Binnemans K, Jones P T, Blanpain B, Van Gerven T, Yang Y, Walton A and Buchert M 2013 *J. Clean. Prod.* **51** 1
[2] Binnemans K, Jones P T, Blanpain B, Van Gerven T and Pontikes Y 2015 *J. Clean. Prod.* **99** 17
[3] Zheng X P, Lin H K 1994 *Miner. Eng.* **7** 1495
[4] Anacleto N M, Lee H G and Hayes P C 1993 *ISIJ Int.* **33** 549
[5] Derevyankin V A, Gasik M I, Anelok L I, Porotnikova T P and Vukelich S B 1981 *Izv. Vyssh. Uchebn. Zaveden. Tsvetn. Met.* **5** 84
[6] Derevyankin V A, Porotnikova T P, Kocherova E K and Yumasheva I V 1981 *Izv. Vyssh. Uchebn. Zaveden. Tsvetn. Met.* **5** 119
[7] Fang X, Hu L, Tan R, Wang Z Process for utilizing waste smelting dregs of rare-earth ferrosilicon. Patent of China CN1302910A Published July 11 2001 (in Chinese)

[8] Jia Y L, Dong Y C and Qin Z Z 1995 *J. Rare Earth.* **3** 133
[9] Lin H K and Zheng X P 1997 *Miner. Metall. Proc.* **14** 18
[10] Yang X H, Long H, Cheng G G, Wu C C and Wu B 2011 *J. Rare Earth* **29** 1079
[11] Babuponnusami A and Muthukumar K 2014 *J. Env. Chem. Eng.* **2** 557
[12] Crane R, Scott T B 2012 *J. Hazard. Mater.* **211-212** 112
[13] Liang J, Komarov S, Hayashi N and Kasai E 2007 *Ultrason. Sonochem.* **14** 201
[14] Van Herck P, Vandecasteele C, Swennen R and Mortier R 2000 *Environ. Sci. Technol.* **34** 3802
[15] Das B, Prakash S, Reddy P S R and Misra V N 2007 *Resour. Conserv. Recycl.* **50** 40
[16] Das B, Prakash S, Biswal S K, Reddy P S R, Mohapatra B K and Tripathy H K 2002 *Utilization of solid waste materials generated at steel plants by physical beneficiation technique: report submitted to Ministry of Steel, India* (Bhubaneswar: Govt. of India Regional Research Laboratory (CSIR)) p 1
[17] Bandara H M D, Mantell M A, Darcy J W and Emmert M H 2015 *Energ.Technol.* **3** 118
[18] Shen X, Li L, Wu Z, Lü H and Lü J 2013 *Metall. Mater. Trans. B* **44** 1324
[19] Mikhailov I, Mandal A R, Kotov S and Kuznetsov D 2013 *Adv. Electrochem.* **1** 83
[20] Umar M, Aziz H A and Yusoff M S 2010 *Waste Manag.* **30** 2113
[21] Xu L and Wang J 2011 *J. Hazard. Mater.* **186** 256

REM-containing silicate concentrates

V F Pavlov[1,*], O V Shabanova[1], I V Pavlov[1], M V Pavlov[1] and A V Shabanov[2]
[1] Special Design and Technological Bureau «Nauka» of the Krasnoyarsk Scientific Center, SB RAS, Krasnoyarsk, Russia
[2] Institute of Physics, Sibirian Branch, Russian Academy of Sciences, Krasnoyarsk, Russia

[*] E-mail: pavlov@akadem.ru

Abstract. A new method of advanced complex processing of ores containing rare-earth elements (REE) is proposed to obtain porous X-ray amorphous aluminosilicate material with a stable chemical composition which concentrates oxides of rare-earth metals (REM). The ferromanganese oxide ores of Chuktukon deposit (Krasnoyarsk Region, RF) were used for the experiment. The obtained aluminosilicate material is appropriate for treatment with 5 – 15% solutions of mineral acids to leach REM.

1. Introduction

For the experiment the ore samples were taken from the Chuktukon deposit. They mostly have iron-manganese oxide composition (40.0-70% Fe_2O_3 and 1-20% MnO), and contain the phases of monazite (Ce,Ln,Y,Th)HO_4, sodium-cerium bisulfate ($NaCe(SO_4)_2*H_2O$), calcite ($CaCO_3$), manganese sulfate, Ca,Ce-phosphate $Ca0.5Ce_2(PO_4)_3$. The instability of the chemical and phase composition of the ore significantly influences physico-mechanical properties of the materials obtained on its basis. The ores under study are complex, and can hardly be enriched due to the high dispersivity of mineral constituents and mutual coalescence [1-4]. Processing of such ores requires new approaches and technology solutions.

2. Experimental

To stabilize and concentrate the rare metal ore in the silicate part of the melt, a pyrometallurgical method was chosen for deep reduction smelting with the melt separating into silicate and metallic portions. This method is also applicable for complex processing of REM-containing technogenic materials, e.g. red muds of aluminum production [5]. Brown coal was used as a reducing agent. The smelting was performed in the induction furnace KIT-25, as is done in [6, 7]. Charging material was loaded into a graphite crucible, heated during 1 hour to 1500-1550°C and kept for another hour, with the samples being taken every 10 min for chemical analysis.

Content from this work may be used under the terms of the Creative Commons Attribution 3.0 licence. Any further distribution of this work must maintain attribution to the author(s) and the title of the work, journal citation and DOI.
Published under licence by IOP Publishing Ltd

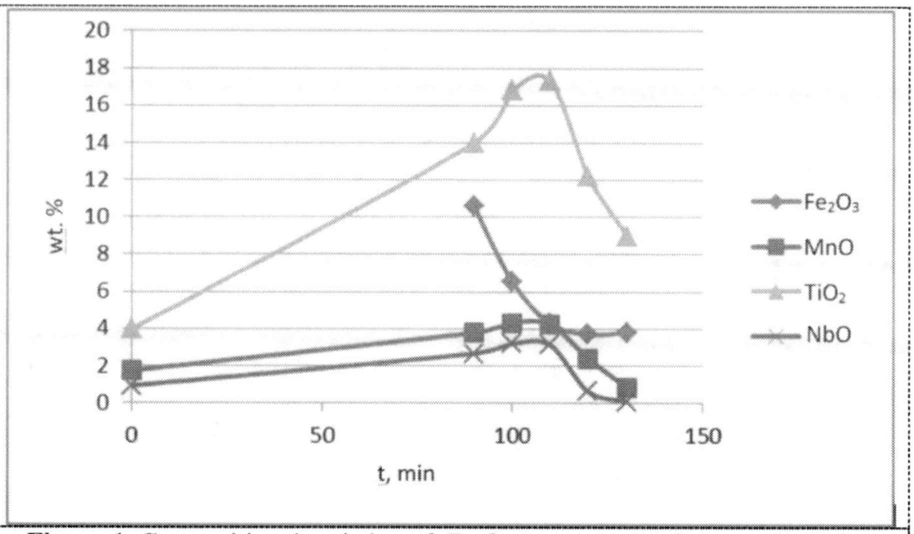

Figure 1. Compositional variation of Fe_2O_3 (curve 1), MnO (curve 2), TiO_2 (curve 3), NbO (curve 4) vs. the reduction time in the silicate part of the melt

The figure 1 shows that upon reducing the ore using brown coal it is mostly iron which is transformed into metallic phase during the first 90 min from switching on the melting furnace. Some part of phosphorus is carried away with the gas phase, the other part is redistributed between the metallic and silicate part of the melt. The composition of the metal sample taken under these conditions is the following, in wt.%: Mg – 0.15; Al – 0.49; Si – 1.02; P – 3.73; S – 0.47; Ca– 0.17; V – 0.17; Mn – 2.13; Fe – 90.5; Nb – 0.26; Cr – 0.21; W – 0.11; Ti – 0.16.

At this stage the main part of titanium, manganese and niobium is left in the slag. The reduction process intensively and deeply proceeds at 1550 - 1600° C, with the time of the melt exposure to this temperature being longer than 110 min (figure 1.). An abrupt decrease in the content of iron, manganese niobium and titanium, which can be seen in the curves in figure 1, evidences their transition into the metallic phase. A synchronous decrease of the content of these elements in the melt evidences their association [8].

The performed preliminary studies of the reduction smelting allow one to assume a possible, almost complete transition of iron, phosphorus, manganese, niobium and titanium into the metallic state. In the silicate part of the melt, the concentration of REM, alkali-earth metals, aluminum and thorium increases.

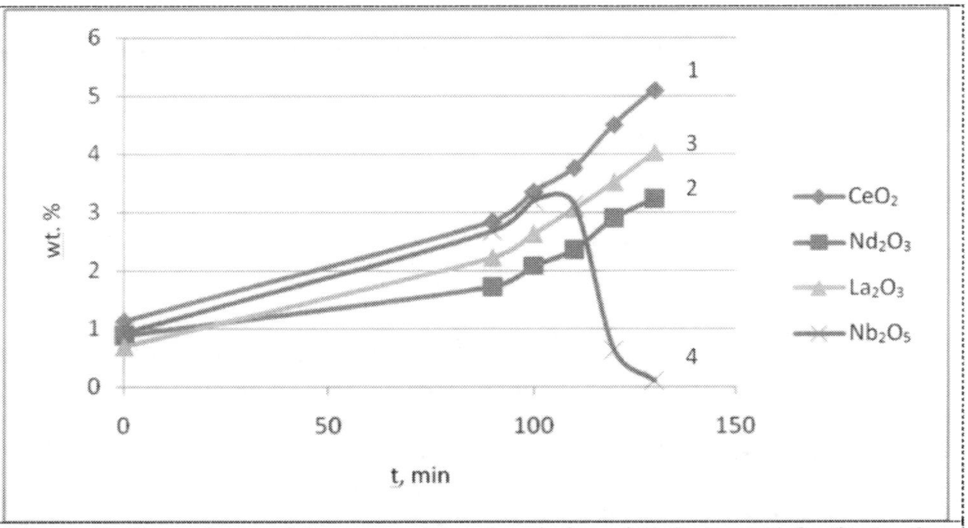

Figure 2. Dependence of the compositional variation of CeO_2 (1), Nd_2O_3 (2), La_2O_3 (3) and Nb_2O_5 (4) on the time of the melt exposure in the process of reduction smelting

The abrupt decrease of the content of niobium oxide in the melt is due to its interaction with silicon carbide and its extraction into the metallic phase.

The silicon carbide formed in the melt by the reaction SiO_2+ 2C =SiC + 2C participates in the reactions of deep reduction of niobium, iron, manganese and titanium. Here, it plays the role of a transporter-agent, resulting in these elements being extracted into the metallic phase, while the silicate part of the melt becomes enriched with rare-earth metals (figure 2, curves 1, 2, 3, figure 3).

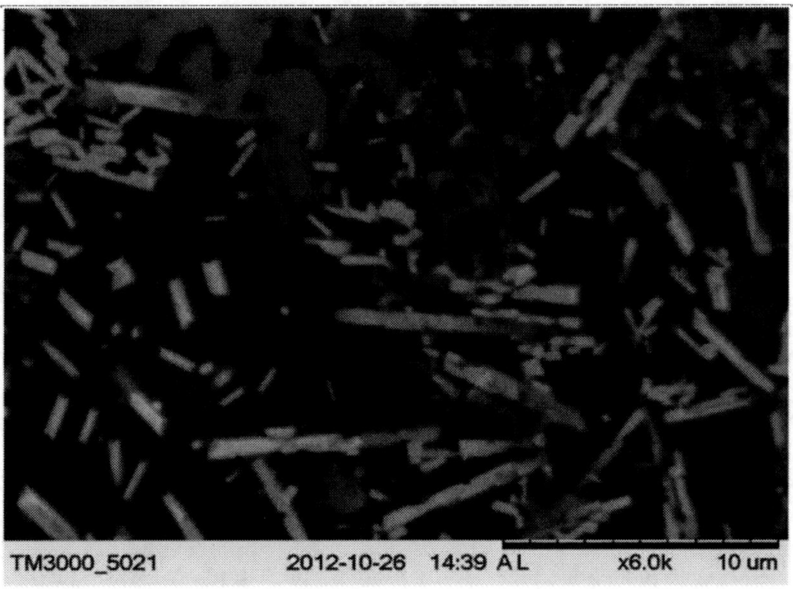

Figure 3. Electron micrograph of the glass sample of the silicate melt

In the figure 3 one can see an aggregation of white bars with the size ranging from 1 to 10 μm. The Rare metal imaging shows that lanthanum is associated with cerium. The detected lanthanum-cerium seeds are in the form of impurity phases in the glass structure [8]. The absence of strong bonds with the glass structure will facilitate their leaching.

The rate of REM leaching is also affected by the specific surface of the glass which increases upon its foaming. It is shown that pouring the melt into water results in the formation of a foamed aluminosilicate material which is presented in figure 4.

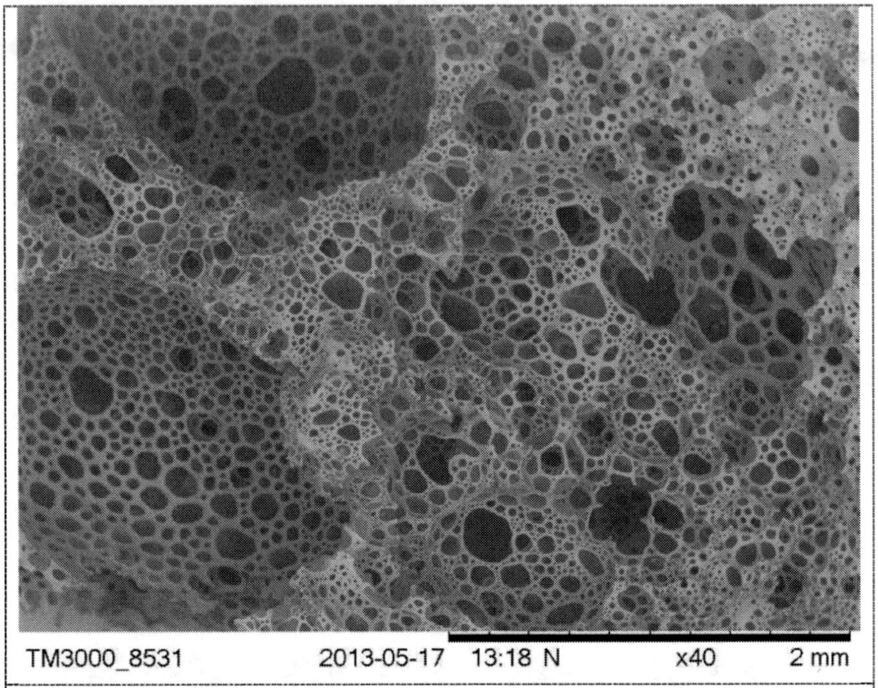

Figure 4. Electron micrograph of the porous aluminosilicate material

The maximum pore size amounts to 150-200 μm. The material represents a sponge composed of numerous porous layers, with the spacing between the layers being approximately equal to the pore diameter. The destruction of the foam layers is due to an abrupt cooling of the melt.

3. Results and discussion

As in the glass, in the foamed material one can observe the REE oxide seeds where lanthanum is associated with cerium and neodymium. The obtained material is appropriate for treatment with 5 – 15% solutions of mineral acids.

The aluminosilicate foam was used to leach REM. The composition of this material in wt. % is the following: Na_2O – 0.29, MgO – 1.98, Al_2O_3 – 9.14, SiO_2 – 26.4, SO_3 – 1.42, Cl – 0.45, K_2O – 0.91, CaO – 45.2, TiO_2 – 3.67, Cr_2O_3 – 1.06, MnO – 0.27, Fe_2O_3 – 0.53, SnO – 0.15, REM oxides – 8.53.

The unground material was treated with sulfuric acid (its concentration being 10%), the insoluble precipitate – mostly gypsum and silica – was filtered and oxalic acid was added to the filtrate for REM to precipitate (figure 5).

Figure 5. Electron micrograph of the REM oxalate crystals

The composition of the precipitated REM concentrate in wt.% is the following: La_2O_3 – 13.8; CeO_2 – 11.7; Nd_2O_3 – 8.7; Y_2O – 2.5; Sm_2O_5 – 2.0; ZrO_2 – 0.5; Nb_2O_5 – 0.2; Cr_2O_3 – 5.3 and others (mostly silica and aluminum oxide) – 55.3. The extraction degree was approximately equal to 60%.

The initial materials and the obtained products were analyzed in the Joint Use Center of The Krasnoyarsk Scientific Center, Siberian Branch of the Russian Academy of Sciences.

4. Conclusion

The present paper considers the peculiarities of the process of pyrometallurgical processing of ores from Chuktukon deposit, i.e. the reduction of iron, manganese, niobium, and titanium oxides. It is shown that the rate of the manganese, titanium and niobium transition into the metallic phase increases only after the formation of silicon carbide in the slag melt. The silicon carbide participates in the transport reactions of the metal reduction. Moreover, the silicon carbide formation is necessary for the pore formation in the slag melt. It has been found that it is possible to obtain a concentrate containing about 40% of oxides of rare earth metals from the silicate part of the melt.

References

[1] Lomayev V G and Lomayeva N V 2006 *Collection «Geology and mineral resources of the Krasnoyarsk Region» Krasnoyarsk, Krasnoyarsk Research Institute of Geology and Mineral Raw Mater.* **7** 20 [in Russian]

[2] Lomayev V G and Kuzmin V I 2004 *Exploration and Protection of Mineral Resources* **6** 23

[3] Lomayev V G and Kuzmin V I 2006 *RF State Registr. 16-01-51/21.* **1** 65

[4] Kuzmin V I 2006 *Nonferr. Metals* **12** 62

[5] Pavlov M V, Pavlov I V, Pavlov V F, Shabanova O V and Shabanov V F 2013 Advanced processing of red muds into the materials of aluminosilicate system concentrating REE *Current problems in mining, production and application of rare-earth elements in Russia: Proceedings of all-Russian conference on rare-earth materials "REM-2013" 20-21 November*

2013 : collection of reports ed B.M. Kerbel.(Seversk: Publishing house of Seversk Technological Institute NRNU MEPhI) p 101 [in Russian]

[6] Pavlov M V 2012 *Chemistry for Sust. Development* **20** 443

[7] Pavlov V.F. 2005 *Physical basis for the technology of obtaining new materials with the desired properties based on the creation of the system of complex use of technogenic and non-ore raw materials SB RAS* Publishing house(Novosibirsk) pp 191 [in Russian]

[8] Shabanov V. F., Pavlov V. F., Kuzmin V.I., Revenko Yu. A. 2012 *Siberian state aerospace university Bulletin Special issue 6 (46)* (Krasnoyarsk) 234 [in Russian]

Synthesis of micro-dispersed zirconium oxide for glass manufacturing

V Goncharuk[1,2], P Starodubtsev[3] and I Maslennikova[1,2]
[1] Far-Eastern Federal University, 8, Sukhanova Str., Vladivostok, 690950, Russia
[2] Far Eastern Branch of Russian Academy of Sciences, 159 Pr. 100-letiya Vladivostoka, Vladivostok, 690022, Russia
[3] Pacific SO Makarov Higher Naval School, 6 Per. Kamskyi, Vladivostok, 690062, Russia

E-mail: gon@ich.dvo.ru

Abstract. A rather simple and original method for processing of zirconium-containing raw material form Algoma deposit (Khabarovsk region, Russia) was suggested, which comprised fluorination of the initial sample with a diluted HF solution followed by the thermal treatment of fluorination products and pyrohydrolysis of zirconium tetrafluoride. Water vapors obtained by hydrogen and oxygen burning in a hydrogen torch as well as by simple evaporation were used for pyrohydrolysis. The feed rate of the water and its temperature were regulated. The temperature of water vapors reached 800–1200 °C. Zirconium dioxide with a purity of 99.97 % or more and a dispersity of 0.1 μm or less was synthesized.

1. Introduction

Zirconium dioxide exhibiting a number of important physicochemical properties such as optical, dielectric, high thermal (melting point is above 2700°C) and mechanical stability, chemical inertness is widely used in practice. It is used in glass manufacturing, constructive ceramic materials, fuel cells, catalyst systems, oxygen sensors, as well as in various areas of microelectronics. Optical properties of fianite single crystals based on zirconium dioxide are close to those of diamond. Recently, zirconium dioxide is widely used in prosthetic dentistry and cosmetology.

Zirconia, which is used in the modern technology, must have a high purity (at least 99.99 %) and dispersity (less 0.1 μm). Conventional methods of zirconium dioxide production from chloride, nitrate and oxalate zirconium solutions comprise the precipitation of its basic salts followed by their one-stage calcination [1]. A technology of alkaline sintering of zircon is based on the process of fusing zirconium-containing raw materials with NaOH and Na_2CO_3 at 450–1050°Cwith subsequent treatment with water [2].

We use fluoride methods for the ZrO_2 production when the zirconium-containing raw material is processed with hydrofluoric acid and ammonium fluorides.

The HF using for zirconium dioxide obtaining was suggested in [3]. It comprised the treatment of zirconium-containing raw material with hydrofluoric acid, aqueous leaching of the reaction products, pulp filtration, neutralization by ammonia with simultaneous precipitation of zirconium hydroxide, separation of the precipitate and its treatment by the reagent with obtaining zirconium-containing solution, evaporation of solution, purification from impurities, and calcination of zirconium-containing

Content from this work may be used under the terms of the Creative Commons Attribution 3.0 licence. Any further distribution of this work must maintain attribution to the author(s) and the title of the work, journal citation and DOI.
Published under licence by IOP Publishing Ltd

precipitate, characterized in that the zirconium-containing raw material was treated by hydrofluoric acid and ammonium fluoride in molar ratio 1-4: 3 per 1 mole of zirconium dioxide and solid ammonium bicarbonate was used as a reagent in an amount of 2.8–3.2 mol per 1 mole of zirconium dioxide. Separation of zirconium from impurities proceeded before evaporation by their successive deposition with the retention of zirconium in solution.

Ammonium hydrogen difluoride (NH_4HF_2) was used for processing of zirconium concentrate consisted of a mixture of baddeleyite, zircon and quartz with the impurity of kaolin $Al_2[OH]_4[Si_2O_5]$ and contained in mass. %: ZrO_2 33–53; SiO_2 40–50; Al_2O_3 0.4–0.6; Fe 0.2–0.4. The method comprised fluorination with NH_4HF_2 at 50–190°C for 3–72 h followed by heating of fluorination product at 330–650°C without the oxygen access for 0.5–3 h with simultaneous condensation of ammonium hexafluorosilicate at 230–270°C and ammonium fluoride at 25–150°C. The latter was returned to the fluorination stage. Desiliconized product was heated in inert atmosphere at 900–1100°C for 4–5 h followed by condensation of zirconium tetrafluoride at 250–850°C and its pyrohydrolysis at 900–920°C for 0.5–1 h with the formation of zirconium dioxide [4].

2. Experimental section

We suggested using the 10–20 % hydrofluoric acid solution for recovering of zirconium-containing raw material. There is no need to raise its concentration as the reaction proceeds exothermally with the isolation of a large heat amount. It is interesting that the quartz dissolution is not observed when a weak hydrofluoric acid solution is used (the dissolution degree is not more than 3% from the initial quantity of SiO_2). The SiO_2 precipitate is removed, and ammonia is added to the solution to separate the iron ions, which occurs at a pH higher than 5. Small excess of ammonium fluoride is needed relative to the reaction stoichiometry:

$$ZrF_4 + NH_4F = (NH_4)_2ZrF_6$$

Ammonium hexafluorozirconate isolates together with a small amount of ammonium hexafluorosilicate. The precipitate is separated, dried and subjected to thermal decomposition at 600–650 °C with the formation of zirconium tetrafluoride and ammonium fluoride. Ammonium fluoride is distilled off in a condenser and recycled. The scheme of micro-dispersed ZrO_2 synthesis is presented in figure 1.

Zirconium concentrate was presented by the Algoma deposit (Khabarovsk region), containing zirconium in the form of amorphous and fine-dispersive gel zircon. It contained in mass. %: ZrO_2 52.1; SiO_2 45.3; Fe 0.16; Al 0.45; W 1.88; Hf 0.95, and impurities Mn 0.5; Cs 0.005; Nb 0.04; Ba 0.028 and Pb, Sb, Sn, Sr 0.001–0.005. 200 g of the initial material were placed into glass-graphite vessel (1000 ml) and 450 ml of 20 % hydrofluoric acid were added. Exothermal reaction proceeded with a slight (40 °C) temperature rise of the solution. The solution was separated from the precipitate by filtration or decantation. Silicon dioxide, some amounts of iron, aluminum and tungsten remained in the precipitate. Zirconium and hafnium dioxides were dissolved quantitatively. Silicon dioxide was insignificantly dissolved (below 3 %). Concentration of the other impurities was 1–2 times lower than in the initial material. Their total content did not exceed 0.1%.

70 g of ammonium fluoride were added to the solution in the second stage. There was a release of a precipitate which was separated from the solution by decantation and dried. As a result, 216 g of the product were obtained, which means virtually full dissolution of zirconium dioxide. All taken NH_4F were spent virtually on the formation of $(NH_4)_2ZrF_6$, and only its part on the dissolved silica.

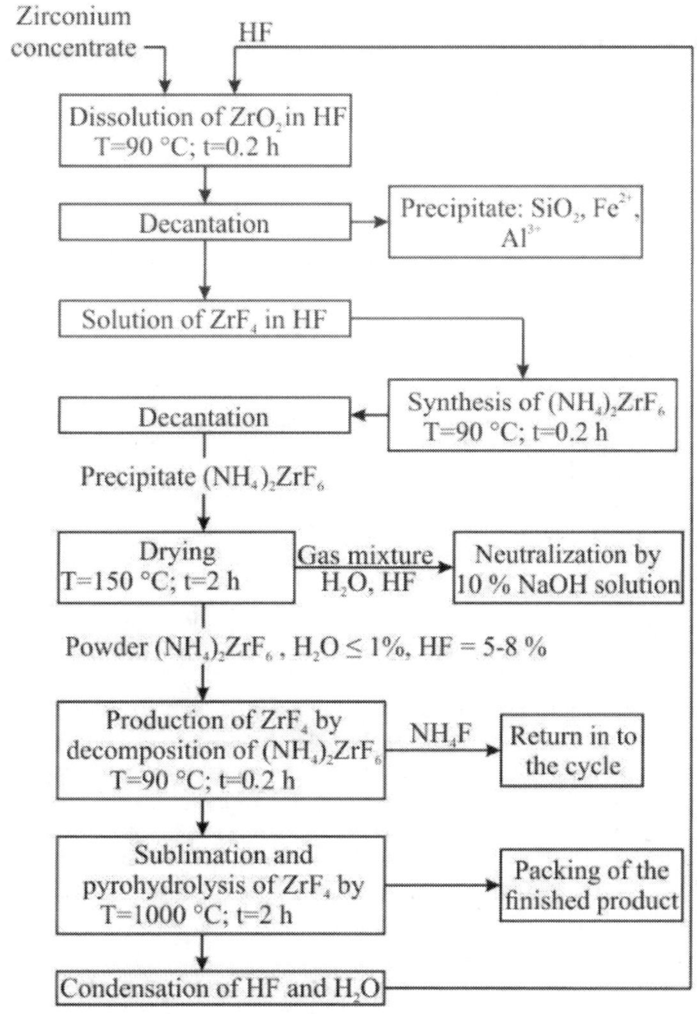

Figure 2. Scheme of micro-dispersed ZrO_2 synthesis

Powdery product was lilac-pink and consisted of $(NH_4)_3ZrF_7$, $(NH_4)_3SiF_7$ and $(NH_4)_2SiF_6$ mixture in accordance with the XRD data. 120 g of the product were heated to 650 °C in a glass-graphite reactor until complete removal of $(NH_4)_2SiF_6$ and NH_4F. The residue mass after 1 h of heating was 36.5 g containing 0.2 % of silicon.

Ammonium fluoride (72.5 g) was condensed at 150 °C and returned in to the cycle. The residue was gray. It consisted of ZrF_4 (98–99 %) and Al, Fe, Hf, Mn, Pb fluoride impurities (1–2 %). This product was placed into a nickel evaporator for ZrF_4 sublimation. Isolated zirconium tetrafluoride was pyrohydrolized in a special device, the scheme of which is given in figure 2.

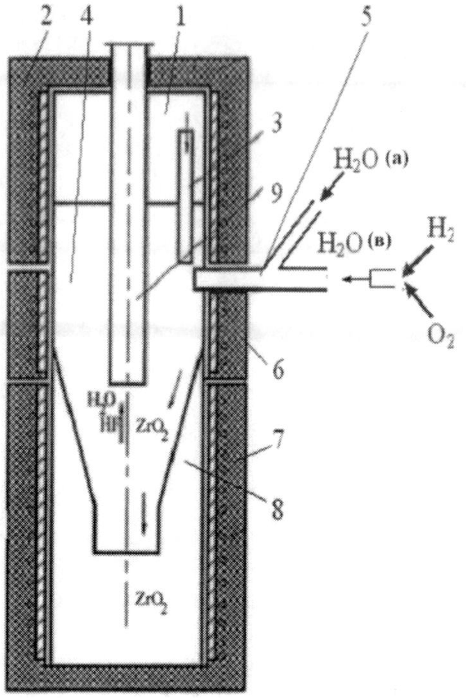

Figure 2. Device for pyrohydrolysis of ZrF_4.
1– vessel for solid ZrF_4,
2– furnace 1, T=900-950°C,
3– pipe for ZrF_4 vapors,
4– reactor,
5– pipes for introduction of water vapors,
6– furnace 2,
7– furnace 3,
8– vessel for storage of ZrO_2,
9– pipe for removing of HF and H_2O vapors.

Conventional water vapors (a) and water vapors obtained from hydrogen burning in oxygen (b) heated to 800–1200 °C came in to the reactor 4 through the connecting pipes 5. Temperature of water vapors was regulated by hydrogen burning rate and by the H_2O(a): H_2O(b) ratio. Reactor was heated to 900–950 °C. During pyrohydrolysis, HF was formed, which came through the pipe 9 in to the apparatus for its trapping and regeneration. The obtained ZrO_2 was gathered in the vessel 8 for their storage. Powder of zirconium dioxide with the particle diameter less than 0,1 μm (100 nm) was produced. Its purity was 99.97 % and the yield was 89.4 %.

3. Conclusion

Using natural zirconium-containing raw material of Algoma deposit (Khabarovsk region, Russia) consisting mainly form amorphous fine-dispersive gel zircon enabled to propose the fluoride method (including dilute solutions of HF and ammonium fluorides) for production of micro-dispersed high purity zirconium dioxide. It provides decreasing number of the stages and operating time and increasing of processing efficiency of raw material.

Acknowledgments
This work was supported by the Ministry of Education and Sciences of the Russian Federation, Agreement # 14.578.21.0024, RFMEFI57814X0024.

References

[1] Vorozhcov A, Zhukov A, Malinovskaya T and Sachkov V 2014 *Synthesis of dispersed metal oxide materials* Volume 2 (Tomsk: NTL)
[2] De Wet W J and Coetzee G H J 2006 *US patent 7063824 B1*
[3] Voskoboinikov N, Skiba G, Kalinnikov V and Nosova L 1998 *RU patent 2116254*
[4] Mel'nichenko E, Epov D, Gordienko P, Shkol'nik E, Nagorskiy L, Kozlenko I and Buznik V 1995 *RU patent 2048559*

Modification of Structural Phase State and Mechanical Properties of Poly-Grained Titanium Alloy Implanted by Aluminum Ions

A V Nikonenko[1], N A Popova[2,3], E L Nikonenko[2,4], M P Kalashnikov[3] and I A Kurzina[1*]
[1]National Research Tomsk State University, 36, Lenin Ave., 634050, Tomsk, Russia;
[2]Tomsk State University of Architecture and Building, 2, Solyanaya Sq., 634003, Tomsk, Russia;
[3]Institute of Strength Physics and Materials Science SB RAS, 8/2, Akademicheskii Ave., 634021, Tomsk, Russia;
[4]National Research Tomsk Polytechnic University, 30, Lenin Ave., 634050, Tomsk, Russia

*E-mail: kurzina99@mail.ru

Abstract. The paper presents TEM analysis of microstructure, phase composition, and mechanical properties of commercially pure titanium. These properties of two types of grains are compared before and after modification of titanium by aluminum ions, namely: large grains (1.4 μm) and small (0.5 μm) grains. The analysis shows that ion implantation results in a considerable improvement of mechanical properties of both large and small grains throughout their implantation depth. However, with increase of the grain size, the stress in the ion-modified surface layer decreases while in the subsurface layer it increases.

1. Introduction

The ion-beam treatment of materials as described in works of Gribkov *et al.* [1] and Kurzina *et al.* [2,3] is currently one of the most intensively developing techniques used for the synthesis of new materials. All experimentally proved models of phase formation at ion implantation are designed basically using coarse-grain metals [1]. However, the decrease of grain sizes, and respectively, the increase of density of grain boundaries and triple points lead to a rapid acceleration of such processes as diffusion, alloying, mixing, formation of secondary phases and defects as observed in works of Gribkov *et al.* [1] and Kurzina *et al.* [2,3], and the emergence of new phases including those not reflected on state diagrams. Thus, a study of phase formation in implanting aluminum ions in titanium materials having various structural conditions (at micro- and meso-levels) is rather relevant. It is possible to assume that physical processes occurred in materials characterized by a heterogeneous condition, i.e. a presence of both small (up to 0.5 μm) and coarse (up to 0.5-5.0 μm) grains or micro- or meso-levels, behave differently as compared to those characterized by a homogeneous condition.
The aim of this study is to analyse and compare the results of research carried out into the microstructure, phase composition, and strength properties of the yield stress, and titanium target in a poly-grained condition before and after aluminium ion implantation. A poly-grained condition refers to a presence of both small (up to 0.5 μm) and coarse (up to 0.5-5 μm) grains in the material workpiece.

Content from this work may be used under the terms of the Creative Commons Attribution 3.0 licence. Any further distribution of this work must maintain attribution to the author(s) and the title of the work, journal citation and DOI.
Published under licence by IOP Publishing Ltd

2. Material and methods

Workpieces of titanium alloy (VT1 0) were used in this experiment. These workpieces were obtained by uniaxial isothermal forging combined with reverse rolling as described in the work of Eroshenko *et al.* [4] and the additional annealing at 500^{0}C. Aluminum ions were implanted in titanium workpieces using the new version of the Mevva (MEtal Vapor Vacuum Arc) ion source Mevva-V.RU described in detail in the work of Gushenets *et al.* [5]. Titanium workpieces were exposed to ion implantation at temperature 623 K, accelerating voltage 50 kV, ionic current density 6.5 mA/cm^2, distance 60 cm from the ion-optical system, implantation time 5.25 h, and fluence $1 \cdot 10^{18}$ ion/cm^2. The microstructure and phase composition were studied with the EM-125K transmission electron microscope (TEM) having the accelerating voltage of 120 kV. A study of the microstructure and phase composition of the implanted materials was carried out for two areas of the modified layer, namely: at a distance of 0-150 nm (surface area I) and ~300 nm (central area II) from the irradiated surface.

3. Results and discussion

Conditions described in the work of Kurzina *et al.* [6] for the alloy preparation provide two types of grains retained after the ion implantation. The size of small grains ranges from 0.1 to 0.5 μm. Their average size is 0.40 μm. The size of coarse grains ranges from 0.5 to 5.0 μm. Their average size is 1.40 μm. The content of small grains is not large and comes to 10%. Small grains are mostly located on triple points of coarse grains as described in the work of Kurzina *et al.* [6].

In this work [6], Kurzina *et al.* observed that the ion implantation lead to a modification of the elementary composition of titanium layers. As is known from the work of Kurzina *et al.* [7], the alloying admixture, namely, aluminum, is distributed non-uniformly both throughout the implanted layers and small and coarse grains of Ti alloy. At first, aluminum forms a solid solution. And the larger the matrix grain the higher the aluminum concentration in the solid solution. Relative to coarse grains, small grains are unified by aluminum. A non-uniform distribution of the alloying admixture along the material results in the different phase composition of small and coarse grains.

Investigations [6] show that at a distance of up to 200 nm from the irradiated surface (area I), both small and coarse grains are not only retained, but the content of small grains even increases by 30% of the total grain content. The average size of both grain types decreases. The both types of grains are also observed in the area II. The content of small grains is the same as before implantation (10%). The average size of coarse grains also remains as before implantation, and the size of small grains is similar to that of the area I.

Aluminum ion implantation in titanium coarse grains results in the formation of secondary phases, namely Ti_3Al and $TiAl_3$ [8], the total number of which in area I is 3% of the material volume, whilst in area II it is 10%. At the same time, Ti_3Al-phase formation is observed throughout the alloyed volume, whilst $TiAl_3$-phase is formed only in the centre of the implanted layer along the grain boundaries. In small grains these phases are not observed. TiO_2 is formed in all grain types and throughout the alloyed layer [8].

Thus, small and coarse grains differ not only by their phase composition but also by their amount. It is stated that, first, the larger grain size the larger phase precipitation. Second, in area II of the ion-alloyed Ti layer, the volume fraction of phase precipitation is always higher. Third, TiO_2 particles precipitate most of all in each area of the ion-alloyed Ti layer and in each grain type.

Also, small and coarse grains differ not only by their qualitative and quantitative phase compositions but also by the defect structure. These differences are retained before and after the ion implantation. Dislocations and dislocation entanglements are present in the material before the ion implantation. Dislocation density ρ increases with the decrease of Ti grain size, namely: in coarse grains $\rho = 1.2 \cdot 10^9$ cm^{-2}, while in small grains $\rho = 2.5 \cdot 10^9$ cm^{-2} (Figure 1*a*, curve 1). After the ion implantation, modification of the dislocation structure is observed in the surface layers. Scalar density decreases, while the excessive density increases. The dislocation structure is mostly cross-linked. In area I, the average value of dislocation density ρ is twice as high in small grains as it is in coarse grains (Figure 1*a*, curve 2). Scalar density of dislocations in coarse grains in area II is over four times as high as it is in area I, while in small grains it increases three times (Figure 1*a*, curve 3). Unlike coarse grains, the average value of dislocation density ρ remains higher in small grains. In the whole material this value is three times higher in area II than in area I. Thus, with distance from the irradiated surface, the scalar density of dislocations both in coarse and small grains of α-Ti phase increases with the different intensity, and is higher than that of the original state in both areas (Figure 1*a*).

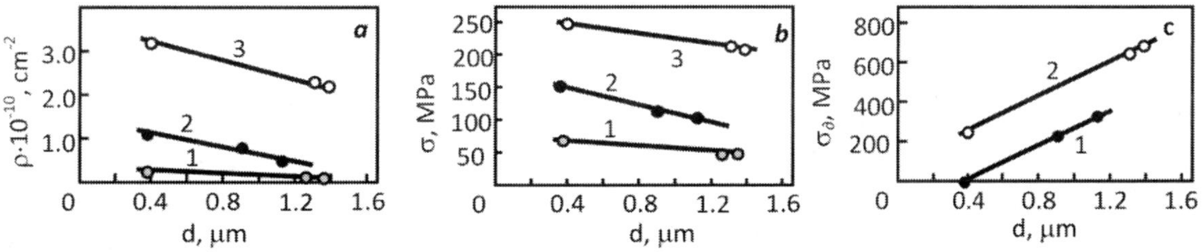

Figure 1. Dependences of scalar density (*a*), shear stresses (*b*), and local stresses (*c*) and the average grain size α-Ti before and after ion implantation: 1 – original Ti specimen; 2 – ion-alloyed layer in area I; 3 – surface layer of the area II

The ion-modified dislocation structure creates internal stresses σ (shear stresses) shown in Figure 1*b*. As this Figure shows, first, with distance from the irradiated surface, the value σ increases throughout the material volume. Second, dependences between the value σ and grain sizes are similar to that of the value ρ. After the ion implantation, stresses increase by 3-5 times with distance from the irradiated surface and relative to the original state. Third, small grains are characterized by stresses higher than that of the coarse grains.

The ion-modified dislocation structure together with the scalar dislocation density is characterized by the high excessive density ($\rho_{\pm}$) which enables internal stresses, i.e. couple or local stresses $\sigma_{f.r.}$. The amplitude of couple stresses both in small and coarse grains differs and also at an average, in each area of the implanted alloy. Ion implantation results in the formation of the elastoplastic couple stresses the amplitude of which significantly increases the value of the yield shear stress. At the same time, in coarse grains the elastic component of the couple stress amplitude is over two times higher than the plastic component. In small grains of area I couple stresses are completely absent, while in area II they are present due to dislocation charges (Figure 1*c*).

The stress yield can be found from equation (1) based on the obtained quantitative structural parameters for coarse and small grains individually. In this equation, contributions made to hardening by forest dislocations and internal fields are quadratic, all other contributions are additive [9]:

$$\sigma = \Delta\sigma_{fr} + \Delta\sigma_{h} + \Delta\sigma_{g} + \Delta\sigma_{O} + \sqrt{\left(\Delta\sigma_{f.r.}^{\ 2} + \Delta\sigma_{f}^{\ 2}\right)}, \qquad (1)$$

where contributions made to hardening by forest dislocations and internal fields are quadratic, all other contributions are additive. This equation covers practically all contributions of resistance to deformation. Here, $\Delta\sigma_{fr}$ is the frictional constraint of dislocations in a crystal lattice of α-Ti phase; $\Delta\sigma_{h}$ is hardening of the solid solution based on α-Ti atoms of alloying elements (Al, C, O); $\Delta\sigma_{f}$ is hardening by forest dislocations which cut off slip dislocations; $\Delta\sigma_{f.r.}$ is hardening by far-ranging stress fields; $\Delta\sigma_{O}$ is hardening by incoherent particles at the dislocation bypass called Orowan mechanism; $\Delta\sigma_{g}$ is hardening due to grain boundaries.

Calculation results are presented in Figure 2. It should be underlined that this type of alloy is practically a single-phase, namely, α-Ti phase in its original state independently of the average grain size. Hence, the contribution to hardening due to the secondary phases $\Delta\sigma_{O}$ is not observed. Hardening by far-ranging stress fields is completely absent also. Components $\Delta\sigma_f$ (the frictional constraint of dislocations in a crystal lattice of α-Ti phase) and $\Delta\sigma_h$ (hardening of the solid solution based on α-Ti atoms of alloying elements Al, C, O) also make minimum contribution to hardening, since according to Auger spectroscopy and the XRD analysis [3], carbon and oxygen are not observed in the solid solution of the original alloys. Therefore, hardening of the original material will be defined by $\Delta\sigma_g$ (hardening due to grain boundaries) and $\Delta\sigma_f$ (hardening by forest dislocations which cut off slip dislocations).

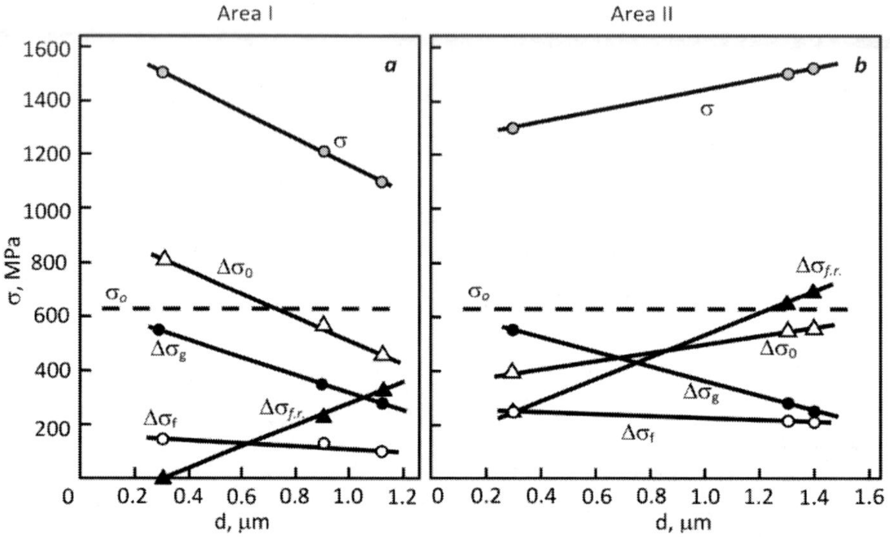

Figure 2. Dependence of the yield stress and certain hardening mechanisms contribution upon α-Ti grain size across the ion-alloyed Ti layer: *a* – area I; *b* – are II; σ_o - yield stress of the original alloy

These factors have different behavior for the ion-modified alloy. Since the ion implantation slightly modifies the grain size in the ion-alloyed layer, the grain-boundary strengthening is also modified. Hardening increases due to grain boundaries of α-Ti phase that are the barriers for the flow propagation. It can be described by the Hall-Petch relation [10]:

$$\Delta\sigma_g = k \times d^{-0.5} , \tag{2}$$

where k is the proportionality factor depending on the material purity, the degree of deformation, the type and the structure of grain boundaries, the boundary hardening by the secondary phase particles, and the grain size [7].

According to Koneva *et al.* [9], forest dislocations play the important part in hardening implanted alloys, both for non-polarized (uncharged) and polarized (charged) dislocation ensembles. A non-polarized dislocation ensemble has no excessive dislocations at $\rho_+ = \rho_-$, where ρ_+ and ρ_- are densities of positively and negatively charged dislocations, respectively. The excessive density of dislocations is then $\rho_\pm = \rho_+ - \rho_- = 0$. In this case, the uncharged dislocation ensemble creates a shear stress (for stress fields generated by the structural dislocation) that can be obtained from [11]:

$$\Delta\sigma_f = m\alpha Gb\sqrt{\rho}, \tag{3}$$

where m is the orientation multiplier; α is the dimensionless factor ranged between 0.05-1.0 depending on the type of dislocation ensemble [9]; G is the shear modulus of the material matrix; b is Burgers vector; ρ is the average value of scalar density of dislocations. In case of a charged dislocation ensemble when the excessive density of dislocations equals to $\rho_{\pm} = \rho_{+} - \rho_{-} \neq 0$, long-range couple stresses occur. TEM provides their detection by bend extinction contours observed in the material structure [11,12]. The value of long-range couple stresses can be obtained from

$$\Delta\sigma_{f.r.} = m\alpha_S Gb\sqrt{\rho_{\pm}}, \quad (4)$$

where $\alpha_S = 1$ is Strunin coefficient [13]. Ion implantation resulted in the accumulation of scalar dislocation density higher than that of the original state of the alloy (Fig. 1*a*). This dislocation structure forms, in turn, higher shear stresses the values of which are also given in Figure 1*b*.

Ion implantation results in polarization of the structural dislocation: bend extinction contours can be observed on TEM images. Thereby, couple (far-ranging) stresses are formed as shown in Figure 1*c*.

As stated above, the ion implantation results in the formation of phase compositions. It is detected that the phase composition depends on the average grain size of the alloy. As is known from the work of Honeycomb [14], the dispersion hardening of the alloy depends on the number of particles, their size, distribution, and distance between them, and also the degree of atomic mismatching between the lattices of the precipitate and the matrix. Mechanisms of dispersion hardening are created for incoherent particles when dislocations bypass precipitates and also for coherent particles when dislocations cut off the particles. In alloys under study, TiO_2 particles are incoherent, whilst particles of phases Ti_3Al, and $TiAl_3$ are coherent. Moreover, mechanisms of dispersion hardening are divided into direct (particles are located inside grains) and indirect (particles are located on grain boundaries and triple points). Direct mechanisms lead to the increase of the yield stress of the alloy, while indirect mechanisms affect its strength. In all cases, hardening can be obtained using the Orowan equation [15]:

$$\Delta\sigma_O = 2\lambda Gb/r, \quad (5)$$

where $\lambda = 0.5$ is the coefficient accounting for the abovementioned cases; r is the distance between particles.

4. Conclusion

Calculations made according to equation (1), showed that both in coarse and small grains, the ion implantation resulted in a considerable hardening of the layer throughout its thickness. However, the effect of grain size on the value σ is different in areas I and II. In other words, nearby the surface of the alloyed layer (area I) the value σ decreases with the increase of the grain size, while in area II it increases. Moreover, the contribution of certain hardening mechanisms to the general alloy hardening differed, and in each case it was different as well.

Acknowledgement

This work was financially supported by Grant N 3.295.2014/K from the Ministry of Education and Science of the Russian Federation

References

[1] Gribkov V A, Grigor'ev F I, Kalin B A and Jakushin V L 2001 *Prospective radiation beam-radiation technologies of metal processing* (Moscow: Kruglii god).

[2] Kurzina I A, Shar keev Yu P, and Kozlov E V 2007 Formation on intemetallic nano-phases at ion implantation *Structure and properties of perspective materials ed I A Potekaev* (Tomsk: Publ YTL) pp 159–195.

[3] Kurzina I A, Popova N A, Kalashnikov M P, Savkin K P, Bozhko I A, Nikonenko E L, Yushkov G Yu, Oks E M, Kozlov E V and Sharkeev Yu P 2011 *Izv. Vuzov. Fizika* **11/3** 112.

[4] Eroshenko A Yu, Sharkeev Yu P, Tolmatsev A I, Korobitsyn G P and Danilov V I 2009 *Advanced Materials* **7** 107.

[5] Gushenets V I, Nikolaev A G, Oks E M, Vintizenko L G, Yushkov G Yu, Ozterhan A and Brown I G 2006 *Review of Scientific Instruments* **77** 6 Art 063301.

[6] Kurzina I A, Kozlov E V, Popova N A, Kalashnikov M P, Nikonenko E L, Savkin K P, Oks E M and Sharkeev Yu P 2011 *Izv. RAN, Seriya fizicheskaya* **76** 11 1384.

[7] Kurzina I A, Kozlov E V, Sharkeev Yu P, Fortuna S V, Koneva N A, Bozhko I A and Kalashnikov M P 2008 Nanocrystal intermetallic and nitride structures formed at ion-plasma exposure (Tomsk: Publ YTL).

[8] Nikonenko A V, Popova N A, Kalashnikov M P, Nikonenko E L and Kurzina I A 2014 *Fundamental problems of modern materials science* **11** 4 437.

[9] Koneva N A and Kozlov E V 2006 Dislocation structure and physical mechanisms of hardening metallic materials *Advanced Materials* (manual) ed D L Merson (Tula: Publisher TSU, MISA) pp 267-320.

[10] Mac Lin D 1965 *Mechanical properties of metals* (Moscow: Metallurgy).

[11] Koneva N A and Kozlov E V 1991 *Izv. Vuzov. Fizika* **3** 56.

[12] Kozlov E V and Koneva N A 2002 *Izv. Vuzov. Fizika* **3** (application) 52.

[13] Strunin B N 1967 *Fizika Tverdogo Tela* **9** 3 805.

[14] Honeycombe R W K 1972 *The plastic deformation of metals* (Moscow: Wiley).

[15] Goldstein M I and Farber V M 1979 *Dispersion steel hardening* (Moscow: Metallurgy).

New technology of extracting the amount of rare earth metals from the red mud

G A Martoyan[1], G G Karamyan[1] and G A Vardan[1]
Ecoatom LLC, 1 Adanai Str., Yerevan, Armenia

[1]E-mail: martoian@yahoo.com

Abstract. The paper outlined the environmental and economic problems associated with red mud – the waste generated in processing of bauxite ore for aluminum production. The chemical analysis of red mud has identified a number of useful elements including rare earth metals. The electromembrane technology of red mud processing with extraction of valuable elements is described. A possible scheme of separation of these metals through electrolysis is also given.

I. Introduction

One of the most urgent problems of resource saving and environmental safety in the mining industry and metallurgy is the disposal of processing waste. Production of aluminum from bauxite ores is accompanied by formation of large quantities of waste in the form of red mud [1]. The production of one ton of commercial aluminum using the Bayer process gives about three tons of red mud. Its chemical composition is determined by the content and properties of leached bauxite, as well as by duration and storage conditions in tailings ponds. They occupy vast areas suitable for agriculture. The leak of red mud caused a real ecological disaster in Hungary in 2010, when in the result of explosion at the aluminum plant the dam holding back a reservoir of these toxic wastes was destroyed. Then the state of emergency for several months was introduced in three areas of the country, also some regions of Croatia have been suffered.

Modern plants for aluminum production do not have the technology of complex processing of bauxite and stored waste in the tailings ponds, considering their further processing as economically not viable, thus creating an environmental hazard due to the high alkalinity and dispersity of the red mud. In Russia it is accumulated more than 100 million tons of red mud.

Table 1 and 2 show the average chemical composition (in wt.%) of red mud produced during the processing of alumina from bauxite rocks of Russian Federation and Iran. Both samples had a moisture content of 40-60%, the alkalinity (pH) - 10-13 and dispersity - 80% fraction of less than 5 microns. As can be seen from the tables, the compositions of both samples are about the same.

As red mud contains a large amount of iron and aluminum, it seems it would be better to turn it into a source of valuable secondary raw material. But processing of red mud in undivided form is difficult task, since it contains too much iron oxide, and it cannot be used as the iron ore as it contains too much aluminum oxide. Moreover, it also contains impurities of sulfur, phosphorus, so it is necessary to find a method of processing based on a significant difference in chemical properties of the compounds of Fe_2O_3 and Al_2O_3, and use it for their separation, or find a way to use this mixture as a raw material to produce the commercial or intermediate products.

Content from this work may be used under the terms of the Creative Commons Attribution 3.0 licence. Any further distribution of this work must maintain attribution to the author(s) and the title of the work, journal citation and DOI.
Published under licence by IOP Publishing Ltd

Figure 1. A typical red mud tailing pond

Table 1. Composition of red mud of Ural plant

Fe_2O_3	CaO	SiO_2	Al_2O_3	MgO	TiO_2	S	P_2O_5	Na_2O
40-55	8-11	5-15	14-16	0.5-1.4	2-5	up to 2	0.2-0.5	up to 2

Table 2. Composition of red mud of Iran plant

Fe_2O_3	CaO	SiO_2	Al_2O_3	MgO	TiO_2	S	P_2O_5	Na_2O
44 - 46	8-11	8-9.5	13-18	0.2-1.6	4-5	3.6	0.2-0.65	2.5-6.5

2. Extraction the amount of valuable and rare earth elements

It turned out that the red mud contains rare earth metals in industrial interest, in particular, scandium and yttrium, as shown in Tables 3 and 4.

We have carried out a study on the extraction of these metals from the bulk of the red mud. In the result, a laboratory method based on the method electromembrane separation was elaborated. The main process steps are shown in Figure 2.

Previously, in a special electrolyzer, the sodium hypochlorite was synthesized. Then, from the slurry tank the red mud was supplied in a glass jar - reactor. A calculated amount of sodium hypochlorite was also poured into this reactor. After a certain time, the dissolved part of the red mud obtained during a mild leaching was selected in a separate reactor and sent to a solution treatment unit. There it was subjected to a special activation that is our know-how. Insoluble part of red mud depleted by rare metals was poured into another slurry tank. Its alkalinity has already been within 7 - 9.

Further, the treated solution was subjected to a electromembrane processing in the system of electrolyzers. As a result, a separate aqueous concentrate of rare, rare earth and other valuable elements were obtained, as well as the NaOH solution, which was directed to the beginning of the process. Solutions were periodically analyzed by the mass spectrometer, and the time of process completion was determined. Simultaneously the chemical composition of the processed red mud has been analyzed (Table 5 and 6).

Table 3. Composition of minor elements in red mud of alumina production in RF

Element	g/kg	Element	g/kg
Sc	0.025060	Cu	0.031695
Ti	31.446005	Zn	0.048848
V	0.566959	Ge	0.004281
Cr	0.255851	Ga	0.035668
Mn	0.226875	As	0.027688
Fe	78.244009	Y	0.053365
Co	0.046470	Mo	0.029679
Ni	0.086609	Cd	0.000808

Table 4. Composition of minor elements in red mud of alumina production in Iran

Element	g/kg	Element	g/kg
Sc	0.019358	Cu	0.068480
Ti	24.298205	Zn	0.235932
V	0.277970	Ge	0.005525
Cr	0.151777	Ga	0.030868
Mn	2.974695	As	0.031339
Fe	78.037918	Y	0.001937
Co	0.078870	Mo	0.002424
Ni	0.303276	Cd	0.001026

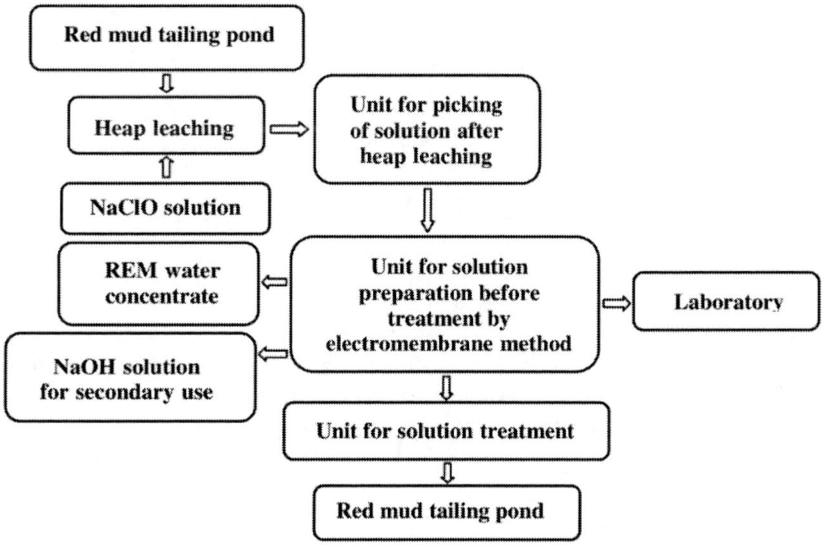

Figure 2. Technological scheme of red mud processing.

Table 5. Composition of red mud of RF after processing

Element	mg/kg	Element	mg/kg
Sc	0.092628	Cu	1.248894
Ti	0.989895	Zn	1.497201
V	2.831367	Ge	0.003759
Cr	0.363909	Ga	0.864186
Mn	0.011424	As	0.714249
Fe	1.405755	Y	0.001866
Co	0.002025	Mo	1.217775

As shown in Tables 4 and 5, the composition of useful elements the red mud significantly decreased after treatment, suggesting that most of them are passed into the solution. The subsequent separation of these elements and the production of metals in its pure form is not the subject of this article, but we consider it necessary to describe the possible technologies of such processing. Various organizations conduct laboratory experiments to extract valuable metals. For example, the pyrometallurgical method - smelting reduction of red mud to obtain a slag rich in valuable elements with further dissolving of slag in respective solutions, its smelting accompanied by separation of solid and liquid fractions, and then the complex processes of cooling and recrystallization [2].

Table 6. Composition of red mud of Iran after processing

Element	mg/kg	Element	mg/kg
Sc	0.136449	Cu	2.043684
Ti	1.462806	Zn	1.629492
V	29.265636	Ge	0.003486
Cr	0.636222	Ga	0.226401
Mn	0.134481	As	1.483803
Fe	5.904255	Y	0.004251
Co	0.004638	Mo	0.092148
Ni	0.022359	Cd	0.000561

Another example is the separation of rare earth elements on the cascade of centrifugal extractors designed by LLC "LIT" of companies group “Skaygrad” (Moscow) [3]. Similar technologies of scandium extraction are developed in National University of Science and Technology (MISIS) [4].

As a possible alternative, we propose the separation of rare, rare earth and other valuable metals using the method of electrolysis based on the difference in electrochemical potentials of metal deposition [5]. A block diagram of this method is shown in Figure 3.

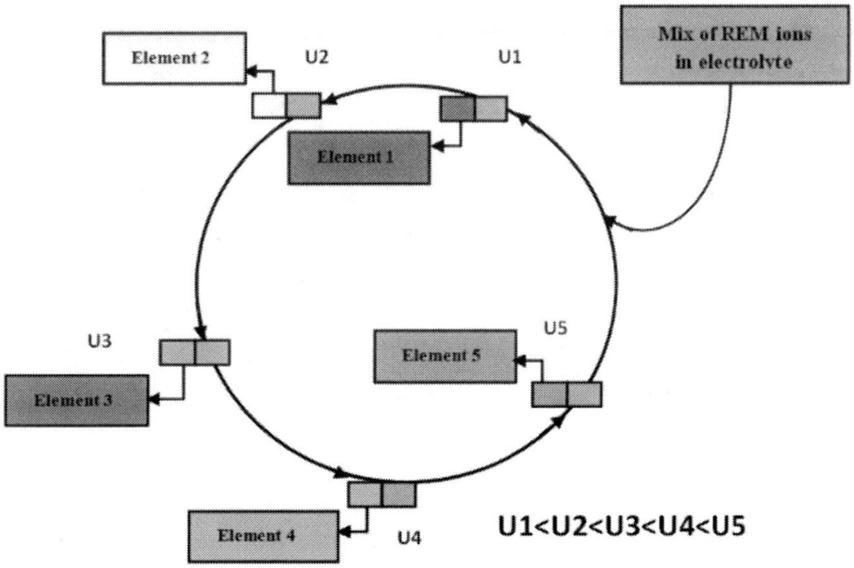

Figure 3. Block diagram of metals separation from the mix of the ions

References

[1] Ritters S K 2014 Making the most of red mud *Chemical and Engineering News* volume 92 Issue 8 pp. 33-35
[2] Golubev A A and Gudim Yu A 2013 Method of Pyrometallurgical Processing of Red Mud Patent WO 2013070121 A1
[3] Abramov A M, Sobol Yu B, Galieva Zh N, Volobuev O I, Solodovnikov A S and Yachmenev A A 2013 Organization of REM production at complex processing of phosphogypsum. *All-Russian Con. on Rare Earth Materials,* Tomsk, RF (in Russian)
[4] Information on http://strf.ru/material.aspx?CatalogId=222&d_no=88685#.VgJYtX3CaKI
[5] Guiragossian Z G, Martoyan G A, Intsheyan S G, Nalbandyan G G and Tonikyan S G 2004 *Proc. of WM'04 Conf., February 29- March 4, Tucson, AZ* WM-4304

IOP Conf. Series: Materials Science and Engineering **112** (2016) 012034 doi:10.1088/1757-899X/112/1/012034

Titanium compounds as catalysts of higher alpha-olefin-based super-high-molecular polymers synthesis

K B Konovalov[1], M A Kazaryan, V N Manzhay and O V Vetrova

Tomsk State University, 36, Lenina avenue, Tomsk, 634050, RF

[1]E-mail: konoko@tpu.ru

Abstract. The synthesis of polymers of 10 million or more molecular weight is a difficult task even in a chemical lab. Higher α-olefin-based polymer agents of such kind have found a narrow but quite important niche, the reduction of drag in the turbulent flow of hydrocarbon fluids such as oil and oil-products. In its turn, searching for a catalytic system capable to produce molecules of such a high length and to synthesize polymers of a low molecular-mass distribution is part of a global task of obtaining a high-quality product. In this paper we had observed a number of industrial catalysts with respect to their suitability for higher poly-α-olefins synthesis. A number samples representing copolymers of 1-hexene with 1-decene obtained on a previous generation catalyst, a microsphere titanium chloride catalytic agent had been compared to samples synthesized using a titanium-magnesium catalyst both in solution and in a polymer medium.

1. Introduction

The first patents that described higher α-olefins polymerization in order to obtain hydrodynamic drag reduction agents for hydrocarbon fluids on Ziegler-Natta catalysts are dated late 1970 till early 1980. It is necessary to note that the catalytic systems which are active with respect to ethylene and propylene, can also be successfully used in higher α-olefins polymerization. One of the first catalysts applied was so called alumothermic titanium trichloride in presence of diethylaluminium chloride (DEAC) as a cocatalyst [1-3]. The activity of alumothermic catalyst was not higher than 2.5 kg of polymer per 1 kg of the catalyst, so its cost made an important piece of the actual cost of the dope. Next generation catalyst of Solvay-type was obtained by titanium tetrachloride reduction with an aluminorganic compound and characterized with much higher activity than previous types [3]. For the moment Solvay-type catalysts are rated as one of the most effective catalysts for obtaining of higher a-olefins super-high-molecular polymers [4]. The Russian-made version of Solvay-type, microsphere titanium trichloride with respect to higher α-olefins was patented in [5] and modified in [6].

It is considered as been proved that the monomer coordination with a transition metal compound is the necessary condition for the catalytic act implementation [7]. The metal organic intermediates being formed take part in the polymerization process, define its mechanism and affect significantly the structure of the polymer being synthesized. Nevertheless, the nature of the active centres and the methods for their action control are not still clear enough.

DEAC, trialkylaluminium, and dialkylaluminium oxide [8] can be used as titanium halogenide activators. The use of alkylalumoxanes in combination with titanium trichloride gives even more

Content from this work may be used under the terms of the Creative Commons Attribution 3.0 licence. Any further distribution of this work must maintain attribution to the author(s) and the title of the work, journal citation and DOI.
Published under licence by IOP Publishing Ltd

IOP Conf. Series: Materials Science and Engineering **112** (2016) 012034 doi:10.1088/1757-899X/112/1/012034

effective catalytic system [9, 10]. Dichloroethane being added to alkylalumoxane as second cocatalyst is forming a synergistic pair with the latter so that titanium trichloride works more intensely [4].

Modern titanium-magnesium catalysts in lower olefins polymerization process show even more high activity than the microsphere titanium trichloride does [7, 11, 12]. Production and application method for titanium-magnesium catalyst in polymerization of higher α-olefins in solution has been patented [13, 16]. In case of propylene polymerization the process stereoselectivity is of significant importance: isotactic polypropylene possesses good working intrinsics. To increase the polymer stereoregularity some compounds having lone electron pairs, so-called electron-donor additives can be used [17]. In production of higher α-olefin-based roentgenoamorphous polymers of superhigh molecular weight the donor additives are obviously of minor effect, as far as the polymer's capability to lower hydrodynamic drag is caused not by the regularity but by the length of the carbon chain. Nevertheless, such donor additives as ketones and monoesters [18], phosphororganic compounds [19], amines, siloxanes [20], and also other classes of compounds [21] are patented.

It should be noted that the following trend appeared in the field under study: the use of titanium trichloride based catalysts for polymerization in the monomer medium and the application of titanium-magnesium catalyst for the poly-α-olefin synthesis in the solvent medium.

2. Experimental Part

The synthesis was performed using microsphere $TiCl_3$, and titanium-magnesium catalyst as a polymerization promoter. Aluminorganic compounds, such as diethylaluminium chloride and triethyl aluminium were used as cocatalysts.

Casting polymerization of 1-octene was performed at 23°C. Into the 600 ml reactor made of polyethylene terephtalate (PETP) equipped with U-shaped three-necked cap, one inlet of which was supplied with a capillary for nitrogen-purging, 440 ml of monomer were loaded. The monomer was purged with dry nitrogen at ambient temperature during 30 minutes. Then, while keeping on purging, 15 ml (0.15 g) of aluminium chloride solution in petroleum solvent and 2 ml (0.08 g) of microsphere $TiCl_3$ preliminarily damped with a small amount of monomer were introduced using a dropping funnel. For a balanced distribution of catalyst in the monomer bulk, nitrogen was barbotated for 15 to 20 minutes until the thickening of the reaction mass, and then the stirring was terminated.

2.1. Analytic procedures

We had determined medium viscosity molecular weights of the synthesized poly-1-octene samples.

2.1.1. Determination of intrinsic viscosity of poly-1-octene

Determination of intrinsic viscosity of 1-octene polymer solutions obtained under different conditions was performed using Ubbelohde viscosimeters, capillary diameter of 0.54 mm.

The viscosimeter was mounted in a thermostat and kept at 20°C (determination temperature) for 10 to 15 minutes. Then the outflow duration was determined for pure tetrahydrofuran and for different concentration solutions. Then relative, corrected, and specific viscosity values were determined.
Intrinsic viscosity [η] (limiting viscosity number) is the limiting value of specific viscosity, when the solution concentration is approaching to zero. The intrinsic viscosity value was calculated using the linear regression equation.

2.1.2. Poly-1-octene average molecular weight determination

The medium viscosity molecular weight (M) was calculated using Mark-Kuhn-Houwink equation:

$$[\eta] = KM^{\alpha}$$

where α is the constant for a given pair 'polymer-solvent' at a given temperature.
Constants K and α values for the system poly-1-octene – tetrahydrofuran are shown in Tab. 1.

Table 1. Viscosimetry Constants K and α Values for Polymer-Solvent Systems Studied [22]

Polymer	Solvent	Temperature, °C	$K \cdot 10^5$	α
Poly-1-octene	tetrahydrofuran	25.0	41.4	0.625

3. Results and discussion

The figure 1 shows dependences of intrinsic viscosity on the conversion depth for 1-octene using titanium trichloride (figure 1, curve 1) and titanium-magnesium catalyst (figure 1, curve 2). As it can be seen, the change of intrinsic viscosity value confirms that the polymer of a greater molecular weight was being formed on titanium trichloride when the monomer conversion is deeper.

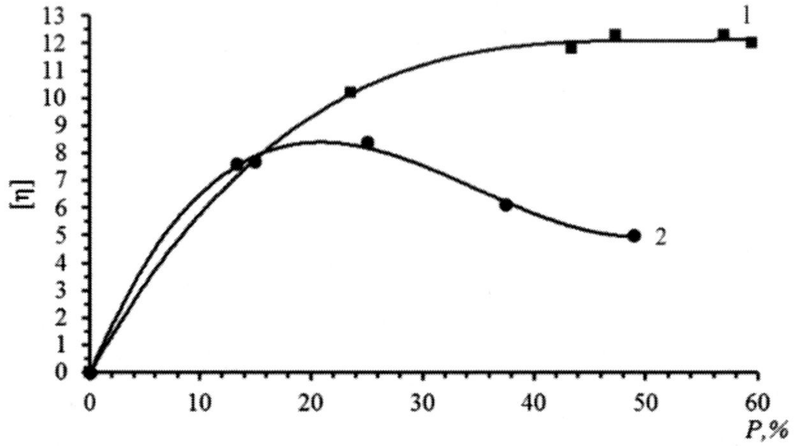

Figure 1. Dependences of intrinsic viscosity value on the conversion depth for 1-octene using microsphere $TiCl_3$ (1) and titanium-magnesium catalyst (2)

The Figure 2 shows dependencies of medium viscosity molecular weight for poly-1-octene on the conversion depth of 1-octene in casting polymerization using titanium trichloride (figure 2, curve 1) and titanium-magnesium catalyst (figure 2, curve 2).

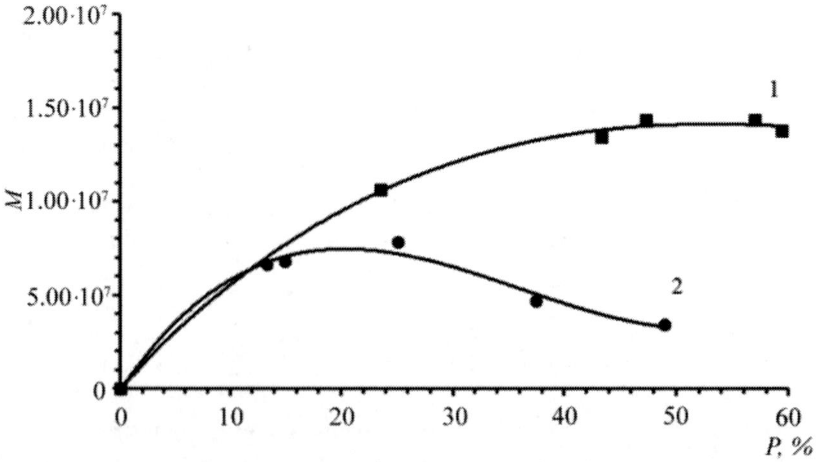

Figure 2. Dependencies of molecular weight value on octene polymerization depth using microsphere $TiCl_3$ (1) and titanium-magnesium catalyst (2)

The data presented show that the catalytic system 'titanium-magnesium catalyst – triethylaluminium' at conversion depth of 30 per cent when polymerization runs in the monomer medium produces the polymer of the molecular weight up to 7.5 million. Further polymerization on that catalytic system would not be expedient as far as low-molecular (ballast) polymer is formed; the latter deteriorates the final product capability to lower hydrodynamic drag in a turbulent flow of oil and oil products. During poly-1-octene synthesis runs on the catalytic system 'microsphere titanium trichloride – diethylaluminium chloride' the final product molecular weight grows up to 50 per cent conversion, after that it represents 'a plateau', the molecular weight being about 14.5 million.

Acknowledgments
The study was sponsored by the Ministry of Education and Science of the Russian Federation, Agreement No. 14.575.21.022, Unique Identifier RFVEFI57514X0022. The authors would like to appreciate Yu. V. Suleymanova for useful consultation she had made.

References

[1] Pat. 4289679 USA
[2] Pat. 4358572 USA
[3] Pat. 4415714 RU
[4] Pat. 2232165 RU
[5] Pat. 2075485 RU
[6] Pat. 2238282 RU
[7] Krencel B and Nehaeva L 1990 Russ. Chem. Rev. **12** 2034.
[8] Pat. 4845178 USA
[9] Pat. 2277103 RU
[10] Pat. 6162773 USA
[11] Sinn H, Kaminsky W, Vollmer H and Woldt R 1980 *Angew. Chem. Int. Ed. Engl.* **19** 390
[12] Babushkin D, Semikolenova N, Zakharov V and Talsi E 2000 *Macromol. Chem. Phys.* **201** 558
[13] Pat. 2171817 RU
[14] Pat. 2221813 RU
[15] Pat. 2152404 RU
[16] Pat. 2230074 RU
[17] Chirkov N 1976 *Polymerization on Complex Organometallic Catalyst* (Moscow: Chemistry)
[18] Pat. 4493904 USA
[19] Pat. 4433123 USA
[20] Pat. 5028574 USA
[21] Pat. 4945142 USA
[22] Zhiqiang F, Linxian F and Shilin Y 1992 *Acta Polymerica Sinica.* **5** 577

Prospects of lithium enrichment on ^{7}Li isotope by method of controlled ions electro-migration

G A Martoyan[1], M M Kalugin[2], A V Gabrielyan[1] and A G Martoyan[1]

[1] SI "Ecoatom" LLC, Yerevan, 1 Adanai str., Armenia
[2] SIC "PolyTechnologies" LLC, Sosnoviy Bor, Russia

[1]E-mail: martoian@yahoo.com

Abstract. This paper deals with a new electro-membrane method of enrichment of ^{7}Li isotope. The data are presented on the importance and application fields regarding the use of ^{7}Li isotopes. Existing methods and criteria of separation of lithium isotopes are discussed. The principle of new technology, regimes of enrichment experiments, and analysis details of obtained products are briefly described.

I. Introduction

Lithium is a rare element scattered in the Earth's crust; its production, processing and purification are quite costly processes. Li is the lightest metal in nature. Natural lithium consists of two stable isotopes ^{6}Li (7.52%) and ^{7}Li (92.48%). One of the main sources of primary raw materials is lithium carbonate (Li_2CO_3). It is known that for industrial production of high purity $LiOH{\cdot}H_2O$ the most preferred method is hydrometallurgical electromembrane technology of conversion of primary lithium carbonate into solution of LiOH with simultaneous removing impurity ions Na^{+}, Ca^{2+}, Mg^{2+}, Al^{3+} and anions SiO_3^{2-}, Cl^{-} available in original Li_2CO_3.

In recent years, lithium has been widely used not only in the nuclear industry, but to regulate the water chemical regime, and also in lithium-ion batteries. However, it should be noted that almost all the important application of lithium is associated with isotopes, and a high level of isotopes enrichment is required up to the mono-isotopic purity of one of its stable isotopes.

^{6}Li isotopes are used:

- For production and replenish the fuel tritium 3H in power CTF reactors (controlled thermonuclear fusion).
- In the industrial production of tritium 3H isotope in nuclear reactors. At irradiation of natural mixture of lithium isotopes with neutrons the nucleus of ^{6}Li, capturing thermal neutrons, decay into helium and tritium (6Li + n → 4He + 3H + 4.78 MeV).

^{7}Li isotopes are used:

- In the development of a wide range of electrical rechargeable batteries, including for portable devices, transportable power sources of electric vehicles and stationary storage of electricity from renewable energy sources to support the utility grid during periods of peak load.
- In nuclear power: the addition of $^{7}LiOH{\cdot}H_2O$ allows to adjust the water-chemical balance of the coolant (pH-factor) in the 1st loop of reactors (boiling and with pressurized water).

Content from this work may be used under the terms of the Creative Commons Attribution 3.0 licence. Any further distribution of this work must maintain attribution to the author(s) and the title of the work, journal citation and DOI.
Published under licence by IOP Publishing Ltd

- Also in water treatment equipment (as part of ion-exchange membranes) of coolant of 1st loop of PW type reactors.
- In medicine: pharmaceuticals on the basis of lithium carbonate ($Li_2 \cdot CO_3$) enriched by the ^7Li isotope (rarely by the ^6Li) are used as antidepressants/anxiolytics.

2. Existing technologies of ^6Li and ^7Li isotopes production

For separation of lithium isotopes most often are used the processes of ion exchange, chemical exchange, electrolysis and recently also electro-migration methods by use of ion-exchange membranes [1-3].

As a basis of widespread methods of industrial separation of lithium isotopes is isotope exchange reaction between the solution of lithium compounds and mercury-lithium amalgam. The method is not without a number of drawbacks, the main of which are:

- toxicity and large amounts of mercury, participating in the process
- amalgam trend to decomposition in aqueous solutions
- formation of dangerous mercury-containing waste
- high energy consumption

Therefore the search and development of new, environmentally less hazardous methods of separation of lithium isotopes is of undoubted practical interest.

The main criteria for evaluation of the process prospects of isotope separation are the followings.

Separation factor. Separation (enrichment) factor r is the ratio of the relative concentration of extracted isotope in the final product (after enrichment) relative to its concentration in the initial (row) mixture of isotopes:

$$r = \frac{n_1'/n_2'}{n_1/n_2}$$

were n_1 and n_2 are concentrations of isotopes with masses m_1 and m_2 before separation, whereas n'_1 and n'_2 are corresponding concentrations after separation.

Productivity - speed of obtaining the desired product.

Safety – level of negative impact on the environment and people health.

Downloading - original amount of the substance contained in the separation unit. Startup time is the time from start-up of separation unit until achieving stationary conditions.

Efficiency - efficiency of use of the initial quantity of raw material.

Prime cost - financial costs for producing the mass unit of product.

As already has been noted, now for separation of lithium isotopes the popular is method based on isotope effect which arises during chemical exchange of lithium between lithium amalgam and aqueous solution of lithium compounds.

In this regard it seems interesting is the possibility to enhance the efficiency of isotope exchange [4] with consequent (simultaneous) slowing of reaction of amalgam decomposition which requires its periodic regeneration. This can be realized, for example, by applying electrical potential difference between contacting liquid phases of lithium amalgam and lithium hydroxide circulating countercurrent in conditions of maintaining the high values of ratio between the speed of the interphase chemical exchange of lithium and amalgam decomposition. It should be mentioned that such approach allows to combine the processes of ion and chemical exchange, electrolysis and electro-migration thereby strengthening their joint action.

3. Modification of amalgam method for ^6Li и ^7Li isotope production

3.1. Short description of the process

a) Lithium hydroxide $LiOH \cdot H_2O$ in aqueous solution was preliminarily subjected to a thorough electromembrane cleaning from impurities (Na, Al, Mg, etc.). It is known that lithium ions in solution are characterized by the highest electrode potential, -3.045 V (due to their small size and high degree of hydration). For comparison, the electrode potential of Mg and Na are -2.37 V and -2.71 V, respectively. This property of lithium ions allows to remove from working compartment of electrodialyzer practically all known impurity cations and anions and to obtain lithium containing solution of very high purity, up to 4-5 (N). Then the purified solution of $LiOH{\cdot}H_2O$ is sent to the amalgam production unit.

b) In this unit the purified solution of $LiOH{\cdot}H_2O$ is fed into the working chamber (2) of the electrolytic cell shown in Figure 1. This is a known effective electrolytic method of producing an amalgam [5], which was proposed earlier for extraction of metals from liquid radioactive waste (LRW). The resulting amalgam is sent to the unit for separation of isotopes ^6Li and ^7Li.

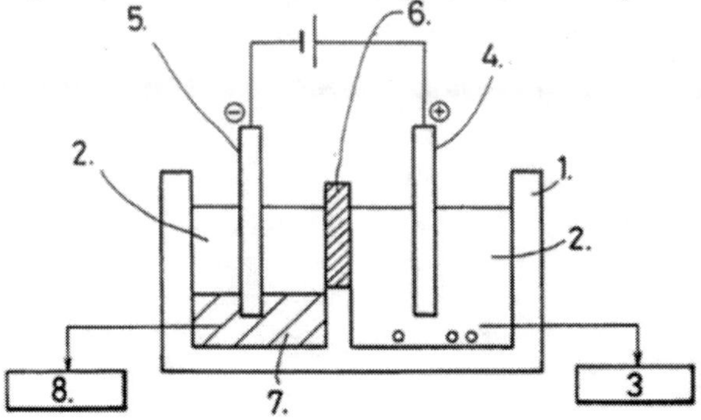

Figure 1. Electrolytic method for obtaining amalgam. 1 – electrolytic cell, 2 – solution of LiOH, 3 – residue, 4 – anode, 5 – cathode, 6 – diaphragm, 7 – mercury, 8 – amalgam.

c) Experiments were conducted in the separation unit [6] shown in the Figure 2 which presents an electrolytic cell with the following operating parameters: I = 150mA, U = 10 V, S = 100cm^2, $V_{mercury}$ = 50cm^3, $V_{solution}$ = 300 cm^3 and lithium content of 15g. Lithium containing solution is pumped in a closed loop.

Under the influence of an applied electric voltage between electrodes, cations of ^7Li and ^6Li isotopes migrate in solution with different speeds.

The spatial separation of isotopes in the cell volume occurs as follows. More movable and light $^6Li^+$ ions will migrate faster through the membranes and bind to mercury amalgam, while the main part of $^7Li^+$ ions is "lagging behind". Some part of cations of iotope $^7Li^+$ having time to migrate through the membrane are raised on the upper surface of the mercury cathode under the influence of the eddy currents generated on the side surface of the mercury amalgam.

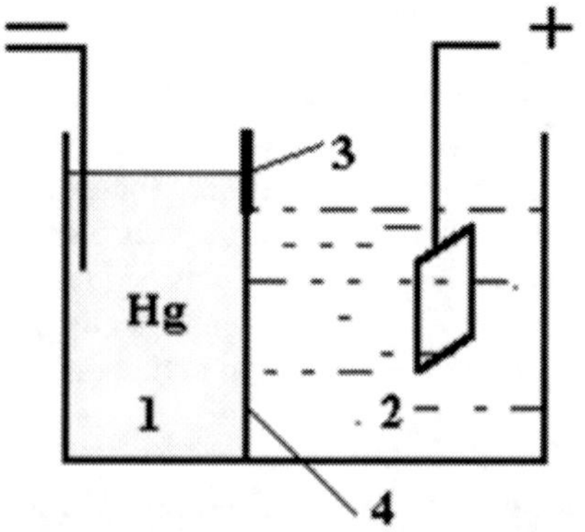

Figure 2. Scheme of electrolytic cell for separation of lithium isotopes. 1 – mercury, 2 – LiOH solution, 3 – dielectric, 4 – cation-exchange membrane.

Thus in this case the enrichment on lithium isotopes is controlled (simultaneously and interconnected) by three different factors: electro-migration through an ion exchange membrane (the probability of passing ^6Li isotopes) is high; the binding of isotopes in amalgam (the probability of binding ^6Li isotopes in mercury cathode) is high; and finally, Eddy currents - the ^7Li isotopes most likely are extracted from the side surface of mercury cathode on its upper surface.

We think that ^6Li isotopes are enriched in the bulk of mercury cathode, while ^7Li ions accumulated on the top surface of the mercury cathode and in the working chamber.

After three-hour operation, samples were taken from the anode (working) compartment, from the solution over the mercury cathode and from the solution after amalgam decomposition.

3.2. Method of measurement of $^7Li/^6Li$ ratio

Analyzes to determine the isotope ratio were conducted by the mass spectrometer ICP-MS ELAN 9000, which gives a good opportunity to identify rare and trace elements and isotopes in a wide range with accuracy up to ng/l.

In experiments the standard solutions of ^6Li, ^7Li (Perkin Elmer) were used.

As matrix solution the bidistilled 2% nitric acid in deionized water purified in apparatus Easypure II (18.2 MOm·cm) was used. For plasma formation and spraying the argon of 99.998% purity was used. Samples were injected into the system by a peristaltic pump in a cross-flow nebulizer, wherein their conversion in aerosol was occurred. Parameters of the analysis are shown in Table 1.

Table 1. Parameters of analysis by mass spectrometer ICP-MS ELAN 9000.

Parameter	Value
Discharge power	1400 W
Consumption of plasma generating gas	15 l/min
Consumption of atomizing gas	0.86 l/min
Scan mode	Peaks determination
Replicates (number of parallel measurements)	6
Integration time	5000 microsecund
Solution feed rate	1.3-1.4 ml/min
Input cones	Nickel

Analyses were carried out as follows: blank – standard – sample respectively. To reduce the "memory" effect the capillaries were washed with a 3% nitric acid during about 5-7 minutes. Results of ^7Li / ^6Li ratio were compared with the ratios in standard solutions and initial solution of lithium hydroxide respectively.

3.3. Preliminary results

^7Li isotope separation coefficient r measured according to procedure described above is equal to 1.178, confirming the prospects of this method of enrichment of lithium isotopes.

Experiments simulating the cascading process (when solution enriched by 7Li in first cell was used as a n initial for second cell) have shown that values of separation coefficient can be kept on the level achieved in the first cascade.

Estimation of scalability made on the basis of performed experiments show that this method is able to provide the production of isotope ^7Li enriched up to 99.95% or higher in a continuous mode.

4. Conclusions

Thus, the proposed electromembrane method of isotopes separation is characterized as following:

- it is a modified version of the amalgam method of lithium isotope enrichment with a broad new opportunities for controlling the parameters;

IOP Conf. Series: Materials Science and Engineering **112** (2016) 012035 doi:10.1088/1757-899X/112/1/012035

- it allows to expect the obtaining of enriched lithium isotopes or mono-isotopic compounds in solution (hydroxides);
- it is characterized by high efficiency and profitability due to low power consumption for isotope separation;
- it has the prospect of scaling up to production on industrial scale;
- it is environmentally more attractive, more technological than existing methods for the separation of isotopes of lithium.

References

[1] Hoshino T. et al. 2011 Basic Technology for 6Li Enrichment Using an Ionic-Liquid Impregnated Organic Membrane, *J. of Nuclear Materials,* **417**, issues 1-3, 696-699

[2] Kitazawa S, Kimura K, Yano H, Shono T 1985 Lithium-selective polymeric membrane electrodes based on dodecylmethyl-14-crown-4, *Analyst* 110, 295–299

[3] Hoshino T, Nakamichi M 2012 Development of fabrication technologies for advanced breeding of functional materials for DEMO reactors, *Fusion Eng. Des.* 87, 486–492

[4] Abroskin I E et all 2001 Method and installation for enrichment of lithium-7 isotope *Patent of Russia* RU 2216391

[5] Otsuka K, Takahashi Y, Tamai H 1990 Electrolytic treatment of radioactive liquid waste to remove sodium *US Patent* No: 4,931,153

[6] Martoyan G, Intsheyan S, Tonikyan S, Demirchyan M and Guiragossian Z 2003 Electrolytic Method of the Extraction of Metals *Patent Application* WO 2004/005586

Structure and phase composition of the superalloy on the basis of Ni-Al-Cr alloyed by Re and La

E L Nikonenko[1,2,4], N A Popova[1,3], N A Koneva[1] and E V Kozlov[1]
[1]Tomsk State University of Architecture and Building, 2, Solyanaya Sq., 634003, Tomsk, Russia
[2]National Research Tomsk Polytechnic University, 30, Lenin Ave., 634050, Tomsk, Russia
[3]Institute of Strength Physics and Materials Science SB RAS, 8/2, Akademicheskii Ave., 634021, Tomsk, Russia

[4] E-mail: vilatomsk@mail.ru

Abstract. Qualitative and quantitative studies of the structure, phase composition, morphology of phase of the high-rhenium alloys additionally doped with La were carried out by TEM and SEM methods. The alloy was obtained by directional solidification method. It was shown that introduction of Re and La to an alloy leads to formation of new phases: β and χ, which bring serious irregularities in the structure of qvazicuboids of γ′-phase.

1. Introduction

Level requirements for materials with the required performance characteristics constantly increase. One of the promising areas of the search for new high-temperature materials is creation of alloys containing intermetallic phases [1,2]. Superalloys on the basis of (γ+γ′) are example of phases where γ-phase is the disorder solid solution with fcc structure and γ′ is the ordered phase with superstructure of $L1_2$. Now superalloys are created on the basis of the Ni-Al alloy alloyed by various elements. Lately, rhenium and lanthanum have been used in alloying superalloys. This work presents investigations of the composition, morphology, and particles size of phases formed due to these chemical elements. Localization of new phases and their distribution in the superalloy is also described in this paper.

2. Material and methods

The purpose was to qualitative and quantitative studies of the structure, phase composition, morphology complex alloyed rhenium alloy additionally doped by La. The alloy was obtained by directional solidification. The main elements of the alloy: Ni – ~70 at.%, Al – ~17 at.% Cr and ~5 at.%. Mo, W, Ta, Ti, Co, Co were the main alloying elements with total elemental content ~ 7 at.%. Re content was 0.4 at.% and La – 0.006 at.%.

The main methods of research of the alloy structure were TEM and SEM [3,4].

3. Results and discussion

The superalloy phases can be classified into primary and secondary phases observed by TEM. This classification is based on the volume fraction of the phases, their role in the alloy [2,5,6]. As we see from Table 1, γ′- и γ are the main phases. As a rule, they form the main structure of an alloy practically in all super alloys. In this alloy they are as quasi cuboids of γ′-phase divided by γ-phase

Content from this work may be used under the terms of the Creative Commons Attribution 3.0 licence. Any further distribution of this work must maintain attribution to the author(s) and the title of the work, journal citation and DOI.
Published under licence by IOP Publishing Ltd

layers. Other phases are secondary. Volume fractions of all phases in the alloy also are given in the Table 1 below. From the table is clearly that the volume fraction of γ′-phase is high. The volume fraction of γ-phase is following in size. β-phase is next (table 1). This symbol is defined β-NiAl phase with a lattice parameter a = 0.288 nm having the CsCl structure, which it can transform into a tetragonal phase type $L1_0$ [7]. The following phase is χ-phase (Table 1). It is topological close-packed phase or in other words it is Franck-Casper's phase [8]. It occurs due to the Re presence in the alloy. Actually, interaction of Re with Mo and W as well as with Co and Al leads to appearance of χ-phase. This phase is formed at annealing.

Table 1. The phase composition of the alloy and quantitative characteristics of the phases

Phases	Type of crystalline lattice	Space group	Crystalline lattice parameters, нм	Volume fractions of phases, ±1%
γ′	cubic lattice.	Pm3m	a = 0.3568-0.3575	85.60
γ	cubic lattice	Fm3m	a = 0.3569	8.00
β	cubic lattice	$Pm\bar{3}m$	a = 0.288	5.00
χ	cubic lattice	$I\bar{4}3m$	a = 0.957-0.960	1.40

It is known that the structure of superalloys is the mixture of (γ+γ′)-phase [2,5,6]. This mixture has morphologically correct crystallographic orientation and generates quasi monocrystalline structure. Part of β-phase and χ-phase completely is formed due to presence of these active phases. Formation of these two phases brings serious changings in structure of quasi cuboid of γ/γ′. Because La and Re are presented only in the local areas and they are not allocated uniformly in the whole volume of the alloy, the alloy was broken only a part of the quasi cuboids. The studies allowed to submit a scheme of the test sample of the alloy in Figure1.

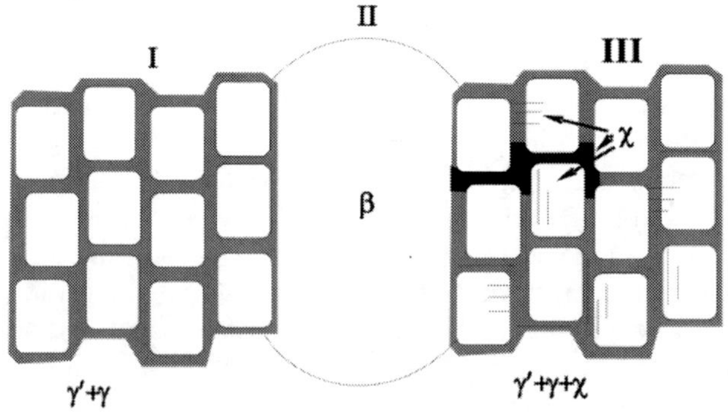

Figure 1. Schematic imagination of the structure of the alloy. Three morphologically-phase states: I - ideal structure of qauzi cuboids (γ′+γ); II-section II of β -phase; III - qauzi cuboids (γ′+γ), containing layers χ-phase

Table 2. Structure –phase state of investigated alloy

State	Volume fraction of state	Phase composition	Sizes of quazi cuboids, nm	Particles size of χ- phase, nm
I	0.65	γ′+γ	320 × 440	-
II	0.05	β	-	-
II	0.30	γ′+γ+χ	320 × 440	30 × 240

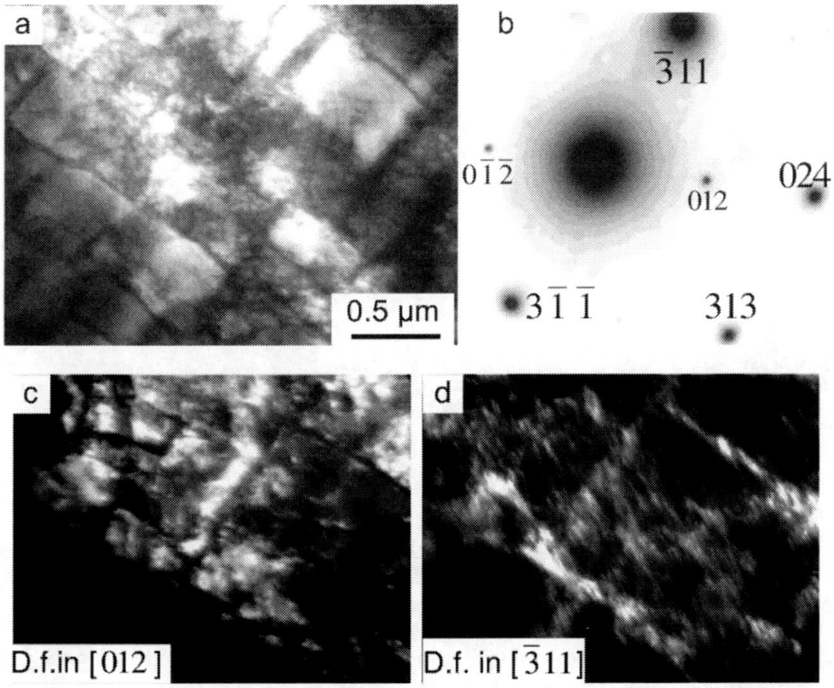

Figure 2. TEM image of quasi cuboid structure (γ′+γ) in alloy (structure-phase state I): a – bright-field image; b – microdiffraction patterns; c,d – dark-field images, obtained correspondingly in superstructure reflex [012] and main reflex [$\bar{3}11$]

It clearly shows three different morphologically-phase states: I, II and III. They are also shown in Table 2. The scheme of Figure 1 clearly shows that the state I is an ideal structure (γ+γ′) with a slightly anisotropic deformation. State III along with quasi cuboids layer contains χ-phase, which can be located in γ′-, and in γ-phase. The orientations of these layers are parallel to the cubic directions of γ′ and γ-phases, as well as the orientation of layers γ-phase, its volume fraction is close to 5%. In state III quasi cubbies are anisotropny also. Figure 1 and 2 shows results of TEM of I and III state their microdiffraction patterns and dark field. Dark field imagers are obtained in the main (Fig. 2d) for the first state and superstructure reflexes (Figure 2c). It is possible to identify γ and γ′-phases: the first layers in Figure 2d, the second it is quasi cuboid (in Figure 2c can see that layers are in unreflection position). In state III local volume fraction of γ-phase is closed to 5%.

State II is characterized by absolutely destroyed structure of quasi cuboid. There is tree component solution of $NiAl_2Re$ in the volume of part II. The last expression it is chemical formula of three component of β-phase. In Figure 3 we can see that layers of γ′ and χ are parallel. These phases have cuboid crystalline lattice but some size effects lead to crystallography disorientation. The lattice parameters of these phases differ enough [9] (a_γ = 0.3569 nm; a_χ = 0.9570 nm). We can see that direction [002]γ′ is almost parallel to [$32\bar{3}$]χ direction. In other words, the fact that the difference in the lattice parameters of these phases, the tendency to minimize of the elastic energy led to the formation of crystallographic orientation with the above orientation relationship and parallel layers χ- and γ-phases. In this case, the elastic stress fields are missing. This is indicated by absence on TEM mages of the extinction contours bending-torsion. Thus, γ- and γ′-phase coherently and almost stress-free contact with the χ-phase.

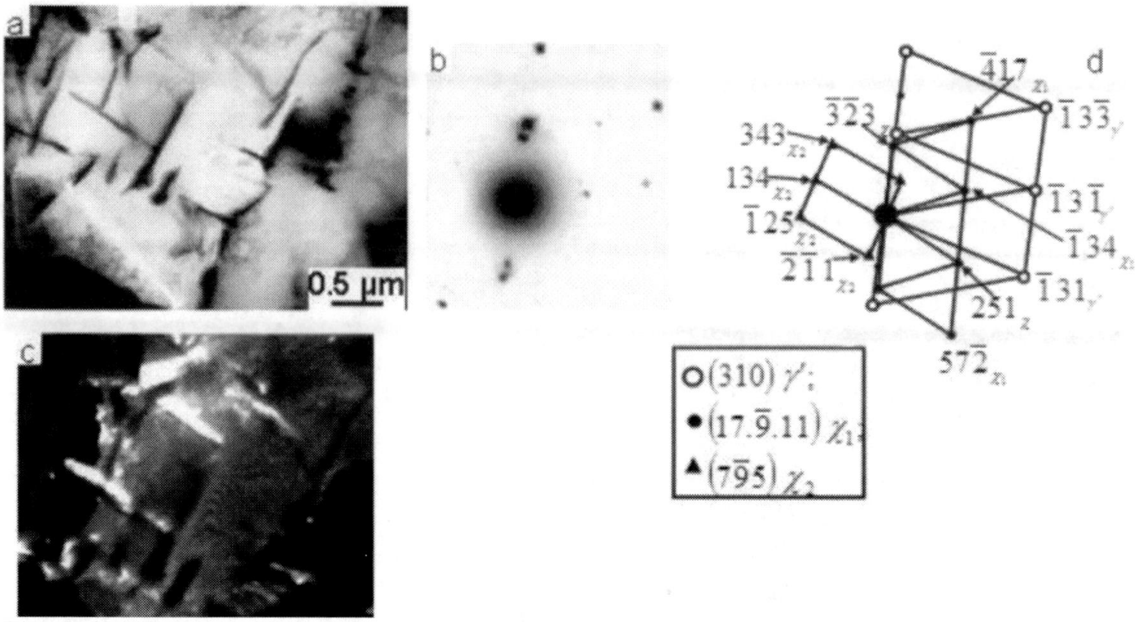

Figure 3. TEM image of structure-phase state III: a – bright-field image; b – microdiffraction patterns; c – dark-field images, obtained in reflex $[\bar{3}\bar{2}3]_{\chi}$ and d - scheme indexing (b)

We will consider contact of β-phase and the massive block of γ/γ'-phases (Figure 4) now. The interface unit and β-phase flows toward the [011] γ'-phase. Orientation ratio β-phase and γ/γ'-phase: $[110]_{\beta}$ II $[111]_{\gamma'}$. This typical ratio for mutual phase transformation of fcc → bcc, for example, in the steels. It is characteristically, the extinction contours bending-torsion near the β – γ/γ' not observed (see Figure 4a). This means that β-, γ- and γ'-phase coherently related.

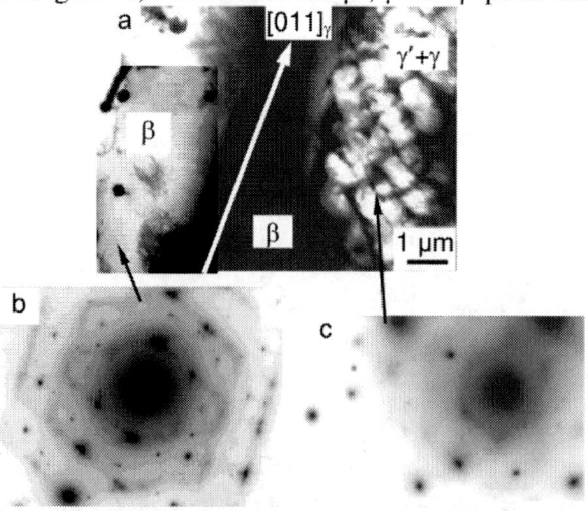

Figure 4. TEM image of fine structure of alloys: a – junction of two structurally-phase states (γ'+γ) and (β): b – micro diffraction pattern obtained from state II (there are only reflexes β-phase); c – micro diffraction pattern obtained from the state I (there are only reflexes γ/γ' mixture)

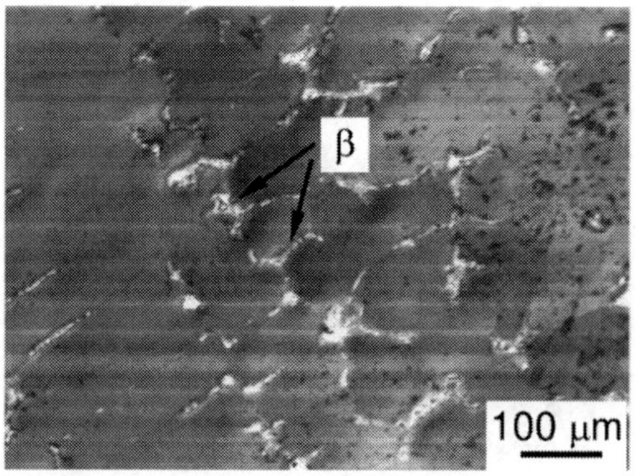

Figure 5. Image of the structure of the alloy obtained by SEM. Black arrows indicate examples of β-particle phase, a white arrow on the (b) - direction of periodicity

The microdiffraction pattern presented in Figure 4b, attract special attention. It characterizes structure of β-phase. Characteristic bands stretching between reflexes indicate micro segregation in this phase [6]. Recall that the chemical formula of this phase is close to the composition of $NiAl_2Re$. Presence of diffuse bands connecting reflexes could indicate firstly, about a nonstoichiometric composition against formula indicated above, secondly, about the presence of micro- segregation in them. It is not surprising that after annealing this phase disappears. The image of structure of an alloy obtained by the SEM method at smaller magnitude is submitted in Figure 5. Most of the volume of the alloy is quasi cuboids, whole and broken, the contrast from which due to the relatively small magnitude is different. Stratification on Re is well visible thanks to precipitation of β-phase which is mainly stabilized by this chemical element. As ratio

$$\frac{P_{V_{Al}}}{P_{V_{Ni}}} = 0.23\,,$$

(where P_{V_i} – volume fraction of the corresponding element), it indicates nearly stoichiometric composition A_3B (A – Ni, B – Al. This means that the amount of β-phase should be few, and that takes place actually. Particles of β-phase are well visible in Figure 5. The total share of γ-phase in an alloy makes the size equal 0.08, γ′-phase – ~0.86, ratio $\Delta\gamma/\Delta\gamma' = 0.09$. It is close to the ratio of the components B and A (A = 0.82, B = 0.18). This is the sum of the ratio of the elements forming γ′- and γ-phase in the alloy under study.

Thus, we note that quasicuboides of g and g¢-phase form 95% of its volume, while 65% are ideal quasicuboides and 30% are those containing the c-phase. Finally, 5% of the volume takes the bcc-based b-phase that is an ordered ternary $NiAl_2Re$.

Acknowledgments
This work was financially supported by Grant N 3.295.2014/K from the Ministry of Education and Science of the Russian Federation

References

[1] Greenberg B A and Ivanov M A 2002 *Intermetallic Ni_3Al and TiAl: microstructure, deformation behavior* (Ekaterinburg: RAS Publ.)
[2] Kozlov E V, Popova N A, Nikonenko E L, Koneva N A, Sizonenko N R, Buntushkin V P and Kolobov Yu R 2006 *Bulletin RAS Physics* **70** **7** 1117
[3] Hirsch P, Howie A and Nicholson R 1968 *Electron Microscopy of Thin Crystals* (Moscow: Wiley)
[4] Gouldsteyna J and Yakovitsa H 1978 *Practical Scanning Electron Microscopy* (Moscow: Wiley).
[5] Kozlov E V, Nikonenko E L, Koneva N A, Popova N A, Sizonenko N R, Kolobov Yu R, Grabovetskaya G P and Buntushkin V P 2005 *Bulletin RAS Physics* **69** **7** 1116
[6] Koneva N A, Nikonenko E L, Popova N A, Sizonenko N R, Kolobov Yu R, Buntushkin V P and Kozlov E V 2005 *Fundamental Problems of Modern Mater.* **2** **4** 32
[7] *Diagrams of Binary Metallic Systems 2000 (Directory* Vol 3) ed N P Lyakishev (Moscow: Mechanical Engineering)
[8] Bonar A A, Velikanova T J and Danilenko V M 1991 *Stability of the Equilibrium Phase and the Phase in the Alloys and the Transition Metals* (Kiev: Naukova Dumka)
[9] *Diagrams of Binary Metal Systems* 1996 *(Directory* Vol 1) ed N P Lyakishev (Moscow: Engineering)

Effect of preparation conditions on physic-chemical properties of tin-doped nanocrystalline indium oxide

T D Malinovskaya*, V I Sachkov, V V Zhek and R A Nefedov
Siberian Physical-Technical Institute, National Research Tomsk State University, 1, Novosobornaya sq., Tomsk, 634050, Russia

*E-mail: malino@sibmail.com

Abstract. In this paper the results of investigation of phase formation and change of concentration of free electrons (Ne) in indium tin oxide system during heat treatment of coprecipitated hydroxides of indium and tin from nitric and hydrochloric solutions and also, for comparison melts of salts nitrates by an alkaline reactant (NH_4OH) are considered.The performed investigation allowed to set the optimal condition of preparation of polycrystalline tin-doped indium oxide with maximal electron concentration.

1. Introduction

There is a growing interest in tin-doped indium oxide (ITO) nanoparticles mostly due to the possibilities of producing highly active catalysts and highly sensitive gas sensors [1].

Tin-doped indium oxide is n-type semiconductor with high electrical conductivity varying in a wide range in dependence on preparation conditions and type of the material (single crystals, ceramics, thin films) [2].

However, it should be noted that most of the publications devoted ITO thin films [3]. And many of the problems of practical use of sol-gel synthesis technology of ITO dispersed materials can not be solved due to the lack of comprehensive studies of the synthesis conditions and the electronic properties of the materials obtained. For example, in work [4] mainly deals with the effects of synthesis temperature and calcinations temperature on morphology and phase structure of ITO particles, but no information about charge carriers concentration. In [5] has information about charge carriers concentration, but examines the impact of her only sintering temperature and Sn concentration.

This work presents the results of study of coprecipitation of hydroxides of indium and tin from nitric or hydrochloric acidic solutions by NH_4OH reactant and following annealing of coprecipitation products, the effect of conditions on the phase composition and major charge carriers concentration (Ne) in prepared nanocrystalline materials. Ne was determined by plasma resonance wavelength from IR-spectra of diffusive reflection [6].

2. Materials and methods

On the basis of EPR, X-ray diffraction and differential thermal analysis the phase formation regularities were established in In-Sn-0 system in dependence on starting solutions type, medium pH value, doping addition (tin) concentration, temperature and annealing time. It was shown that independently on used reactants nature the closest interaction of co-precipitating hydroxides takes place in the medium with pH = 6-8 (figure 1).

Content from this work may be used under the terms of the Creative Commons Attribution 3.0 licence. Any further distribution of this work must maintain attribution to the author(s) and the title of the work, journal citation and DOI.
Published under licence by IOP Publishing Ltd

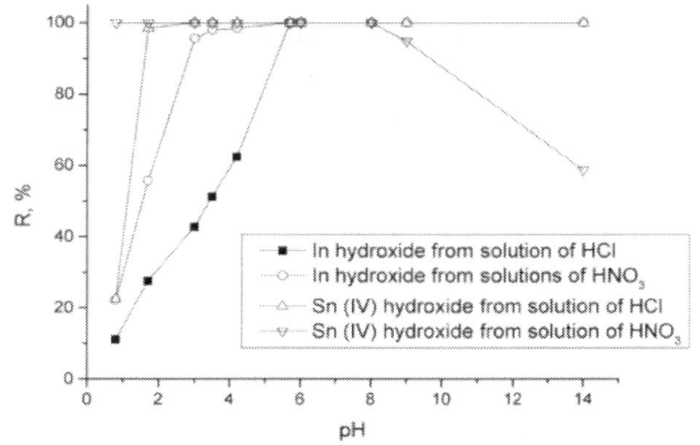

Figure1. The dependence of a degree sedimentation R on pH of solutions

Co-precipitation products annealing leads to the formation In_2O_3 defect structure of cubic modification with enlarged interplanar distances in comparison to standard values, or two-phase mixture of this phase and tin oxide of tetragonal modification is formed (figure 2, 3, 4).

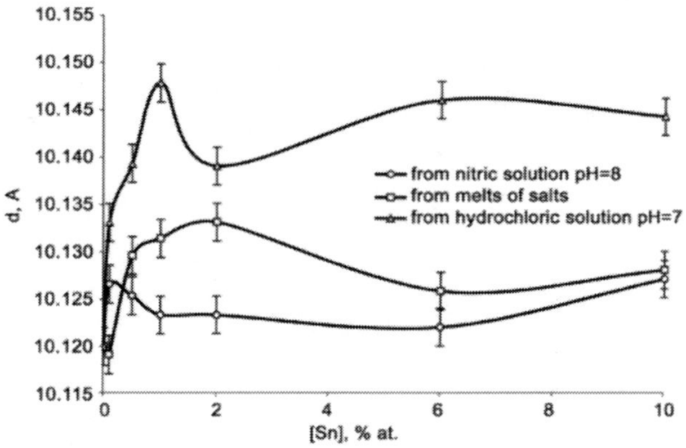

Figure 2. Lattice parameter change in indium oxide as a function of tin content (T=800 °C)

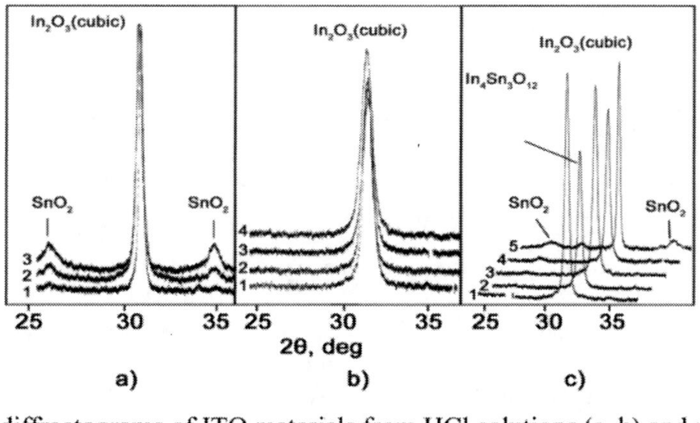

Figure 3.X-ray diffractograms of ITO materials from HCl solutions (a, b) and melts of salts (c) containing 1-1, 2-2, 3-6, 4-10 atom % tin at different temperatures: a, b-900 °C, c-1100 °C

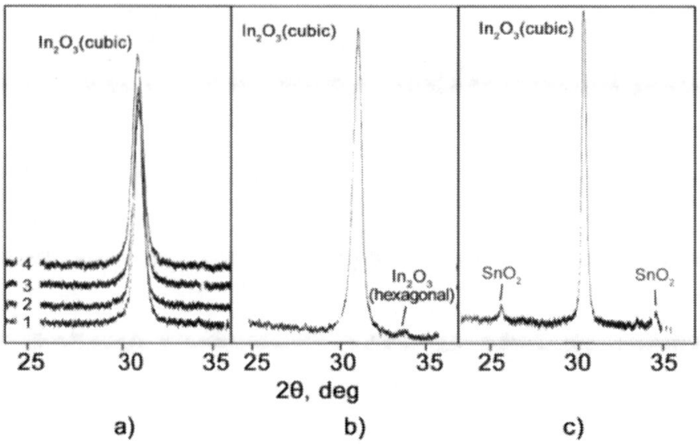

Figure 4.X-ray diffractograms of ITO materials from HNO3 solutions containing 1-1, 2-2, 3-6, 4-10 atom % tin at different temperatures: a, b- 900 °C, c-1100 °C

3. Results and discussion

The dependence of the change of free electrons concentration on content of tin in indium and tin oxide (ITO) materials was established (table 1, 2).

Table 1. The electron concentration as a function of tin content in ITO materials prepared from different reagents at T = 900°C and pH = 7.

reagents	C_{Sn}, at. %							
	0.1	0.5	1	2	4	6	10	14
$InCl_3$ $SnCl_4$ NH_4OH	2.6	2.6	2.3	2.2	2.1	2.1	2.1	2.0
$In(NO_3)_3$ $Sn(NO_3)_2$ NH_4OH	1.5	1.7	2.1	2.1	2.1	2.1	2.1	2.1
$In(NO_3)_3$ $Sn(NO_3)_2$	0.5	0.9	1.1	1.2	1.2	1.3	1.3	1.4

Table 2. The electron concentration as a function of heat treatment temperature and tin content in ITO materials.

C_{Sn}, at. %	Temperature of heat treatment, °C				
	300	500	700	900	1000
0.1	2.3	1.6	1.6	2.6	2.6
0.1*	1.7	1.5	1.6	2.3	2.6
2.0	1.1	1.2	1.7	2.3	2.4
10.0	0.7	0.9	2.1	2.1	2.1
10.0*	0.8	1.0	2.1	2.1	2.1

Note: * Samples were kept at this temperature for 60 min

The formations of ITO thus, occurs, apparently, at removal of NO_3- ions, which as determine, from IR-spectrum of co-precipitation In(III) and Sn(IV) hydroxides (figure 5), and the products of their thermal decomposition - nitrogen-containing radicals of a different composition, about that the occurrence on curves DTA for nitrate system of several endothermic peaks (figure 6c) testifies. For systems obtained by zol-gel method from hydrochloric and nitric solutions (figure 6a, 6b) the

endothermic peaks on curves DTA relates to evaporation of water and decomposition of indium hydroxides. Endothermic peak on 380 °C (figure 6 b) relates to removal of NO_3- ions.
According to EPR data the basic intermediate product of thermal decomposition of NO_3--ions stabilized in forming structure of an indium oxide are the radicals NO_2. They formed in the greatest amount after heat treatment at 400 °C. Probably, the formation of nitrogen-containing radicals NO_2 is accompanied by removal of oxygen from the system and, hence, by great deviation from stoichiometry in structure of indium oxide, it results in occurrence of hexagonal modification In_2O_3 at 400 °C. The presence of NO_2 attaches some amorphism to structure of cubic In_2O_3, it results in low values of Ne (figure 7, 8). The spectrum EPR disappears after heat treatment at 600 °C.

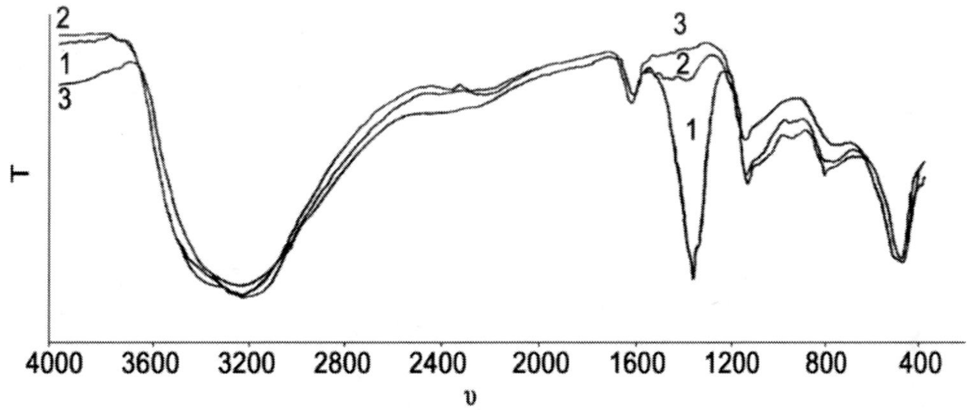

Figure 5. IR absorption spectrum of co-precipitation products from hydrochloric (3) and nitric (1, 2) solutions, where heated at 50 °C, after hot (2) and cold (1) washing procedure

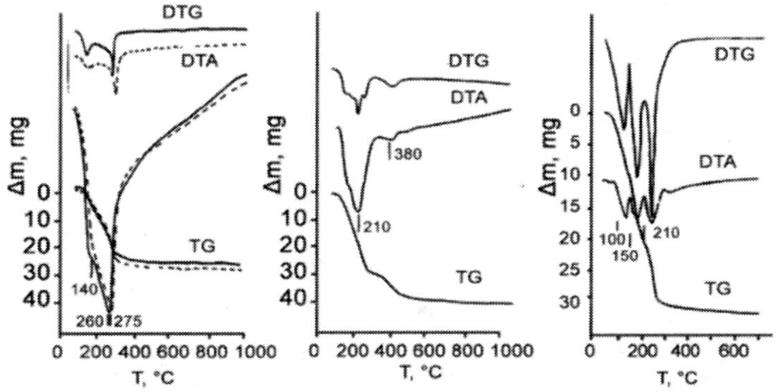

Figure 6. Curves of DTA, TG, DTG for ITO from:
a- hydrochloric solution, b -nitric solution, c - melts of salts

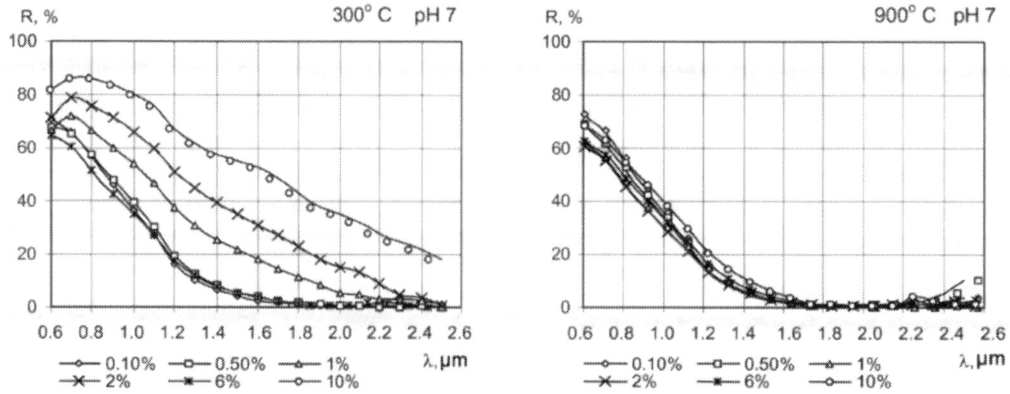

Figure 7. Reflectance spectrum of ITO materials prepared from hydrochloric solutions at different temperature and tin content

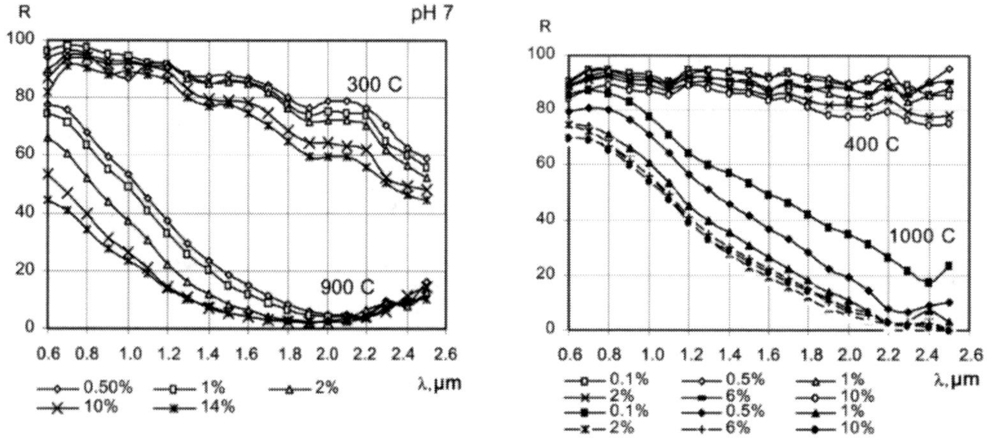

Figure 8. Reflectance spectrum of ITO materials prepared from nitric salts melts (a) and solutions (b) at different temperature and tin content

4. Conclusion

We studied the electronic and spectral properties for ITO material obtained under different conditions of solid phase synthesis. It is shown the upper bounds of mobile electrons concentration are achieved for materials obtained from chloride solutions. At temperatures up to 600 ° C a decisive contribution to the concentration of free charge carriers makes oxygen deficiency of the crystal structure of indium oxide doped with tin, while at higher temperatures, is a determining factor is the concentration of the dopant. The ligation efficiency of indium tin oxide, is greatly reduced with increasing tin concentration higher than 1 at. %.

Acknowledgement

The work was financially supported by project RFMEFI57814X0026 of the Ministry of education and science of Russia.

References

[1] Schmid G, Bumle M, Geerkens M, Heim I, Osemaun C and Sawitowski T 1999 *Chem. Soc. Rev.* **28** 179
[2] Jikang Y, Wanxia T, Erxin N, Guoyou G, Gang Y, Zhe S, Jinghong D, Jiamin Z, Yichun L and

Jia-Hong Y 2014 Adv. Mater. Res. **833** 61

[3] Gary C, Goosey, Marti 2009 Materials used in manufacturing electrical and electronic products n Environmental Science and Technology, 27 (Electronic Waste Management), 40-74

[4] Hanafi Z M, Hammad F F and Youssef A M 2009 Preperation and effects of temperature on nanocrystalline ITO particles synthesized by CO-precipitation with different SnO_2 content *Canadian Journal of Pure and Applied Sciences* **3 2** 839-845

[5] Psuja P, Hreniak D, Strek W 2009 The influence of sintering temperature and Sn^{4+}concentration on electrical and optical properties of ITO nanocrystallites, *Journal of Physics: Conference Series* **146** 012012

[6] Hadj Tahar R B, Ban T, Ohya Y and Takahashi Y1997 *J. Appl. Phys.* **83/4** 2139

Formation Structural Phase Gradients in Rail Steel During Long-Term Operation

Yu F Ivanov[1,2,3]*, K V Morozov[4], O Peregudov[4], V E Gromov[4], N A Popova[5,6], E N Nikonenko[2,5]

[1] High Current Electronics Institute SB RAS, 2/3, Academicheskii Ave., 654052, Tomsk, Russia
[2]National Research Tomsk Polytechnic University, 30, Lenin Ave., 634050, Tomsk, Russia
[3]National Research Tomsk University, 36, Lenin Ave., 634050, Tomsk, Russia
[4]Siberian State Industrial University, 42, Kirov Str., 654007, Novokuznetsk, Russia
[5]Tomsk State University of Architecture and Building, 2, Solyanaya Sq., 634003, Tomsk, Russia
[6] Institute of Strength Physics and Materials Science SB RAS, 8/2, Akademicheskii Ave., 634021, Tomsk, Russia

*E-Mail: yufi55@mail.ru

Abstract. The paper presents results of the structural and phase analysis of the surface layer composition in the type R65 rail steel in its original state and after long-term operation. It is shown that long-term operation of rail steel is accompanied by its structural and phase modification at a depth of not less than 2 mm. The structural elements are detected that can be stress concentrators.

1. Introduction

The reliable operation of railways mostly depends on the quality and durability of rails. The increase of traffic loading and working capacity of railways requires the increase of their reliability and durability and improved requirements for hardness, rolling-contact fatigue resistance, and brittle fracture resistance. Oil quenching and tempering according to the State Standard, is currently one of the widespread methods of thermal strengthening of rails. This method allows producing quenched rails with a uniform sorbite or troostosorbite structure that, in turn, allows the choice of the optimum hardness level [1-3]. It should be noted, however, that carbon steels are characterized by a variety of structures the formation of which strongly depends on thermal treatment conditions and defines the material behavior during its operation [2,4-7].

This work aims at the investigation of the structural evolution of the type R65 rail steel that occurs during its long-term operation.

2. Material and methods

The type R65 rail steel was used for investigations the properties and the elementary composition of which is subject to terms of the State Standard. The steel specimens were cut of the original product that was in service (1000 mn t gross tonnage). All specimens demonstrate the increased wear of gauge corner or the inner edge of a rail the structural phase condition of which was analyzed.

Content from this work may be used under the terms of the Creative Commons Attribution 3.0 licence. Any further distribution of this work must maintain attribution to the author(s) and the title of the work, journal citation and DOI.
Published under licence by IOP Publishing Ltd

Thin films were evaluated using the transmission electron microscopy (TEM), while section etching was carried out by metallography techniques (4% solution of nitric acid and alcohol). Thin films were prepared by electrolytical thinning of plates situated on the railhead running surface and at 2 and 10 mm distance from this surface.

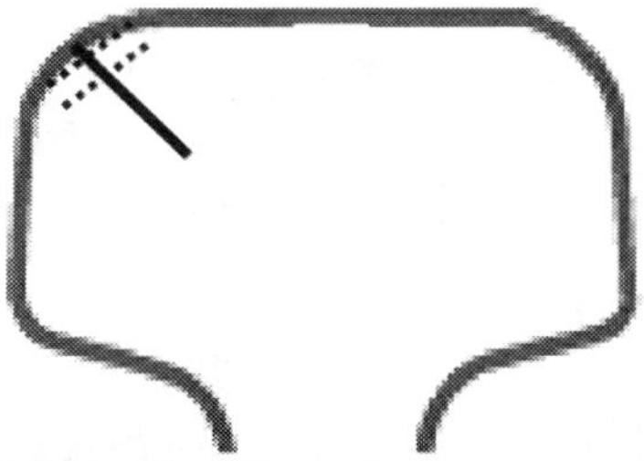

Figure 1. Schematic of specimen preparation for TEM examination. Solid line – direction of gauge corner; dash lines – layers used for thin films.

3. Results and discussion

A visual examination allows detecting arc-shaped discontinuity flaws in steel specimens the original product of which was in service and has 1000 mn t gross tonnage. These discontinuity flaws are observed at 500 μm depth at a certain angle to the surface (Figure 2 *a*) and are filled with corrosion products.

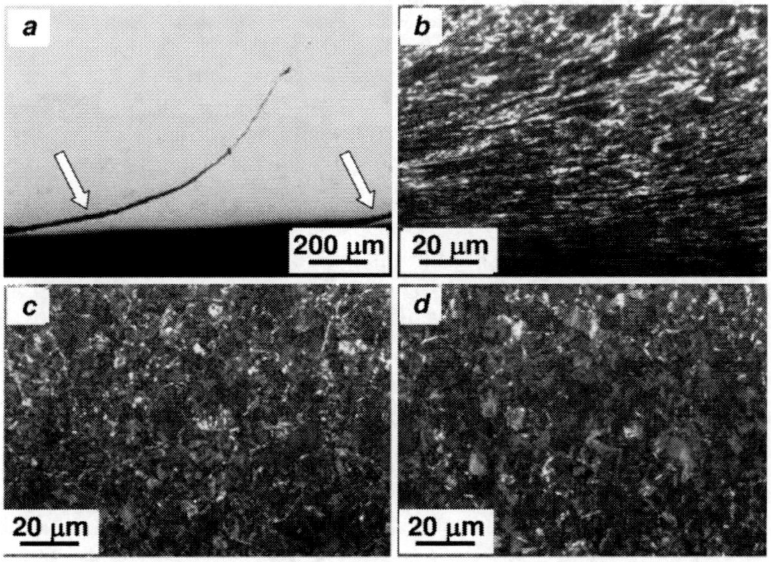

Figure 2. Optical images of gauge corner microstructure (after rail service); *a* – rolling-contact fatigue while in service; *b* – deformation bands on gauge corner surface; *c, d* – steel section microstructure at 2 mm (*c*) and 10 mm (*d*) distance from rail head running surface.

No traces of oxides, brittle-fractured and plastic silicates as well as any defects of metallurgical origin are observed on edges and discontinuity flaw propagations that indicates to the formation of these defects while in operation. Deformation of the metal structure is observed on the surface of gauge corner, grains being oriented at an angle of 30 degrees or less to the surface (Figure 2*b*). The metal microstructure is sorbite after the section etching having the ferrite network fragments (Figure 2*c*). With distance from the center, perlite dispersion insignificantly decreases as well as the amount of ferrite (Figure 2*d*). The grain size in the rail steel is estimated according to the State Standard.

Investigation of the steel failure after the rail operation, shows the formation of the multilayer structure presented in Figure 3a. This layer comprises 20 μm thick surface layer (Figure 3b) and 200-

300 μm thick transition layer (Figure 3a) that transits to the base material. The surface layer has the great number of microcracks, micropores, and shelly spots (Figure 3*b*). The structure of the transition layer is characterized by a relatively small (0.5 ... 1 μm) size of fracture facets (Fig. 3*c*). It is important to note that the interfacial layer that separates the transition layer from base material has no structural defects induced by deformation (micropores, microcracks, etc.

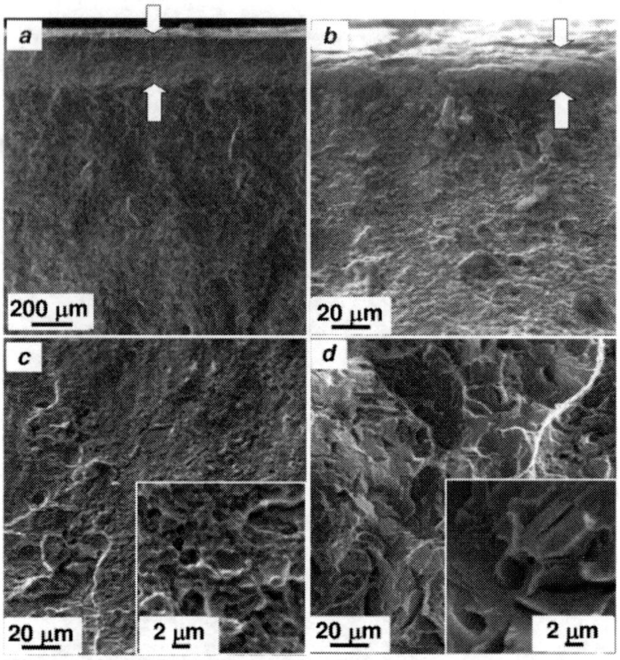

Figure 3. SEM image of multilayer structure: *a* - rail head running surface (upper arrow), transition layer and base material interface (lower arrow); *b* – surface layer; *c* – transition layer; *d* – base material.

TEM investigations of thin films allow detecting the structural phase conditions of the rail steel. As a result, the morphology of the base material at 300 mm distance from the head running surface is detected, namely: lamellar perlite (Figure 4a); grains (regions) of ferritic-carbide mix (Figure 4b). The structure mainly comprises perlite grains the content ratio of which is 0.7 (Figure 4a), while that of ferritic-carbide mix is 0,2; the rest content is free ferrite.

The steel structure significantly depends on its operation. A substructure forms in the surface layer with the fragment size of 500 nm (Figure 5). TEM images show that the obtained layer contains a large number of the second phase inclusions (iron carbides and oxides).

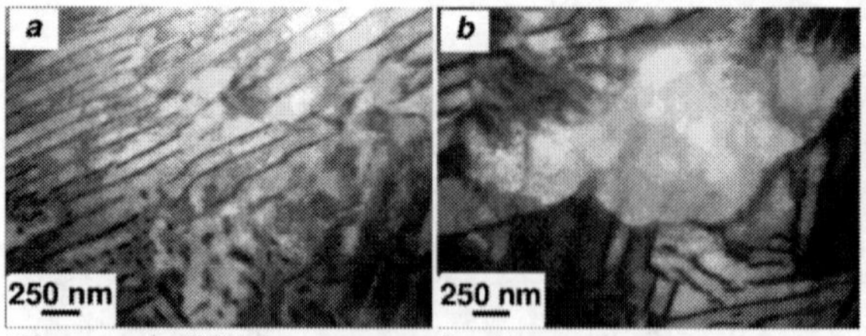

Figure 4. TEM images of rail steel structure: *a* – lamellar perlite; *b* – ferritic-carbide mix.

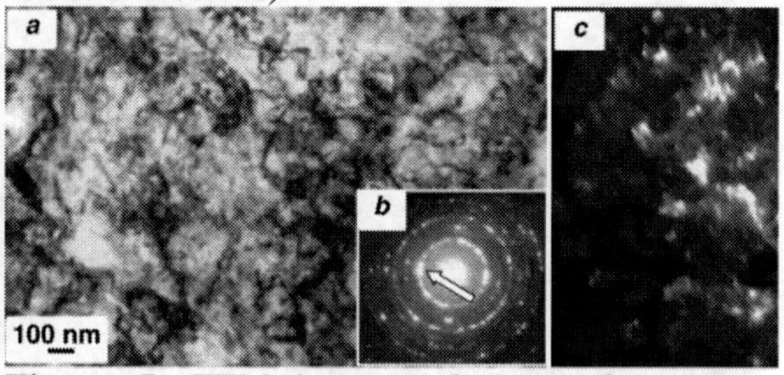

Figure 5. TEM images of the surface layer structure while in service; *a* – bright field; *b* – X-ray diffraction pattern (arrow indicates the dark

field reflection); *c* – dark field obtained in α-Fe reflection [110].

At a distance of 2 mm, the steel structure is similar to that of the base material after its operation. The difference is in the degree of the material deformation. Perlite grains are fragmented and a great number of extinction bend contours is detected (Figure 6a). The regions of free ferrite and ferritic-carbide mix have a higher density of dislocations (Figure 6b).

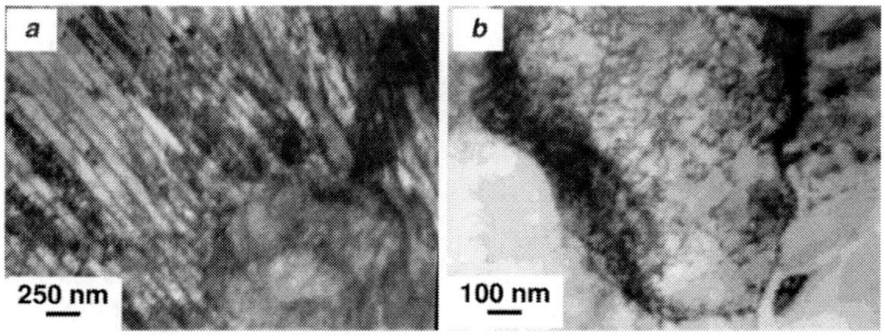

Figure 6. TEM images of base material formed after the rail service. Layer locates at 2 mm distance from the rail head running surface.

4. Conclusion

A study of the structural phase structure of the type R65 rail steel formed as a result of its operation (1000 mn t gross tonnage) was carried by SEM, TEM and optical microscopy investigations. It is shown that the rail operation leads to the formation of a multilayered steel structure. The surface layer 20 μm thick is characterized by micropores, microcracks and shelly spots.

The morphology of the layer located at 2 mm distance from the rail head running surface is similar to that of the base material (steel structure before its operation). The structure is mostly presented by lamellar pearlite, ferritic-carbide mix and free ferrite. They are characterized by dislocation density higher than that of the original state of the structure.

Acknowledgement

This work was financially supported by Grant N 3.295.2014/K from the Ministry of Education and Science of the Russian Federation.

References

[1] Nesterov D K, Razinkova N N and Chernyakova L E 1991 *MiTOM* **4** 45 (in Russian)
[2] Bernstein M L and Rahshtadta A 1983*Metallurgy and heat treatment of steel* (Moscow: Metallurgy) (in Russian)
[3] Vorozhischev V I 2008 *The Composition and Production Technology of High Performance Rails* (Novokuznetsk: Novokuznetsk Printing Plant) (in Russian)
[4] Kurdiumov V G, Utevsky L M and Entin R I 1977 *Transformation in Iron and Steel* (Moscow: Nauka) (in Russian)
[5] Tushynskiy L I, Bataev A A and Tikhomirova L B 1993 *Pearlite Structure and Strength of the Structural Steel* (Novosibirsk: TO Science) (in Russian)
[6] Tushynskiy L I 1990 *Theory and technology of hardening metal alloys* (Novosibirsk: Nauka) (in Russian)
[7] Gromov V E, Yuriev A B, Morozov K V and Ivanov Yu F 2014 *The Microstructure of Hardened Rails* (Novokuznetsk: Inter-Kuzbass) (in Russian)

The preparation of the Ti-Al alloys based on intermetallic phases

N Kosova[1,2,*], V Sachkov[1,3,4], I Kurzina[1], A Pichugina[1,3], A Vladimirov[1], L Kazantseva[1] and A Sachkova[3]

[1] National Research Tomsk State University, 36, Lenina Ave., Tomsk, 634050, Russia
[2] Siberian Research Institute of Agriculture and Peat, Gagarina Str., Tomsk, 634050, Russia
[3] National Research Tomsk Polytechnic University, 30, Lenina Ave., Tomsk, 634050, Russia
[4] Institute for Problems of Chemical & Energetic Technologies of the Siberian Branch of the Russian Academy of Sciences, 1, Socialisticheskaya Str., Biysk, 659322, Altai Region, Russia

[*] E-mail: kosovanatalia@yandex.ru

Abstract. This article deals with a method of obtaining materials in the Ti-Al system. Research was carried out in accordance with the phase diagram of the system state. It was established, that both single-phase and multiphase systems, containing finely dispersed intermetallic compositions of phases Ti_3Al, TiAl and $TiAl_3$, are formed. Additionally, it was found that the pure finely dispersed (coherent-scattering region (CSR) up to 100 nm) intermetallic compound $TiAl_3$ is formed at molar ratio of Ti:Al = 1:3. Experimentally proved the possibility of produce the complex composition of alloys and intermetallic compounds and products based on them.

1. Introduction

The titanium and titanium alloys are widely used in industrial production due to their properties such as: prevalence in nature, low density, high specific strength, corrosion and heat resistance.

It is known, that aluminum is the main alloying element for titanium that is associated with the ability to increase the temperature of titanium allotropic transformation and to enhance the solubility of isomorphous and eutectic-forming β-stabilizers.

The alloys obtained by aluminum doping have the significant advantages over The other alloying components. The density of the aluminum is much less than the density of titanium, so the addition of aluminum increases the specific strength of alloys. Moreover, under enhancing of aluminum content in alloys the heat resistance of the alloys and the elastic modulus are increased, and the tendency to hydrogen embrittlement is decreased. In addition, it makes titanium alloys more sensitive to the salt corrosion and it reduces their technological plasticity [1].

The system Ti-Al is the basis system of the many titanium alloys due to its high functional properties. This system is widely used in aerospace, chemical and petrochemical industries. Alloys of Ti-Al system, based on intermetallic compounds, are more perspective materials for the engines of new generation. The layered metal-intermetallic titanium-aluminum composites, including

Content from this work may be used under the terms of the Creative Commons Attribution 3.0 licence. Any further distribution of this work must maintain attribution to the author(s) and the title of the work, journal citation and DOI.
Published under licence by IOP Publishing Ltd

intermetallic phases, are particularly attractive for industries and scientists. It is related to the fact that these composite materials are able to operate at high temperatures and mechanical loads [2, 3].

Depending on the temperature of the intermetallic interlayers formation under the thermo-power influences on the original layered titanium-aluminum the composite aluminide layers can be produced. They can be singlphasic or biphasic, consisting of an aluminum matrix and the ordered particles of the thermodynamically probable of the aluminide $TiAl_3$ [4].

At the different temperatures of the intermetallic interlayers formation, under thermo-power influences on the original layered titanium-aluminum composite, various types of aluminide layers can be produced. Such as: singl- and biphasic layers

According to the Ti-Al phase diagram, in the system can be up to seven stable intermetallic phases. The most stable intermetallic phases, that increase the physico-mechanical properties of titanium aluminide, are γ-TiAl, α2-Ti_3Al and γ-TiAl + α2-Ti_3Al [4 - 6]. Sometimes in modern diagrams of binary alloys of titanium – aluminum system, not only four well-established phases α2(Ti_3Al), γ(TiAl), $TiAl_2$ and $TiAl_3$, but also Ti_5Al_{11}, Ti_2Al_5 are present [5].

The presence of thermodynamically stable intermetallic phases in composite materials permits significantly enhances the physical and mechanical characteristics of the systems. The direction, connected with the choice of an optimal method of producing composite Ti-Al system based on intermetallic phases, is important [4, 6, 7]. In traditional metallurgy of titanium its alloys with aluminum are prepared by melting the components or by method of powder metallurgy. The common method for obtaining Ti-Al alloys is aluminothermic reduction by out-of-furnace method [8]. The inability to precisely control the parameters of the process during the melting, the high cost of a large number of thermite additives, which consumes aluminum, low productivity of the process all these components are very important disadvantages of the out-of-furnace metallurgy.

The examined method of obtaining the alloys of refractory metal is to compact the mixture of powders of metal hydrides with further dehydrogenation [9, 10]. On the other hand this method has some disadvantages which are connected with the technological peculiarities and the high energy consumptions.

In this paper, a new method of producing complex composite materials of Ti-Al system is proposed. This method allows to prepare the systems with predetermined composition and with thermodynamically stable intermetallic phases of Ti-Al system, according to the phase diagram of the binary system. The preparation of the materials consists of three consecutive stages: obtaining of the titanium hydride; pressing the powders of hydride and aluminum; kilning at the predetermined temperature.

This article presents the data on the obtaining of samples of alloys of Ti-Al under varying the weight ratio of titanium and aluminum (from 55.83 to 100 wt. % Al). The purpose of this work is to obtain the new materials of Ti-Al system, investigate its structure and phase composition.

2. Materials and Methods

The titanium hydride was produced from titanium sponge (CAS 7440-32-6). A weighed amount of the titanium was placed in a quartz boat and heated in a furnace (Nabertherm RS 120/750/13) in the hydrogen stream. The heating rate of the furnace was 10°C / min to 375-450°C with a hydrogen volume flow 500-800 ml/min. This sample was held for 0.5-1.5 h at this temperature, then it was heated to 800-1050 C at the hydrogen volume flow 1000-2000 ml/min, after that it was cooled down to room temperature. The obtained titanium hydride was mixed with a nano-dispersed aluminium powder (according to the manufacturer, the average size of AlNP particles was 115 ± 10 nm, specific surface area – 19.4 ± 3 m^2/g, loading of active aluminium – 80.8 ± 0.6 %) and pressed (HERZOG TP-60) under a pressure 10-20 tones/sm^2. A round plate with a diameter of 20 mm and a thickness of 2 mm was formed. These samples were exposed to the thermo-programmed heat-treating (800-1050 °C) in the quartz boat.

The samples were obtained in a vacuum system. X-ray analysis of the titanium-based composites were carried out with diffractometer Rigaku Miniflex 600 with CuKα-radiation in the range of 10°-90°

(2θ) with a step scan 0.02° and the rate of registration - 2 deg/min. Identification of the diffraction peaks, calculation the areas of coherent scattering (CSR) were carried out via JSPDS database.

3. Results and Discussions

A series of samples, under variation of the ratio of the elements, were obtained by the method described above (Table 1). The choice of components and composition of the mixture corresponds to the phase diagram of Ti-Al.

Table 1. The composition of the alloys of the system Ti-Al

Sample	Mixture, wt.%		Molar relation in mixture
	Ti	Al	
1	100	0	Ti
2	44.17	55.83	5Ti : 11Al
3	36.72	63.28	Ti : 3Al
4	34.31	65.69	3Ti : 10Al
5	23.5	76.5	2Ti : 11Al
6	0	100	Al

Figure 1 (Ti-Al phase diagram) shows which phases are to be expected at equilibrium for different combinations of aluminium content (expressed in atomic %) and temperature (in °C). For farther investigation and for synthesis were chosen composition that marked at the Figure 1: samples N1, 2, 4, 5, 6 were exposed to annealing at temperature of 1050° C, sample N3 – at 800 °C.

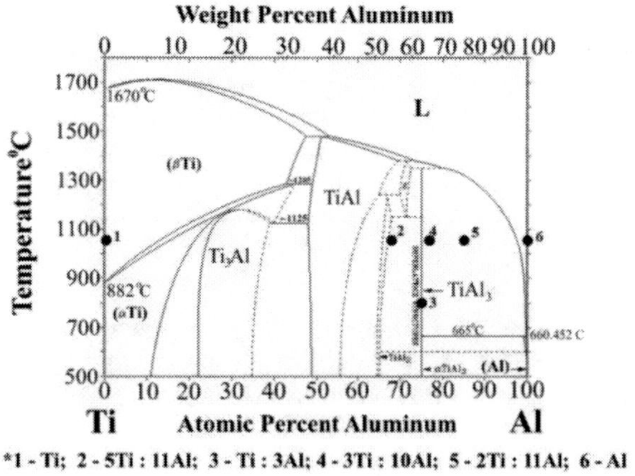

Figure 1. Ti-Al phase diagram. The synthesized systems: *1* – Ti, T_{an} =1050 °C; *2* - 5Ti: 11Al, T_{an} =1050 °C; *3* - Ti: 3Al, T_{an} =800 °C; *4* - 3Ti: 10Al, T_{an} =1050 °C; 5 - 2Ti: 11Al, T_{an} =1050 °C; *6* – Al, T_{an} =1050 °C.

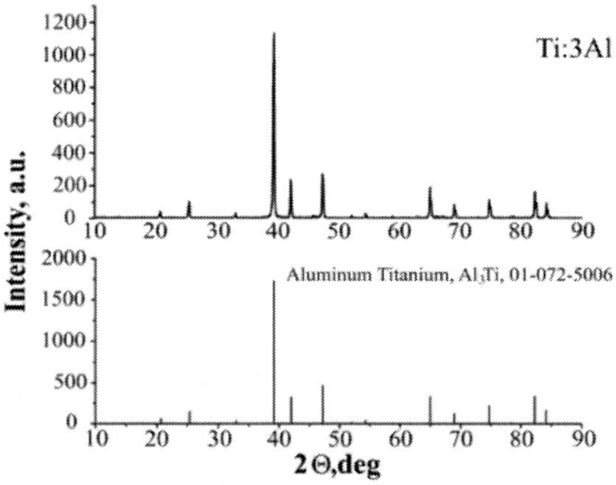

Figure 2. X-ray diffraction of Ti-Al alloy at the molar ratio Ti : 3Al. T_{an} = 800 °C.

Using this method the sample was obtained in a molar ratio Ti: 3Al. $TiAl_3$ compound was obtained under synthesis conditions described earlier, and a yield was about 100% (Figure 2). According to the X-ray phase analysis it is a single-phase product $TiAl_3$ (PDF № 01-072-5006) with dimetric system and lattice parameters: a = b = 3.8503 nm and c = 8.6091 nm. CSR calculation showed that there are particles with size from 39 to 74 nm in this sample.

According to the data presented on Figure 3 the sample with molar ratio Ti:Al=3:10 is a mainly (99 %) stable AlTi3 phase presented (PDF № 00-052-0859). It has dimetric system and lattice parameters: a = b = 5.7968 nm, c = 4.6567 nm. The CSR are from 8 to 24 nm.

The sample that was synthesized at a molar ratio Ti:Al=2:11, according to XRD, contains the next mixture of phases: 16 wt. % $TiAl_3$ (PDF number 01-079-5718) with dimetric system (a = b = 3.7994 nm and c = 8.4566 nm), 18 % mass. Ti_3Al (PDF number 00-052-0859) with hexagonal system (a = b = 5.8083 nm and c = 4.6595 nm), 64 wt. % Ti (PDF number 00-044-1299) with hexagonal system (a = b = 2.9703 nm and c = 4.6837 nm) and 2 % mass. $AlTi_2$ (PDF number 00-029-0095 with hexagonal system (a = b = 2.9947 nm, c = 13 5861 nm). At the molar ratio Ti:Al=2:11 the sample is presented as the mixture that contains: 58 wt. % $TiAl_2$ (PDF № 01-072-9142) with tetragonal system (a = b = 3.7994 nm, c = 24.2955 nm), 40 wt. % Ti_3Al_5 (PDF № 03-065-9789) with orthorhombic system (a = 4.0322 nm, b = 3.9135 nm, c = 4.0483 nm) и 2 wt. % (PDF № 01-075-9784) with hexagonal system (a = b = 2.9905 nm, c = 13.5660 nm). It should be noted that the all phases correspond to the state diagram. They are characterized by the parameters of the crystal lattice according to the theoretical values. According to the analysis of XRD spectra it was show that the all phases are characterized by CSR to 100 nm and the low degree of stress. This fact shows that by this method the difficult composition, containing dispersed intermetallic phases of Ti-Al system, could be obtained. The presence of Ti_3Al and TiAl is associated with thermodynamics of the phases formation. The formation of these phases is characterized by a minimum of Gibbs energy [11].

Figure 3 demonstrates the distribution the phase composition of the Ti-Al system by varying the mass content of aluminum and the annealing temperature.

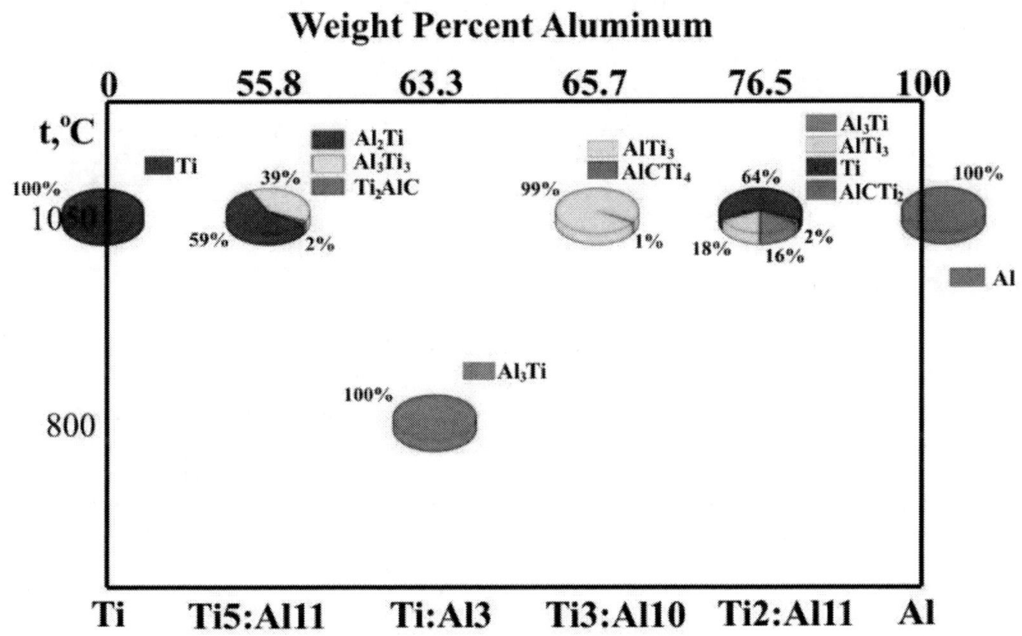

Figure 3. Phase composition of the Ti-Al system (Weight present aluminum) at different mass content of aluminum and annealing temperature.

As Figure 3 illustrates different systems are formed at the various content of aluminum (in wt.%). So, when the content of aluminum to 60 wt. % mainly the phase Ti_3Al is formed; at the 64-66 wt.% - single-phase system; at 80 wt.% – multiphase system. The obtaining these phases does not converge with the data the phase diagram of Ti-Al. Therewith, the formation of Ti_3Al and TiAl phase composition is the most thermodynamically advantageous, because they have minimal Gibbs energy.

So, the proposed technology of obtaining the alloys from the metals powders could be used for synthesis of materials with specified composition.

4. Conclusion

In current work the new method of obtaining of materials based on the Ti-Al system was suggested. Materials containing from 55.83 to 100 wt.% of aluminum were obtained. The phase composition of the samples was investigated. The formation of intermetallic phases of Ti_3Al, TiAl, $TiAl_3$ was established. Both single-phase and multiphase intermetallic systems were prepared by this method. It was determined that pure finely dispersed (CSR up to 100 nm) intermetallic compound $TiAl_3$ is formed at molar ratio of TiAl=1:3.

The study promotes application of this method to obtain of high strength alloys and functional materials of new generation.

Acknowledgements

This work was supported by the federal target program «Research and development on priority directions of scientific-technological complex of Russia for 2014-2020» (RFMEFI57814X0002).

References

[1] Kurzina I A, Popova N A, Nikonenko E L, Kalashnikov M P, Savkin K P, Sharkeev Yu P and Kozlov E V 2012 *Bulletin of the Rus. Acad. of Sci.: Physics* **76** 64

[2] Fleischer R L, Dimiduk D M and Lipsitt H A 1989 *Ann. Rev. Mater. Sci.* **19** 231

[3] Vecchio K S Synthetic Multifunctional Metallic-Intermetallic Laminate Composites *J. Miner., Met. and Mater.* 2005 **3** 25

[4] Welsch G, Boyer R and Collings E W *Mater. Properties Handbook: Titanium Alloys ASM International*, ed. Lampman S, 1993 p 1176

[5] Appel F, Öhring M, Paul J D H and Lorenz U Proc. of the 2nd Int. Symp. «Structural Intermetallics», eds. Hemker K J et al. 2001 *Minerals, Metals & Mater. Soc.* 63

[6] Advanced Light Alloys and Composites. *Series: NATO Sci. Partnership Sub-Series: 3* 1998 **59** 532

[7] Imayev V M, Imayev R M and Oleneva T I 2011 *Let. Mater.* **1** 25

[8] Huang R, Bai C, Lv X and Liu S 2012 Preparation of Titanium Alloy from Titania-bearing Blast Furnace Slag, in 3rd Int. Symp. on High-Temperature Metallurgical Processing (eds Jiang T, Hwang J-Y, Masset P, Yucel O, Padilla R and Zhou G), (John Wiley & Sons, Inc., Hoboken, NJ, USA) doi: 10.1002/9781118364987.ch10

[9] Swainson I P, Dolukanyan S K, Aleksanyan A G, Shekhtman V Sh, Mayilyan D G and Yonkeu A′ L 2010 *Canadian Journal of Physics* **88** 741

[10] Dolukhanyan S K, Aleksanyan A G, Ter-Galstyan O P, Shekhtman V Sh, Sakharov M K and Abrosimova G E 2007 *Rus. J. Physical chemistry B* **1** 563

[11] Kurzina I A, Kozlov E V, Sharkeev Yu P, Ryabchikov A I , Stepanov I B, Bozhko I A, Kalashnikov M P, Sivin D O and Fortuna S V 2007 *Surf. Coat. Technol.* **201** 8463

IOP Conf. Series: Materials Science and Engineering **112** (2016) 012040 doi:10.1088/1757-899X/112/1/012040

Study of rhenium production process by hydrogen reduction of rhenium hexafluoride

V I Vybyvanets, A V Kosukhin[1], A V Cherenkov and G S Shilkin
Federal State Unitary Enterprise "Scientific Research Institute Scientific Industrial Association "LUCH", Zheleznodorozhnaya Str. 24, Podolsk, Moscow Region, 142100, Russia

[1]E-mail: kosuhin_artem@mail.ru

Abstract. Study of heterogeneous reduction of rhenium hexafluoride with hydrogen was carried out. The dependence between the rate of rhenium deposition from mixture of rhenium hexafluoride and hydrogen and temperature and rhenium hexafluoride concentration has been studied. A study of structure of various rhenium precipitates was carried out. Microhardness of rhenium after deposition, and after one hour annealing at temperatures 1200 and 1800 °C was carried out.

1. Introduction

Gaseous rhenium hexafluoride reacts with hydrogen as follows:

$$ReF_{6(g)} + 3H_{2(g)} \rightarrow Re_{(s)} + 6HF_{(g)}$$

Thermodynamic analysis shows that the process proceeds with heat liberation (~ 288 kJ / mol), and it becomes possible even at room temperature.

The minimum temperature at which the rhenium metal is obtained inside the copper tube (hot-well reactor) is 200 °C under atmospheric pressure [1,2,3]. The bulk of the metal was in powder form. With increasing temperature, the reaction has homogeneous-heterogeneous nature, a compact layer of rhenium forms together with the powder on the surface of the substrate.

Minimum powder formation is observed at 250 °C and ratio of H_2: ReF_6 = 25. These conditions provide a 95-100% conversion of ReF_6 to metal, including 70-75% - to compact. The density of the latter with thickness of 0.25-0.5 mm is 87-89% of the theoretical and the average microhardness (under a load of 1 ~ N) is 9100 MPa. The decrease of the system pressure to 267 Pa and the increase of ratio H2: ReF6 to 50-400 [4,5] do not result in formation of rhenium precipitates with density close to the theoretical value.

Plastic coatings with thickness 0.025-0.050 mm inside tube with diameter 19 mm and length 250 mm were obtained at a temperature of 600 °C, ratio of H2: ReF6 = 12.5, pressure of 1.3 kPa and flow rate of the gas mixture of 270 cm^3 / min. Rhenium deposition rate under these conditions is 0.013 mm / h and completeness of ReF_6 recovery is 25%.

The coating has a columnar structure. X-ray studies showed that absence in it of any structure.

In [6,7] the dependence of rhenium deposition rate on the content of rhenium hexafluoride in gas mixture and deposition temperature was studied. Deposition of rhenium was carried out in the temperature range of the substrate (400-1500 °C) at a total pressure of 133 Pa, hydrogen flow rate of $5*10^{-7}$ m^3 / s and rhenium hexafluoride flow rate $(0.3\text{-}1.1)*10^{-7}$ m^3 / s. Kinetic data at different temperatures are placed on one straight line that indicates on diffusive character of the reaction. Investigation of the effect of the substrate temperature on the deposition rate shows that there is a

Content from this work may be used under the terms of the Creative Commons Attribution 3.0 licence. Any further distribution of this work must maintain attribution to the author(s) and the title of the work, journal citation and DOI.
Published under licence by IOP Publishing Ltd

weak dependence of the rhenium deposition rate on the temperature. At high temperatures, a large role is played by homogeneous nucleation.

Kinetic regularities of heterogeneous recovery of rhenium hexafluoride by hydrogen have not been investigated in these studies.

The purpose of this work is the study of heterogeneous recovery of rhenium hexafluoride by hydrogen.

2. Experimental part

Study of heterogeneous reduction of rhenium hexafluoride with hydrogen was carried out using the apparatus (figure 1) consisting of a reaction chamber (1), rhenium hexafluoride feeder (5), bottle gas purifier (8), gaseous components dosing units (3, 4), condensers (6). In cases when the experiments are carried out at lowered pressure, a forepump (9) and optical manometer (10) for pressure measurement inside the chamber.

The reaction chamber (figure 2) is a vertically arranged cylinder with water cooled jacket (1) and two removable flanges (4) which also function as mixers. Copper current leads were electrically insulated from the reaction chamber with fluoroplastic and rubber seals. To counterbalance substrate temperature elongation, one of the current lead was fitted with a movable electric contact. Hydrogen was purified from oxygen, moisture and dust particle impurities using a bottle gas purifier (dew point 70 °C).

Deep purification of rhenium hexafluoride was done using highly efficient rectifying columns [8].

Rhenium hexafluoride and hydrogen were dosed using a flow meter including a divided capillary and a differential manometer. The gas mixture consisting of rhenium hexafluoride and hydrogen were fed through the top of the reaction chamber. Unreacted rhenium hexafluoride and hydrogen fluoride produced by the reaction were collected in two tandem condensers cooled by liquid nitrogen.

Pressure in the system was maintained within (1÷1.5) kPa and measured with an optical manometer.

The average composition of the gas mixture fed into the reaction chamber was calculated according to known hydrogen consumption and loss of weight of the rhenium hexafluoride containing vessel. Temperature was measured with chromel-alumel thermocouple. Copper pipes with diameter 3 mm were used as substrates.

Rate of rhenium deposition was determined by change of precipitate diameter before and after the process.

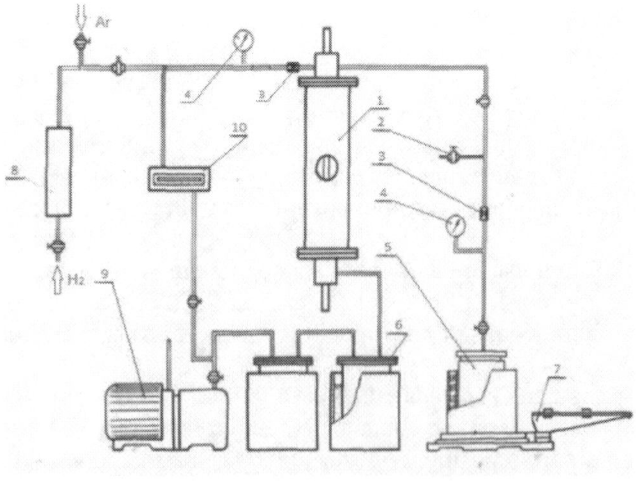

Figure 1. Diagram of the apparatus for rhenium precipitates production: 1 – reaction chamber; 2 – valve;

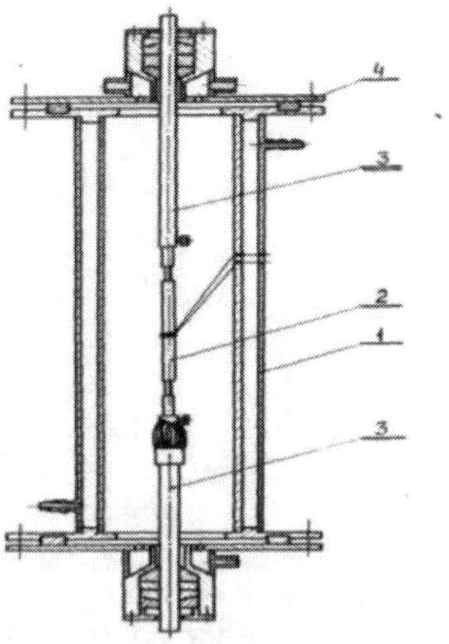

Figure 2. Diagram of the reaction chamber for rhenium precipitates

3 – capillary; 4 – differential manometer; 5 – vessel with ReF_6; 6 – condenser; 7 – scales; 8 – bottle gas purifier; 9 – forepump; 10 – optical manometer.

production: 1 – body; 2 – substrate; 3 – current lead; 4 – flange.

3. Results and discussion

Experimental results of study of temperature dependence of rhenium deposition obtained in this work are presented in figure 3.

As it can be seen from the data presented, the initial temperature of rhenium deposition is 350 °C, which is higher, than the temperature obtained in another work [1]. This result is probably related to the fact that the authors of the aforementioned work activated the gas mixture by heating it (rhenium deposited inside a tubular substrate). As the temperature rises above 350 °C, rhenium deposition rate increases exponentially, however at temperature over 450 °C reaction between rhenium hexafluoride and hydrogen proceeds throughout the whole volume with formation of fine-dispersed powder.

The results presented in figure 3 were obtained for rhenium hexafluoride concentration in gas phase equal to 0.015 molar fraction. Increase of rhenium hexafluoride content in gas phase results in rhenium deposition rate growth without changing its functional dependence; however that leads to decrease of volume reaction initiation temperature.

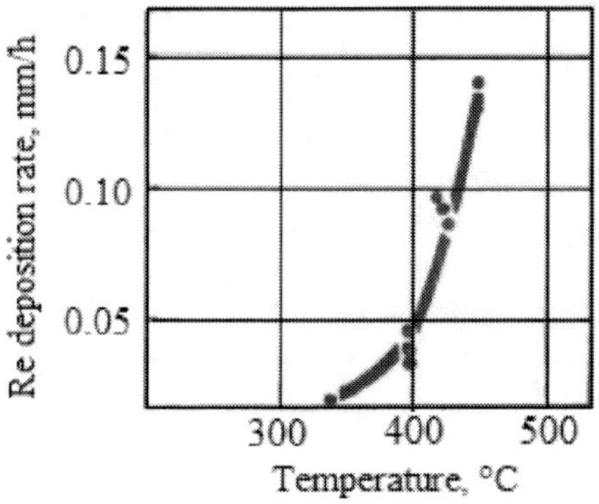

Figure 3. Dependence of rhenium deposition rate from substrate temperature

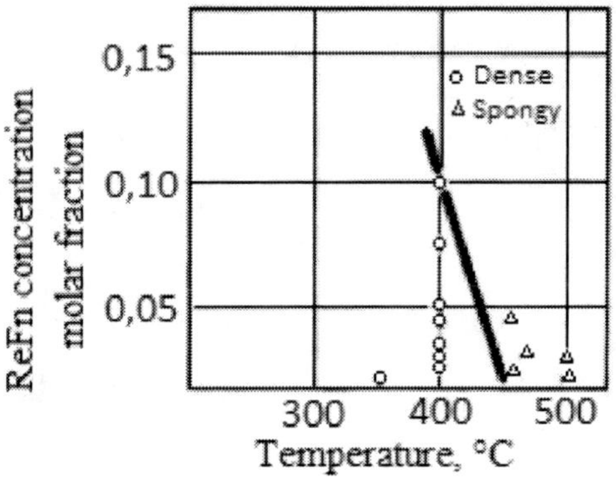

Figure 4. Conditions of formation of various rhenium precipitates depending on the parameters of the process

To obtain high-quality rhenium precipitates on large diameter substrates it is necessary to increase the gas flow velocity, and at higher temperatures the gas mixture should be fed in such way, that it would get simultaneously to entire reactive surface of the substrate (e.g. gas mixture feeding through a muzzle towards the whole generatrix of the rotating product).

The structure of rhenium precipitates was studied using metallographic slices with 10÷500-fold magnification, and the microhardness was measured using PMT-3 device with 100 g load.

The analysis of precipitates structure has shown that three types of precipitates form within the studied intervals of temperature and rhenium hexafluoride concentration in gas phase. These precipitates types are columnar, laminated and loose spongy. The conditions of these precipitates formation are presented in the figure 4.

In the area of relatively low temperatures and rhenium hexafluoride concentrations dense columnar precipitates form (figure 5a). Increase of the temperature results in formation of laminated precipitates (figure 5b). Further increase of the aforementioned parameters is attended by the reaction proceeding throughout volume and at the near-surface layer of the substrate with simultaneous heterogeneous reaction running on the substrate as well. These conditions result in formation of loose spongy precipitates which are a mixture of powder and solid rhenium. (figure 5c).

IOP Conf. Series: Materials Science and Engineering **112** (2016) 012040 doi:10.1088/1757-899X/112/1/012040

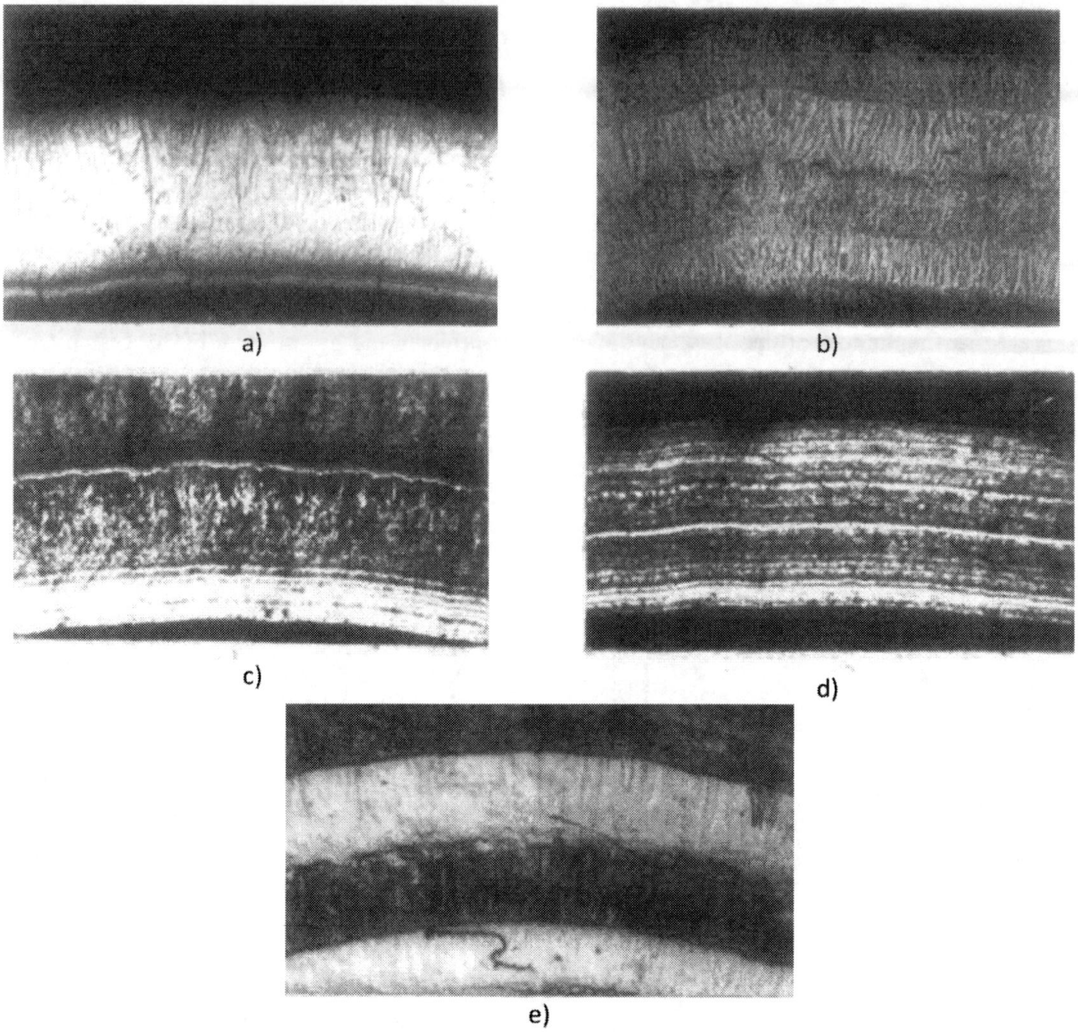

Figure 5. Structure of rhenium precipitates (T = 400 °C, C_{ReF6} = 0.02 molar fraction): b) T = 420 °C, C_{ReF6} = 0.015 molar fraction; c) T = 450 °C, C_{ReF6} = 0.05 molar fraction; d) T = 450 °C, C_{ReF6} = 0.02 molar fraction; d) T = 400 °C, C_{ReF6} = 0.05 molar fraction

Laminated precipitates form within the dense-to-spongy structure transition area. On the one hand, in this area homogeneous reaction starts to run, while a slight change of the process parameters results in prevalence of heterogeneous or homogeneous reaction. Apparently, that circumstance may explain varying thickness of layers of the laminated precipitates (figure 5d). At the same time, laminated precipitates with greater thickness of layers form inside the area where dense precipitates crystallize (Fig. 5e). Such lamination is related to instability of the process parameters. Therefore, production of dense precipitates with homogeneous structure requires automatic maintenance of the parameters at a preset level.

Initially, microhardness of obtained rhenium precipitates was (600-650) kg/mm^2. After one hour annealing at 1200 and 1800 °C, obtained levels of microhardness were (400-450) kg/mm^2 and (300-350) kg/mm^2 respectively.

4. Conclusion

The dependence between the rate of rhenium deposition from mixture of rhenium hexafluoride and hydrogen and temperature and rhenium hexafluoride concentration has been studied. It has been demonstrated, that at rhenium hexafluoride concentration 0.015 molar fraction within temperature range (350 – 400) °C the reaction proceeds at the surface of the substrate, while at 400 °C it runs both at the surface and throughout the whole volume. Increase of rhenium hexafluoride concentration in the gas mixture lowers the region where heterogeneous reaction runs.

Conditions of formation of different types of rhenium precipitates were defined. Microhardness of rhenium after deposition – (600 – 615) kg/mm^2, while after one hour annealing at temperatures 1200 and 1800 °C– (400 – 450) kg/mm^2 and (300 – 350) kg/mm^2 respectively.

References

[1] Hoertel F W and Donaldson J C *Electrodeposition and Surface Treatment* 1973/1974 vol 2 pp 327-341
[2] Donaldson J C, Hoertel F W, Cochran A A 1968 *J. Less-Common Metals* **14(1)** 93-101
[3] Hoertel F W, Donaldson J C 1967 *Nucl. Sci. Abstr.* **21(11)** 1919
[4] Weinberg A F, Lindgren J R, Elsener N B, Mills R G 1965 *Nuclear Application* **1(6)** 581-583
[5] Federer J, Schaffhouser A, Leitten C 1965 *U.S. Report of Investigation* 7
[6] Lakhotkin Yu V and Krasovsky A I 1979 *Protective coatings* 50-53
[7] Lakhotkin Yu V and Krasovsky A I 1983 *Metals Magazine (in Russian)* **1** 24
[8] Vybyvanets V I, Kosukhin A V, Cherenkov A V and Shilkin G S 2015 *Proc. of the XVth All-Russian Conf. "Ultra-Pure Substances and Materials. Production, Analysis, Use"*, Nizhny Novgorod, Russia.

IOP Conf. Series: Materials Science and Engineering **112** (2016) 012041 doi:10.1088/1757-899X/112/1/012041

Investigation into Rhenium Fluorination Process

V I Vybyvanets, A V Kosukhin[1], A V Cherenkov and G S Shilkin
Federal State Unitary Enterprise "Scientific Research Institute Scientific Industrial Association "LUCH", Zheleznodorozhnaya Str. 24, Podolsk, Moscow Region, 142100, Russia

[1]E-mail: kosuhin_artem@mail.ru

Abstract. The process of rhenium and fluorine interaction in the interval of temperatures from 200 °C up to 600 °C has been investigated. Influence of fluorine charge, fluorination square and temperatures on the process efficiency and fluorine use efficiency have been determined. Presence of two areas of process behavior, namely kinetic and diffusion has been demonstrated. It was shown that in the course of fluorination a mixture of hexafluorides and heptafluorides was formed. Dependence of rhenium fluorides composition on temperature has been determined

1. Introduction

A promising method of obtaining tungsten-rhenium alloys is vapor deposition, namely joint hydrogen reduction of metal hexafluorides [1,2].

Tungsten hexafluoride is currently produced on an industrial scale. The main technological production method is direct interaction of elemental fluorine with tungsten powder [3].

Information about the fluorination process of rhenium powder is contradictory. The first studies of the interaction of rhenium with elemental fluorine were published in [4]. The authors determined the initial reaction temperature (125 ° C) at which there was formation of a gaseous product and condensates of deep purple and green colors. They concluded that it was rhenium hexafluoride on the basis of chemical analysis of the obtained product. By optimizing the installation of fluorination, the authors of [5, 6] obtained a more pure rhenium hexafluoride due to pre-treatment of fluorine and improvement of technological methods. The obtained compound in the liquid and solid phases had pale yellow color. The melting temperature of ReF_6 (18.8 °C) and dependence of its density on temperature were determined. The authors did not find rhenium heptafluoride in the obtained product.

The authors of [7], exploring the interaction process of rhenium with fluorine, determined that a mixture hexa- and heptafluorides were formed at temperature of 400 ° C. The studies on the effect of fluorination temperature of rhenium on the composition of formed fluorides are given in [8]. It was shown that with increasing temperature the mixture is enriched with rhenium heptafluoride. Attempt of quantitation of rhenium hexa- and heptafluorides and fractional separation of the mixture were unsuccessful.

This work was carried out to study the fluorination process of rhenium metal powder.

2. Plant, Source Materials and Experimental Technique

A technological process of rhenium fluorides synthesis includes: fluorine making, a rhenium powder fluorination, and condensation of the made fluorides (figure 1).

Content from this work may be used under the terms of the Creative Commons Attribution 3.0 licence. Any further distribution of this work must maintain attribution to the author(s) and the title of the work, journal citation and DOI.
Published under licence by IOP Publishing Ltd

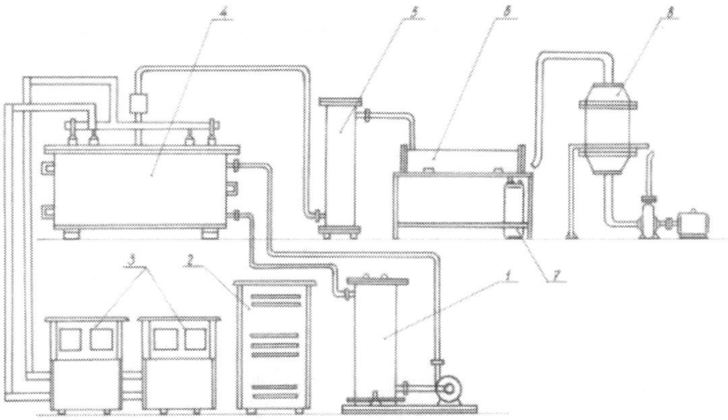

Figure 1. Laboratory-scale plant for rhenium fluorides making: 1 – Thermostat; 2 – Autotransformer; 3 – Rectifier; 4 – Electrolyzer; 5 – Column; 6 – Fluorination Reactor; 7 – Capacitor; 8 – Filter.

Fluorine was produced by electrolysis of potassium bifluoride melt. Fluorine was purified from fluoric hydrogen by its passing through the grained fluoric sodium heated up to 100 °C. Rhenium fluorination was performed in a horizontal cylindrical reactor made of nickel with 100 mm in diameter and 1,000 mm in length (figure 2).

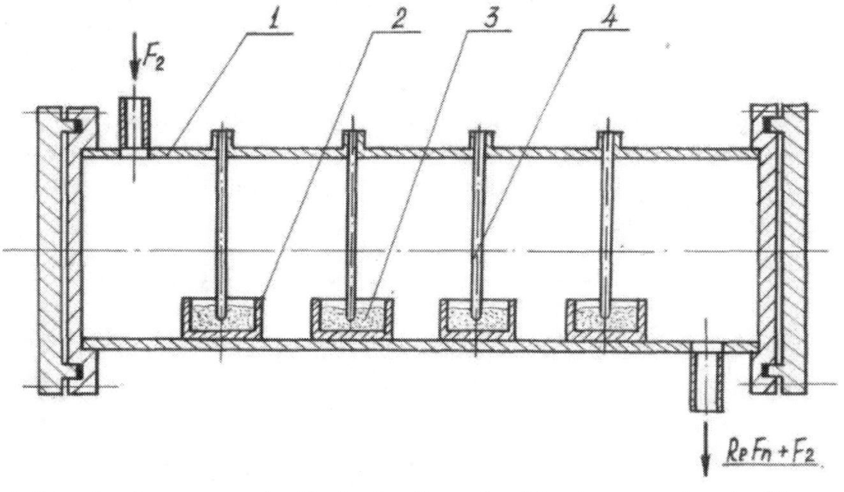

Figure 2. Fluorination reactor 1 – Body; 2 – Boat; 3 – Rhenium powder; 4 – Thermocouple cover

Fluorine charge was specified by current loading on an electrolyzer and changed from 50 up to 200 g/hour (figure 3).

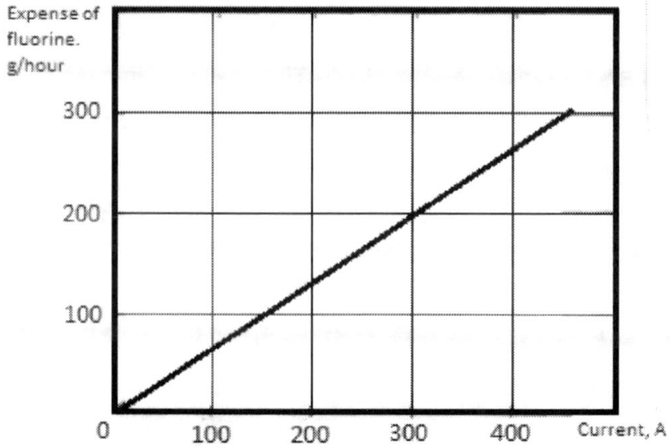

Figure 3. Dependence of fluorine charge on current loading on electrolyzer

The process temperature was controlled by changing the power of the radiator placed outside the fluorination reactor and measured by a chromel-alumel thermocouple placed into a nickel cover. One to four boats having a reaction surface square of ~ 1 dm^2 each were loaded into the fluorination reactor. Formed reaction products were trapped in a tank made of stainless steel cooled by liquid nitrogen. Upon the process completion condensed fluorine was distilled out of the tank, and analysis sample was selected for determination rhenium and fluorine content in it. Powdery metallic rhenium of P-1 TU 48-4-195-87 brand was used as a feed stock. For purifying from oxides and moisture the rhenium powder was exposed to heat treatment in hydrogen at a temperature of 1000 °C for 2 hours.

3. Influence of fluorine charge on fluorination rate and its efficiency

Figure 4 shows dependence of fluorination rate and also output by fluorine on its charge at the process temperatures from 200 to 600 °C and fluorination square of 1 dm^2.

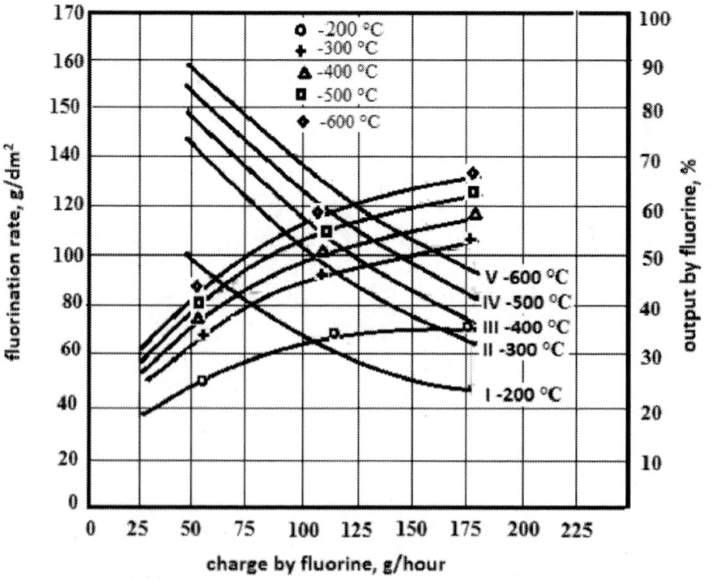

Figure 4. Dependence of fluorination rate output by fluorine on its charge

In the field of low charges of fluorine a fluorination rate varies in proportion to the quantity of fluorine carried. In this case the output by fluorine is 50 % at 200 °C, and 90 % at 600 °C. At increase in the charge of fluorine (more than 75 g/hour) a fluorination rate is slowing down, and this effect occurs in a greater degree at a temperature of 200 °C. At large fluorine charges a fluorination rate is almost constant, and output by fluorine sharply decreases from 90 % to 40 % at 600 °C, and from 45 % to 18 % at 200 °C. The specified decrease in the output by fluorine is connected with a high linear speed and short time of its contact with a rhenium powder.

Analysis of the dependences given in figure 4 shows that at increasing in the fluorination process efficiency by 10 g/hour dm^2 the output by fluorine decreases by 10 % at the average for all range of temperatures investigated.

From the above-stated follows that an optimal mode of the process for the fluorination square of 1 dm^2 is a temperature of 300 – 400 °C and fluorine charge is equal 100-120 g/hour, at which the efficiency of 95 g/hour is reached at the output by fluorine of 50-55 %. Increase in temperature up to 600 °C leads to insignificant increase in the process efficiency and output by fluorine, but in this case the fluorination reactor material stability decreases with respect to fluorine.

4. Influence of reaction surface square on process rate and its efficiency

One of the ways to raise the fluorination process efficiency is expansion in a reaction surface. Figure 5 shows dependences of fluorination rates and output by fluorine on the reaction surface at temperatures of 200, 300, 400, 500 and 600 °C and constant charge of fluorine equal to 200 g/hour.

It can be seen from the given data that expansion in the reaction surface from 1 to 4 dm2 leads to decrease in a specific rate of fluorination; however the process efficiency rises from 60 to 130 g/hour for 200 °C and from 80 to 200 g/hour for 600 °C. And for a temperature of 200 °C the most appreciable raise in the process efficiency is observed at expansion of the square of fluorination from 1 to 3 dm2. For temperatures of 300, 400, 500 and 600 °C expansion of the fluorination square leads to uniform raise of fluorination rate.

From the above-stated follows that different character of the process is observed at a fluorination temperature of 200 °C at changing the square of fluorination. At transition from one area of a reaction to another one a fluorination rate sharply changes. This conclusion is also confirmed by dependence of a fluorination rate on the temperature in Arrhenius coordinates.

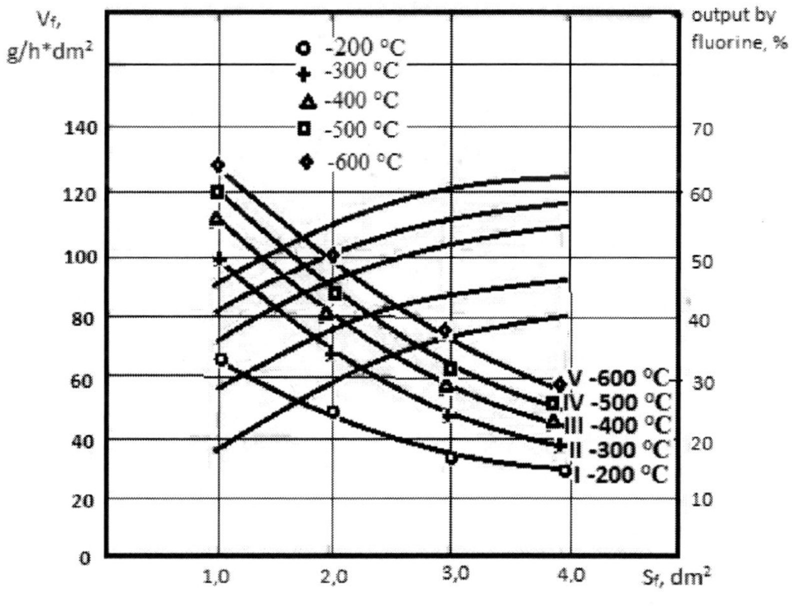

Figure 5. Dependence of fluorination rate and output by fluorine on reaction surface size

Analysis of the dependences given in figure 4 shows that optimal conditions for the process are the following: a temperature of 350-400 °C, the square of fluorination of 4 dm2. At these parameters the fluorination rate of 160-170g/hour is reached at the output by fluorine equal to 45-55 %. At a temperature of 600 °C and fluorination square of 4 dm2 the process efficiency and output by fluorine increase slightly in comparison with normal conditions, 200 g/hour and 62 % respectively.

Based on experimental dependences of fluorine charge and reaction surface square influence on a fluorination rate which are given in figures 4 and 5, a dependence of a fluorination rate and output by fluorine on the process temperature was made at constant fluorine charge of 200 g/hour (figure 6).

From the results given in figure 6 it can be seen that influence of process temperature on the fluorination rate at various reaction surfaces is not the same. So, at the rhenium powder square of 1 dm2 up to a temperature of 300-350 °C a proportional increase in the fluorination rate can be observed, and above 350 °C – insignificant increase. These dependences are less typical at a reaction surface of 2 and 3 dm2, and they are practically absent at a reaction surface of 4 dm2. Processing of the integrated data on a rhenium fluorination rate in Arrhenius coordinates given in figure 7 demonstrates a different character of the process.

Transition from one area of the process to another is observed at the temperature change, and at expansion of the reaction surface a transition point on the curves moves to the area of lower temperatures in which a dependence of the fluorination rate on temperature becomes more apparent, and a limiting factor of the process is chemical interaction of rhenium powder with fluorine. The activation energy in this area is E = 25 kcal/mole.

In the area of higher process temperatures (300 °C) the fluorination rate slowdown can be observed that is typical for transition from a kinetic to diffusion area. The temperatures of the process transition from a kinetic area to diffusion one at different fluorination squares and fluorine charge of 200 g/hour, which were determined in accordance with the data in figure 7, are given below.

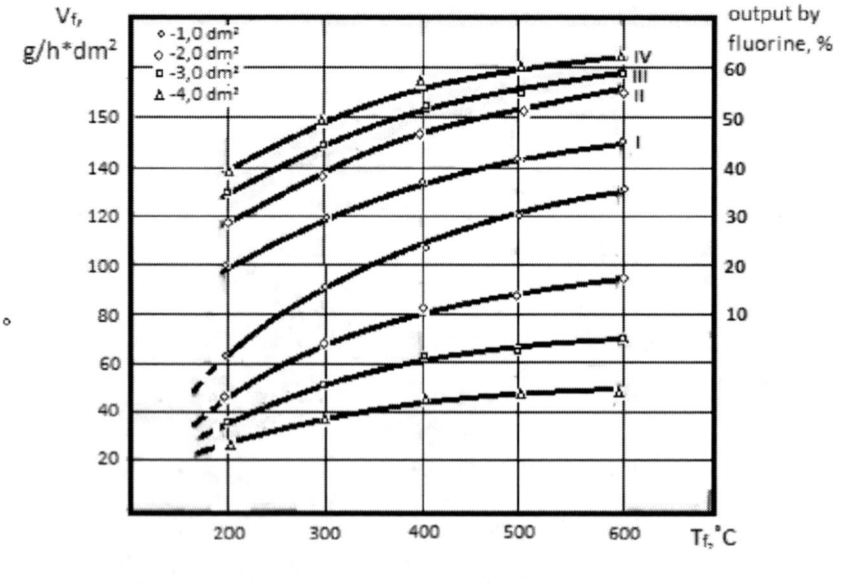

Figure 6. Dependence of fluorination rate and output by fluorine on process temperature

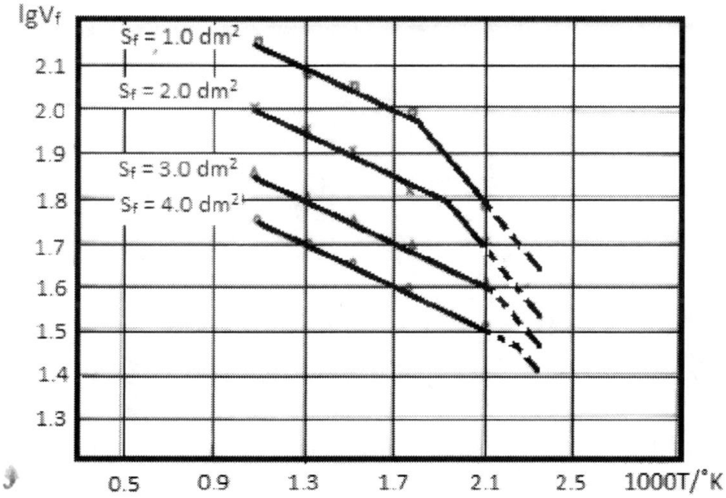

Figure 7. Dependence of fluorination rate on temperature

Table 1 - The temperatures of the process transition from a kinetic area to diffusion one at different fluorination squares

S, dm^2	1	2	3	4
T tr, °C	280	240	190	160

The curve made in accordance with the presented data (figure 8) separates the process areas and demonstrates different areas of the process going.

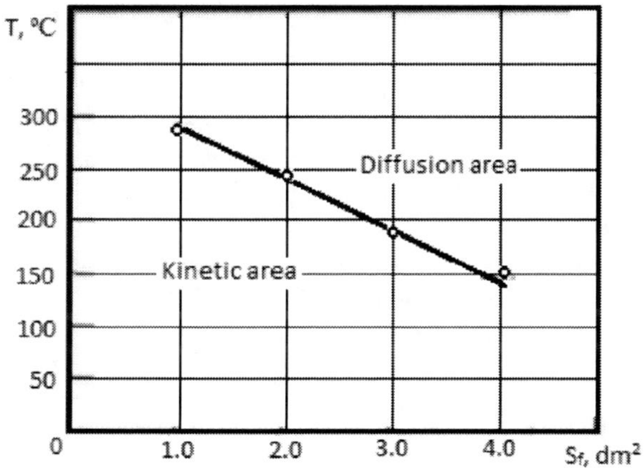

Figure 8. Different areas of process going

5. Determination of rhenium fluorides chemical composition

Rhenium fluorides were analyzed for rhenium and fluorine content in them under a procedure of specifications TU 2134-035-076.22928-2003; according to them rhenium was determined by a gravimetric method, and fluorine – by an ionometry method.

The gained analysis results showed that rhenium fluoride in all samples consisted of a mixture of hexafluorides and heptafluorides (figure 9). Increase in the fluorination temperature leads to more rhenium content in the mixture of heptafluorides.

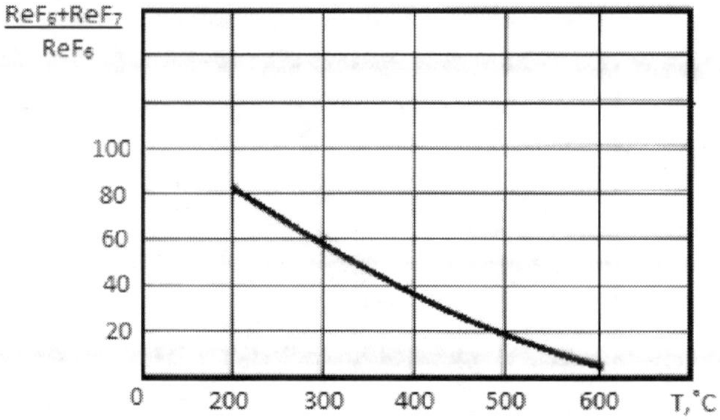

Figure 9. Dependence of ReF_6 monofraction content in mixture on process temperature

6. Conclusion

The process of rhenium and fluorine interaction in the interval of temperatures from 200 °C up to 600 °C has been investigated.

Influence of fluorine charge, fluorination square and temperatures on the process efficiency and fluorine use efficiency have been determined. The maximal process efficiency was 210g/hour at a temperature of 600 °C and fluorine charge of 200 g/hour; in this case the fluorine use efficiency was 60%. Apparent activation energy of the process was 25kcal/mole.

Presence of two areas of the process going and areas of their existence has been demonstrated. Dependence of rhenium fluorides composition on temperature has been determined. It was shown that in the course of fluorination a mixture of hexafluorides and heptafluorides was formed. Rise of the process temperature leads to increase in rhenium heptafluoride content in the mixture and on the contrary temperature decrease leads to increase in hexafluoride content.

References

[1] Korolev Y M, Stolyarov V I 1981 *Reduction by hydrogen of fluorides refractory metals* (Moscow: Metallurgy)
[2] Lahotkin Y V, Krasovsky A I 1989 *Tungsten-rhenium coatings* (Moscow: Science)
[3] Tungsten hexafluoride (technical). Technical specifications 2154-035-07622928-2003.
[4] Ruff O, Kwasnik N 1932 *Zeitshzift fur anarganische und allgemeine chemie* vol 209 pp 113-122
[5] Ruff O, Kwasnik N 1933 *Angew. Chemie* **46 (48)** 739
[6] Ruff O, Kwasnik N 1934 *Angew. Chemie* **47 (26)** 480
[7] Malm J, Kwasnik N 1960 *Journal of the American Chemical Society* **82**
[8] Malm J 1961 *Journal Nucl. Chemie* **20** 189

Production and investigation into properties of high-pure rhenium fluorides

V I Vybyvanets, A V Kosukhin[1], A V Cherenkov and G S Shilkin
FSUE "Scientific Research Institute Scientific Industrial Association "LUCH",
Zheleznodorozhnaya Str. 24, Podolsk, Moscow Region, 142100, Russia

[1]E-mail: kosuhin_artem@mail.ru

Abstract. Two products of synthesis were obtained: synthesis product at a fluorination temperature of 300°C, synthesis product at a fluorination temperature of 600°C. Two source products were separated and rectified during this work. Monofraction of rhenium hexafluorides and heptafluorides were obtained. The temperatures of liquids in the system ReF_6 - ReF_7 were determined. Liquid-vapor equilibrium in the system was investigated. Viscosity and surface-tension at the liquid-to-vapor boundary are measured for liquid ReF_6 and ReF_7.A promising method of tungsten-rhenium alloys making is a gas-phase precipitation, namely joint hydrogen reduction of these metals fluorides.

1. Introduction

Higher volatile rhenium fluorides are of great interest for the production of high- purity precipitates of metallic rhenium and its alloys with tungsten by hydrogen reduction of their fluorides [1]. However, in the technical literature data on the synthesis and physico-chemical properties of rhenium hexa - and heptafluorides practically are not available or are contradictory.

The main technological production method is a direct interaction of elemental fluorine with rhenium powder [2,3]. Studies on the effect of fluorination temperature on the composition of the formed fluorides showed that at low temperatures the mixture enriched with rhenium hexafluoride and at high temperatures vice versa. The attempt to separate a mixture of fluorides by fractional distillation ended unsuccessfully. The studies on the chemical separation of fluorides mixture showed that at the interaction of mixture with fluorine at temperature of 400 °C rhenium heptafluoride is formed, and rhenium hexafluoride is formed at interaction with metallic rhenium.

Quantitative determination of rhenium hexa- and heptafluorides in their mixture was unsuccessful. Chemical analysis method is not allowed to solve this problem due to the insufficient accuracy of concentration measurement of fluorine and rhenium. The study of the spectrum of infrared radiation of fluorides was inconclusive due to the proximity of their energy levels.

The purpose of this work is obtaining and investigation of the properties of higher rhenium fluorides.

2. Phase equilibrium in ReF6-ReF7 system

Liquidus temperatures in the system were measured by a visually- polythermal method [4]. The obtained data are given graphically in figure 1. Rhenium hexafluoride and heptafluoride formed a

Content from this work may be used under the terms of the Creative Commons Attribution 3.0 licence. Any further distribution of this work must maintain attribution to the author(s) and the title of the work, journal citation and DOI.
Published under licence by IOP Publishing Ltd

IOP Conf. Series: Materials Science and Engineering **112** (2016) 012042 doi:10.1088/1757-899X/112/1/012042

diagram of a eutectic type with a degenerated eutectic from ReF6 side. Polymorphic transformation of ReF7 was revealed in the system at a temperature of ~ 24.5°C.

Solution heat ofReF7 in ReF6 calculated from a tangent of a liquidus line slope in coordinates of lgN - 1/T (where N – a mole fraction of ReF7) is 13.2 kcal/mole. Melting temperature of pure ReF6 and ReF7 is respectively 18.5°C for ReF6 and 48.3°C for ReF7.

Liquid-vapor equilibrium in the system was investigated by an ebulliometric method by boiling points within a range of the pressures exceeding normal (from 760 to 1.085 mm Hg). The temperature was measured by thermometers with a division value of 0.1°C. The measurements data are given in figures 1 and 2. It follows from the experiment data that ReF6-ReF7 system is practically ideal.

In the assumption that the system is ideal the components relative volatility is 3.5 at a normal pressure from the side of pure ReF6, and 4.2 from the side of pure ReF7. It specifies that separation ReF6 and ReF7 by a rectification method does not cause difficulties. The obtained data on boiling temperatures of pure fluorides were 33.4°C for ReF6, and 72.2°C for ReF7.

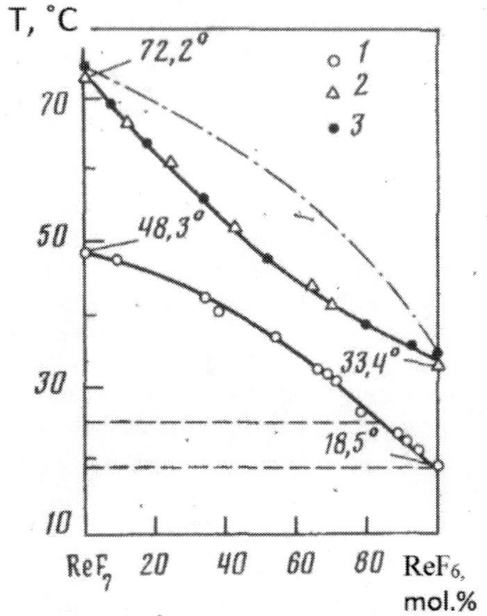

Figure 1. Crystal -liquid and liquid-vapor equilibria (at 760mmHg) in ReF_6-ReF_7 system: 1 - Liquidus temperatures under experimental data; 2 – Boiling temperatures of pure ReF_6 and ReF_7 and their mixture under experimental data; 3 – Data on the liquid line calculation in the assumption that the system is ideal; a dash-dotted line is a result of the vapor line calculation in the assumption that the system is ideal.

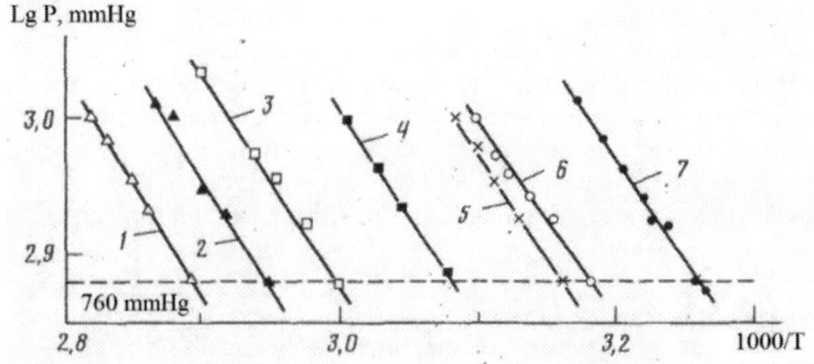

IOP Conf. Series: Materials Science and Engineering **112** (2016) 012042 doi:10.1088/1757-899X/112/1/012042

Figure 2. Dependence of ReF_6 and ReF_7 and their mixes saturated vapors pressure on temperature: 1 – pure ReF_7; 2 – 11.9 mol.% ReF_6; 3 – 24.5 mol.% ReF_6; 4 -43.0 mol.% ReF_6; 5- 64.3 mol.% ReF_6; 6 -69.2 mol.% ReF_6; 7 – pure ReF_6.

3. Synthesis of rhenium fluorides

Rhenium hexafluoride (ReF6) and rhenium heptafluoride (ReF7) used for researches were synthesized by fluorination of a rhenium powder of brand Re-0. The impurity composition of a source powder is given in Table 1.

For removing oxides and moisture, a rhenium powder was subjected to thermal treatment in a hydrogen flux at a temperature of 900-1000°C for two hours. Rhenium fluorination was performed in a horizontal cylindrical reactor made of nickel with a diameter of 100 mm and length of 1m. Two products of synthesis were obtained: synthesis product at a fluorination temperature of 300°C, synthesis product at a fluorination temperature of 600°C.

When the end product should be ReF6 and fluorination were too deeply with formation of significant amounts of ReF7 in the mixture, the situation can be corrected by adding metal rhenium to the mixture with conversion of superfluous ReF7 into ReF6. In paper [4] it was pointed to ease of ReF7 interaction with metal rhenium already at the temperatures close to its melting temperature (~50°C). We have specified a kinetic characteristic of the process.

4. Separation and purification of rhenium hexa - and heptafluorides by rectification

Investigation into liquid-vapor equilibrium in the ReF6-ReF7 system shows that their mutual separation and purification by a rectification method can be performed very effectively.

Two source products were rectified during this work. The rectification was carried out at atmospheric pressure with the use of a plate- and- sieve columns made of quartz glass with a diameter of 30 mm.

During the first experiment a synthesis product at a fluorination temperature of 300°C was rectified with the use of a column with 15 real plates. Selection of fractions lasted for 8.5 hours. The product mainly consisted of ReF6. Five fractions of purified ReF6 (1-5 fractions) were obtained. The fractions were selected with reflux ratios (R): fractions 1-5 at R from 70 to 60; fractions 6-8 at R from 150 to 200; fraction 9 at R = ~ 20; and fraction 10 at R close to 1.

During the second experiment a synthesis product at a fluorination temperature of 600°C was rectified with the use of a column with 25 real plates. Selection of fractions lasted for 7 hours. The product mainly consisted of ReF7. Four fractions of purified ReF7 (4-7 fractions) were obtained. The fractions were selected with reflux ratios (R): fraction 1 at R =200; fractions 2-3 at R from 90 to 70; the other fractions at R from 15 to 10.

The temperature of distillation curves are graphically shown in figure 3.

The results of analysis on the impurities content in the source sample and purified fraction are given in Table 1.The analysis was made by chemical- spectral method after the samples dissolution in the deionized water.

Gas chromatography analysis of rhenium fluorides purified by rectification showed that they did not contain any volatile impurities at a sensitivity of 10^{-3} volume percents.

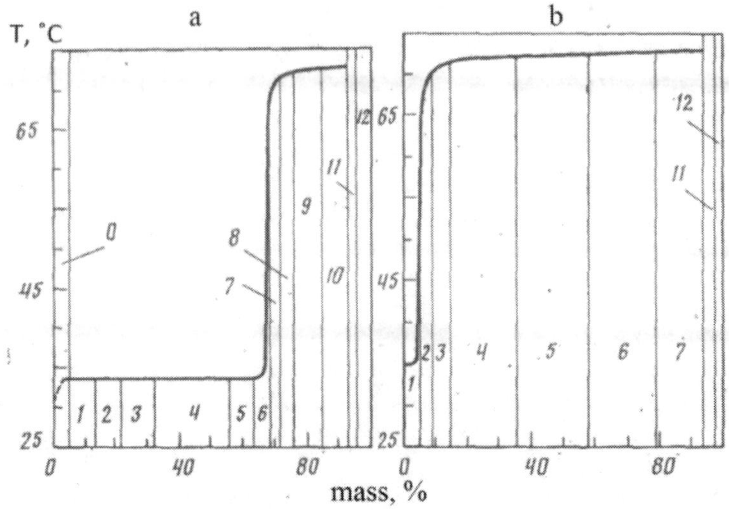

Figure 3. Temperature curve of rhenium hexa - and heptafluorides distillation: a mass percent of fractions from the initial loading is given on the X-axis, and fractions selection temperature is given on the Y-axis. The head fraction is presented by a zone zero; vat residue is presented by zone 11, and losses – by zone 12; *a*- Rectification of a synthesis product at a fluorination temperature of 300°C; *b*- Rectification of a synthesis product at a fluorination temperature of 600°C.

Table 1. Impurity composition of rhenium powder, source fluorides mixture and rhenium hexafluoride purified by rectification (10^{-4} mass %).

Impurity	Re powder of brand Re-O	Source fluorides mixture ReF_6+ReF_7	Purified fraction ReF_6
Al	5	0.6	0.05
Fe	7	0.1	0.05
K	10	<0.01	<0.01
Ca	10	2	0.2
Cu	3	2	0.1
Na	3	0.5	<0.05
Ni	2	<0.04	<0.01
Si	30	0.2	0.01
Mo	3	1	<0.2

5. Density, viscosity and surface tension of liquid rhenium hexafluoride and heptafluoride

In this work, density, viscosity and surface tension at the liquid-to-vapor boundary were measured in temperature range from fluorides melting point to the temperatures exceeding their normal boiling point for 20-40°C.

All physical and chemical studies were carried out inside evacuated and sealed off devices made of quartz glass. Each property was measured with three different devices. Density was measured with pyknometers-dilatometers according to techniques described in papers [6-8]. Viscosity was measured with capillary efflux viscometer [6, 9]. Design of the viscometers used was presented in [10]. Surface tension was measured using liquid capillary ascension technique, similar to paper [11].

Table 2. The experimental data on density, viscosity and surface tension were processed using least square method and formulated using respective interpolation equations

Compound	Equation	Range of measurement, °C	Standard deviation of measurement
ReF_6	ρ_l=3.717-0.00587*t, g/cm^3 $Lg\eta_l$=-0.945+261.4/t, cP σ=25.0-0.136*t, dyne/cm	18-76 20-76 20-76	$\Delta\rho_{st}$=0.0025 $\Delta\eta_{st}$=0.04 $\Delta\sigma_{st}$=0.1
ReF_7	ρ_l=3,909-0,00488*t, g/cm^3 $Lg\eta_l$=-1.359+446.7/t, cP σ=28.8-0.086*t, dyne/cm	50-96 51-96 52-96	$\Delta\rho_{st}$=0.002 $\Delta\eta_{st}$=0.005 $\Delta\sigma_{st}$=0.3

Regarding the density of liquid, our data correlate well with the data [12] presented for one temperature $\rho_l(ReF_6)$=3.58(22°C) g/cm^3 and $\rho_l(ReF_7)$=3.65(52°C) g/cm^3.

Processing of experimental data on liquid ReF_6 and ReF_7 viscosity in 1/ρ-1/η coordinates showed that liquid ReF_6 and ReF_7 are non-associated liquids.

6. Process operations in production of ReF_6 monofraction

The studies performed made it possible to develop the following sequence of process operations to produce ultra-pure rhenium hexafluoride monofraction:

1. Production of rhenium powder by hydrogen reduction of ammonium perrhenate at temperature (800÷850)°C:
$$2NH_4ReO_4 + 7H_2 \rightarrow 2Re + 2NH_3 + 8H_2O$$
2. Two-hour purification of rhenium powder from oxides and moisture in hydrogen flow at temperature (900÷1000) °C;
3. Fluoridation of rhenium powder:
$$2Re + 7F_2 \rightarrow 2ReF_7$$
$$Re + 3F_2 \rightarrow ReF_6$$
4. Transformation of ReF_7 into ReF_6:
$$6ReF_7 + Re \rightarrow 7ReF_6$$
5. Purification of ReF_6 from attending impurities by rectification

7. Conclusions

1. Higher rhenium fluorides were synthesized.
2. Separation and purification of rhenium hexafluoride and heptafluoride were performed
3. Phase equilibria in rhenium hexafluoride and heptafluoride system were studied
4. Density, viscosity and surface tension of liquid rhenium hexafluoride and heptafluoride were measured.
5. A process for ultra-pure rhenium hexafluoride monofraction production was developed.

References

[1] Lahotkin Y V, Krasovsky A I 1989 *Tungsten-rhenium coatings* (Moscow: Science) pp 1-158
[2] Malm J 1961 *J. Nucl. Chemie* **20** 189
[3] Zaitseva T S, Karelin A I, Lapin P V 1978 *Synthesis of high rhenium fluorides*
[4] Niselson L A and Perekhrest G L 1958 *Inorg. Chem. J.* **3(9)** 2150
[5] Dudin A S, Vovna V I, Rakov E G and Lopatin S N 1978 *Higher Ed. Institutions News: Chemistry and Chem. Technol. (in Russian)* **21(11)** 1564
[6] Pugachevich P P, Niselson L A, Sokolova T D and Anurov N S 1963 *Inorg. Chem. J.* **8(4)** 791
[7] Niselson L A and Sokolova T D 1962 *Inorg. Chem. J.* **7(12)** 2653
[8] Niselson L A, Sokolova T D and Stolyarov V I 1967 *Phys. Chem. J.* **41(7)** 1654
[9] Niselson L A, Sokolova T D and Stolyarov V I 1965 *Phys. Chem. J.* **39(12)** 3025

IOP Conf. Series: Materials Science and Engineering **112** (2016) 012042 doi:10.1088/1757-899X/112/1/012042

[10] Niselson L A and Tretiakova K V 1969 *Phys. Chem. J.* **43(8)** 2172
[11] Niselson L A and Sokolova T D 1965 *Inorg. Chem. J.* **10(7)** 1516
[12] Malm J G and Selig H 1961 *J. Jnorg. Nucl. Chem.* **20** 189

IOP Conf. Series: Materials Science and Engineering **112** (2016) 012043 doi:10.1088/1757-899X/112/1/012043

Structural Transformations in heat resistant coatings containing rare earth elements

N I Afanasiev[1*], O K Lepakova[1] and N I Kosova[2]

[1]Tomsk Scientific Center SB RAS, Tomsk, 634021, Pr. Akademichesky 10/4, Russia
[2]Tomsk State University, Tomsk, 634050, Lenin ave. 30, Russia

*E-mail: af42@yandex.ru

Abstract. Degradation of two-layered coatings and ZhS6U alloy microstructure were studied during long-term processes of high temperature annealing and creeping. It was shown that yttrium and zirconium oxides are promising as protective coatings for heat resistant nickel based alloy.

1. Introduction

One of the main targets of material science today is providing adequate environmental protection of heat resistant alloys. This target can be solved by application of protective coatings. Under influence of high temperature chemical composition and structure of coating and alloy may significantly change due to diffusion transfer processes. Component diffusion of protection system into alloy decreases its strength and creep resistance; diffusion of alloy components into coating leads to decreasing its heat resistance. Moreover, diffusion of impurities fluxes in interfacial area has an impact on recrystallization, creeping and freeness in grain boundaries, and pore formation [1].

Study of mechanism of coating structure degradation is very important for developing of optimal protection systems and techniques that can be used to coat alloy surfaces. Also, it is valuable to study the influence of coatings on mechanical properties of heat resistant alloys.

Ni(Co)-Cr-Al-Y–based heat resistant coatings [2] are used to protect alloy-treated heat resistant nickel based composites. Studies have shown [3] that these coatings do not prevent from full penetration of oxygen to alloy surface which may lead to internal oxidation of alloy under coating and deteriorating its protective properties.

2. Experimental section

In this work, double-layer coatings were used to prevent internal oxidation of alloys. On treated surface of ZhS6U (C-0.18, Cr-9.0, Co-4.8, W-10.3, Mo-1.5, Al-5.4, Ti-2.6, Ni-base, Nb-1.0, Zr-0.04, B-0.02) alloy the first (internal) layer consisting of Ni-20%Cr-13%Al-0.5%Y (weight %) was coated. Composition of the second (external) layer is 40%Y_2O_3-40%ZrO_2-10% Cr_2O_3-10% Al_2O_3 ceramics. The thickness of internal coating was 100 microns and external coating was 50 microns. Coatings were studied using high energy vacuum plasma spraying techniques [4]. The method of vacuum plasma technology for high-energy (MVPT) developed VIAM. Use a vacuum arc ion energy 30-150ev. The cathode is made of an alloy coating. Method of vacuum-plasma technology of high

Content from this work may be used under the terms of the Creative Commons Attribution 3.0 licence. Any further distribution of this work must maintain attribution to the author(s) and the title of the work, journal citation and DOI.
Published under licence by IOP Publishing Ltd

energies (VPTHE) developed in VIAM. Used vacuum arc ions with energy of 30-150ev. The cathode is made of alloy coating.

Mechanical tests on stretching and creeping were conducted at temperature of 1273 K and vacuum pressure of 10^{-3} kPa in air atmosphere. Flat samples were used with size of working heads $2*3*20$ mm^3. Structural and chemical transformations at surface layers of alloy and in the coatings were studied during annealing and creeping processes by X-ray structure analysis, X-ray spectrum analysis, optical and electron microscopies.

3. Results and discussion

Figure 1a,b presents general view of two-layered coating and surface layer of ZhS6U alloy at different time of annealing with air at temperature of 1273 K. Long-term annealing leads to the structural degradation of ceramic coating. It was observed that after 50 hours of annealing, coating is becoming crumbled; pores and microfractures are formed, deteriorating and layer destruction take place.

On surface of ZhS6U alloy and less on Ni-20%Cr-13%Al-0.5%Y alloy coating formation of oxide scale was observed which is crumbled and exfoliated with increase of annealing time. After 100 hours of annealing with air nickel-based coating became thin in 10-15%, but still remained indiscrete and preserved its protective properties. Both between coatings and internal layers of alloy diffusion zones were formed (figure 1). Dependence of diffusional zone from annealing time complied with parabolic law. Growth rate of diffusional zone in the process of creeping is 3-4 times less than at annealing without loading. This is probably connected with compressed tensions that appeared in cross-section of sample during the process of creeping.

ZhS6U alloy has dendritic structure (figure 1a). Predominant strengthening phases are γ' (Ni_3Al) and carbides Cr_7C_3 (figure 1e). Volume fraction of γ'-phase consists of 55%. Particles with size of 0.1-0.3 microns have cubic shape (figure 1e). Large particles of γ'-phase and γ'/γ –eutectic are observed with size of up to 10 microns. At all modes of thermal treating topological close-packed phases in alloys were not observed. After 4 hours of annealing at temperature of 1273 K size of grains in alloy is 30-40 microns. Carbide particles have shape of plates and bars with length of 15 microns and thickness up to 5 microns (Fig.1e). Coalescence of grains and phases in alloy is observed during annealing and creeping at 1273 K. After 50 hours of annealing size of grains increases twice as much, and size of particles γ'-phase reaches 0.8-1 microns.

After treating, both coatings have ultrafine-grained structure with size of grains and phases of 0.2-0.3 microns. Grained structure in alloy Ni-20%Cr-13%Al-0.5%Y has high thermal stability. Annealing during four hours at temperature of 1273K leads to increasing of average size of grains in this layer up to one micron (figure 1*d*), and after 50 hours up to 2.2 μm. In 40%Y_2O_3-40%ZrO_2-10% Cr_2O_3-10% Al_2O_3 ceramic coating average size of grains consists of 5 (figure 1*c*) and 15 microns after corresponding thermal treating.

NiAl, Ni_3Al phases and free chrome in the first coating were observed by X-ray structure analyses. Set of detected phases is typical for all studied modes; however, distribution of relative intensity depending on thermal treating is changed. Relative intensity of X-ray lines of Ni Al phase decreases, and Ni_3Al-phase increases. Decreasing of amount of NiAl phase is connected with aluminum consumption for Al_2O_3 oxides and aluminum diffusion into alloy. In Y_2O_3 - ZrO_2 ceramic based coating phases Y_2O_3, Cr_2O_3, ZrO_2, α - Al_2O_3 were found.

Figure 2 shows distribution of alloying elements in coatings and at surface layer of alloy. It is observed higher content of aluminum and titanium between alloy and coating. This is connected with aluminum, diffusion from coating into alloy which lead to increasing of volume fracture of γ'-phase and oxygen penetration into alloy surface with formation of titanium oxides (figure 1*f*).

In alloys with one layered coating oxidation of alloy surface under coating starts after 5-10 hours of annealing and creeping in air at temperature of 1273K. In alloy with two layered coating formation of oxides takes longer time (40-50 hours). During this period of time formation of open-ended pores is occurred in ceramic coating, and oxygen is penetrating to the surface of first coating and diffuses through it. On the surface of ZhS6U alloy under Ni-20%Cr-13%Al-0.5%Y alloy formation of titanium and nickel oxides take place. Structure of oxides is presented in figure 1*d*.

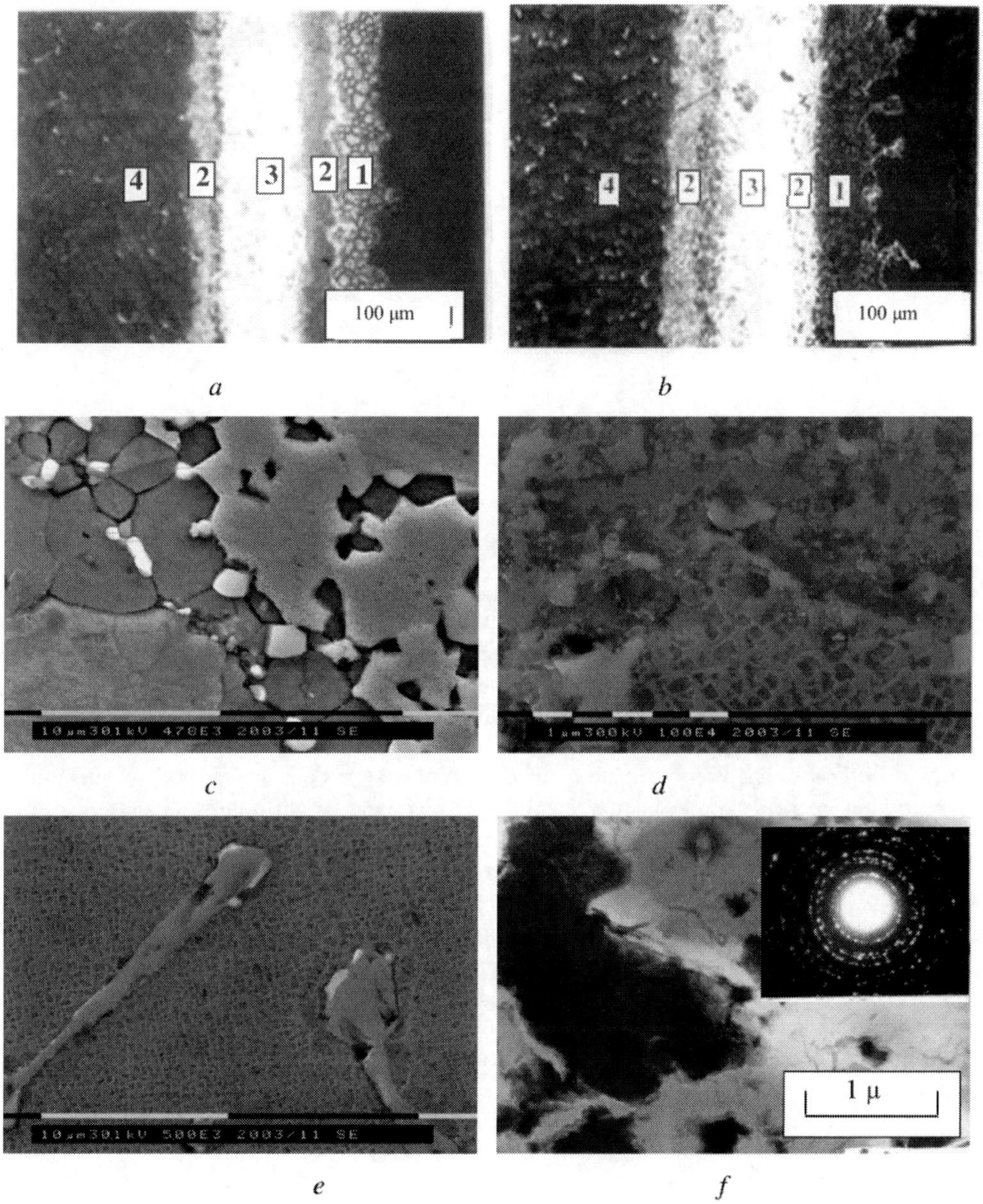

Figure 1. *a*, *b* – two-layered structure of coating after 4 and 50 hours of annealing at temperature of 1273K, 1 – ceramic coating, 2 – diffusion zones, 3 – Ni-20%Cr-13%Al-0.5%Y alloy coating, 4 – alloy, *c* – grained structure of ceramic coating after 4 hours annealing, *d* – carbides and γ'-phase in ZhS6U alloy after 4 hours of annealing, *f* – oxides of Ti and Ni in alloy after coating with microelectronogram.

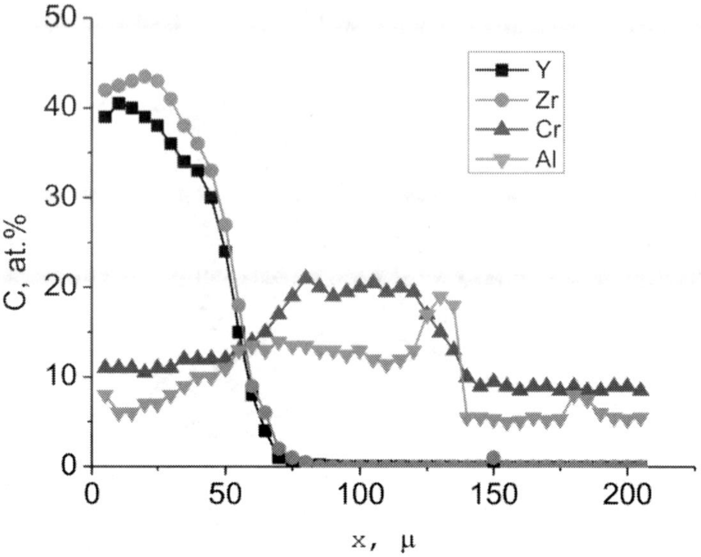

Figure 2. Dependence of element concentration in coatings and in surface layer of ZhS6U alloy from the depth

Ultimate strength of ZhS6U alloy without coating and with one layered and two layered coatings at temperature of 1273 K in vacuum on temporary base of 100 hours without taking into account cross section of covering consists of 170 MPa. Coatings at 1273 K have low ultimate strength at high loading (3-5 MPa), therefore, if take into account cross-section of coating, the strength of samples is decreased [5]. Tests with air at the same conditions lead to decrease of time when samples without coating start destructing up to 50 hours and with one-layered coating from Ni-20%Cr-13%Al-0.5%Y alloy up to 80 hours. Difference in time, when samples start destructing both in air and in vacuum, decreases when size of sample is increasing. It is connected with abnormal demonstration of scale factor at long term tests in corrosion conditions. Additional coating on the base of Y_2O_3 – ZrO_2 ceramics allow increasing the time till destruction of samples up to 90 hours.

Y_2O_3 – ZrO_2 - ceramic based coatings have lower heat resistance in comparison with coating from Ni-20%Cr-13%Al-0.5%Y alloy. Coating with thickness 50 microns is completely destructed for 80 hours of annealing in air atmosphere at temperature of 1273 K. During annealing it is observed intensive growth of grains in ceramic coating on internal and external interfacial areas with pore formation. After 50 hours of annealing at 1273 K on boundaries of grains formation of open-ended pores is occurred and oxygen easily penetrates to the first coating. Ni-20%Cr-13%Al-0.5%Y coating remains indiscrete and preserves its protective properties after 100 hours of annealing in air with no load. Additional ceramic coating increases time when ZhS6U alloy sample starts destructing on 10% in the process of high temperature creeping in air.

4. Conclusions

Putting additional ceramic coating with composition of 40% Y_2O_3-40% ZrO_2-10% Cr_2O_3-10% Al_2O_3 on Ni-20% Cr-13Al-0,5% Y coating leads to delay of oxygen penetration to the ZhS6U heat resistant surface for 80 hours. Destruction time of ZhS6U samples with two-layered coating at ultimate strength of 170MPa and 1273K temperature increases from 100 to 150 hours in comparison with destruction time of one-layered Ni-20% Cr-13% Al-0.5% Y samples. During annealing it is observed intensive growth of grains in ceramic coating on internal and external interfacial areas with pore formation. It is not recommended to use ceramics with 40% Y_2O_3-40% ZrO_2-10% Cr_2O_3-10% Al_2O_3 composition as one- layered coating for heat resistant nickel-based alloys. Prior to putting coating on alloy it is necessary to use interface heat resistant coating which contains nickel or cobalt as a base.

References

[1] Kolobov Yu, Valiev R, Grabovetckaia G, Zhiliaev A, Dudarev E, Ivanov K, Ivanov M, Kashin O and Naidenkin E 2001 *Grain Boundaries Diffusion and Properties of Nanostructure Materials* (Novosibirsk: Nauka)
[2] Shalin R 1981 *Heat Resistant Alloys for Gas Turbines* (Moscow: Metallurgy)
[3] Afanasiev N, Bushnev L, Kasymov M, Kolobov Yu, Muboyadzhan S and Pomyalov Ya 1986 *Izvestiya Vuzov: Physics* **12** 109
[4] Burke J and Wise F 1985 *Surface treating and materials reliability* (Moscow: Mir)
[5] Afanasiev N, Bushnev L, Avramchik A 2010 *Physics and chemistry of materials processing* **1** 36

Deformation and Damage Accumulation in a Ceramic Composite under Dynamic Loading

M V Korobenkov[1,3], S N Kulkov[1,2], O B Naymark[4], U V Khorechko[5] and A V Ruchina[5]

[1] National Research Tomsk State University, Tomsk, Russia
[2] Institute of Strength Physics and Materials Science SB RAS, Tomsk, Russia
[3] Institute of Marine Technology Problems FEB RAS, Vladivostok, Russia
[4] Institute of Continuous Media Mechanics UB RAS, Perm, Russia
[5] National Research Tomsk Polytechnic University, Tomsk, Russia

[1]E-mail: korobenkov@ftf.tsu.ru

Abstract. Methods of computer modelling were used to investigate the processes of deformation and microdamage formation in ceramic composite materials under intense dynamic loading. It was shown that there was no damage caused by dynamic compression in the vicinity of phase borders of a nanostructured aluminum oxide matrix and reinforcing particles of tetragonal zirconium dioxide. Also, the local origination of microdamages occurs only in the zones close to micropores.

Keywords: ceramic composites, dynamic exposure, computer modelling

1. Introduction

By the present day, the range of application of structural ceramics combining high hardness, thermal resistance, wear strength and chemical inertness has appreciably expanded. This is due to a successful development of technologies for the production of ceramic composites on the basis of powders with nanocrystallic structure, which enables mass production of materials with augmented strength and crack resistance that are successfully implemented in oil and gas industry, and are used as protective coating. The use of new generations of ceramic materials demanded the creation of adequate models of their mechanical behavior under high dynamic loading. The fracture of brittle ceramic materials usually occurs in dynamic regime and is accompanied by a fast change of the local stress condition. However, the determination of the influence of geometrical factors on the mechanical behavior is a complex challenge.

The available experimental data is not enough for determining the influence of geometrical factors. In this connection, the solution of the problem on predicting the properties of ceramic composites on the basis of computer modelling is a promising task. This is conditioned by the ability of computer modelling of mechanical behavior of heterogeneous condensed media to improve the understanding of regularities describing the deformation, development of damage and fracture of ceramic composites under severe dynamic exposure.

Content from this work may be used under the terms of the Creative Commons Attribution 3.0 licence. Any further distribution of this work must maintain attribution to the author(s) and the title of the work, journal citation and DOI.
Published under licence by IOP Publishing Ltd

The physical and mathematical model for studying the kinetic processes of damage and fracture of composites with an oxide ceramic matrix filled with submicron impurities accounts for the results obtained after investigating the microstructure of materials at different scale levels [1]. The numerical modeling of deformation and fracture was performed with the use of either the discrete-continual particle method or the finite difference method with the accuracy of second order.

2. Computational Model

The model volume of a structural inhomogeneous medium (RVE) is filled with particles of identical effective size *d*. Physicomechanical properties of each particle are set according to its attribution to matrix phase, grain boundary phases (matrix grain-to-matrix grain, matrix grain-to-reinforcing particle), reinforcing particles. The volume of microcracks is filled with pseudoparticles, the mass density is zero. Introduction into the model of the pseudoparticles allows modelling free surfaces in a microcrack without altering the calculation algorithm. The lack of estimated stresses on the surface of a crack will be carried out automatically during the calculation of particle-pseudoparticle couples.

The complete system of equations describing nonstationary loading in relation to model volume includes conservation equations for weight, impulse and energy, and kinematic equations defining ratios, initial and boundary conditions.

As a criterion of brittle local fracture of particles that form composite ceramic material, the following condition was adopted:

$$\sigma_1 \geq \sigma_{tension}, \qquad (1)$$

where σ_1 is the main component of the stress tensor, $\sigma_{tension}$ is the tensile strength of the ceramic matrix condensed phase, grain boundary phases, phase of reinforcing particles, under quasi-static or dynamic deformation conditions.

To evaluate $\sigma_{tension}$, the data from [2, 3] were used.

The present work uses media damage parameter *D* considered as a result of accumulation of damage increments during the microstructure evolution at lower scale level over the time from t_0 to $t_0 + m\,\Delta t$ [4–7]:

$$D = \sum_{i=1}^{m} \frac{\left[\Delta\varepsilon_{eq}^{p}\right]_i}{\varepsilon_f}, \qquad (2)$$

Where $\Delta D_i = \frac{\left[\Delta\varepsilon_{eq}^{p}\right]_i}{\varepsilon_f}$ is the increment of the damage parameter over the deformation time Δt,

ε_{eq}^{p} is the intensity of inelastic deformation,

ε_f is the ultimate strain at the moment of macroscopic fracture.

The criterion of local fracture of a particle of the matrix is as follows:

$$D = 1. \qquad (3)$$

3. Results and Discussion

The modeling results showed that there are local areas of tensile stresses occurring during the propagation of a compression wave in the sample after the heterogeneous media compression with high speed of deformation in the front of shockwaves (Figure 1).

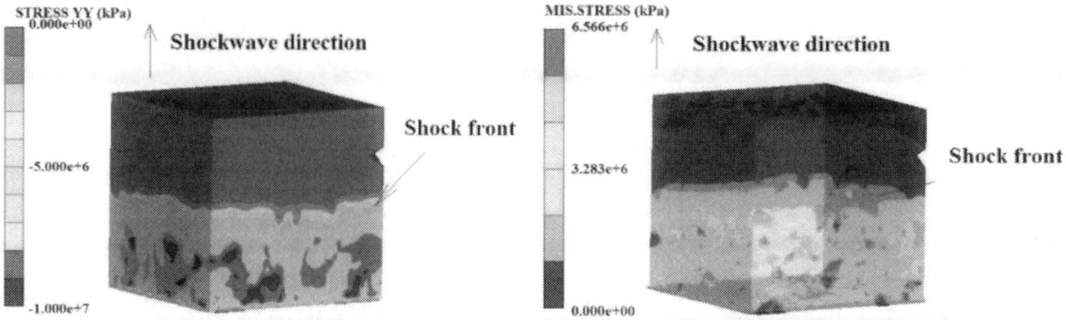

Figure 1. Distribution and strength of shear stresses during the shockwave propagation in Al_2O_3-(15vol.%t-ZrO_2) composite material

In the case of elastic-brittle behavior of ceramic composites, the size of J-integral corresponds to the energy stream flowing to the top of the crack. In this case,

$$J = J_{1C} = G, \qquad (4)$$

where the J_{1C} is fracture toughness, G is the crack energy release rate.

Figure 2 depicts the distribution of specific energy of damage origination in a model volume of Al_2O_3-(15vol.%t-ZrO_2) composite behind the front of the shockwave over consecutive moments of time. The values of specific energy of damage formation calculated in the course of the modelling of dynamic loading of a composite model volume can be used for fracture toughness determination.

Figure 2. Specific work of damage origination behind the front of shock compression

The loss of shear strength of the nanocomposite during high-speed deformation could be caused by the localized propagation of damages or microcracks. The complete loss of shear strength of a model volume will be reached in the case of its fragmentation when the damage parameter averaged over the volume becomes less than criterial value equal to 1.

In the considered Al_2O_3-(t-ZrO_2) model ceramic composites, the local origination of microdamage occurs only in the zones close to microcavities (micropores) in the case of high-speed deformation. There was no damage origination detected under shock compression in the vicinity of phase boundaries of Al_2O_3 nanostructural matrix and the reinforcing particles.

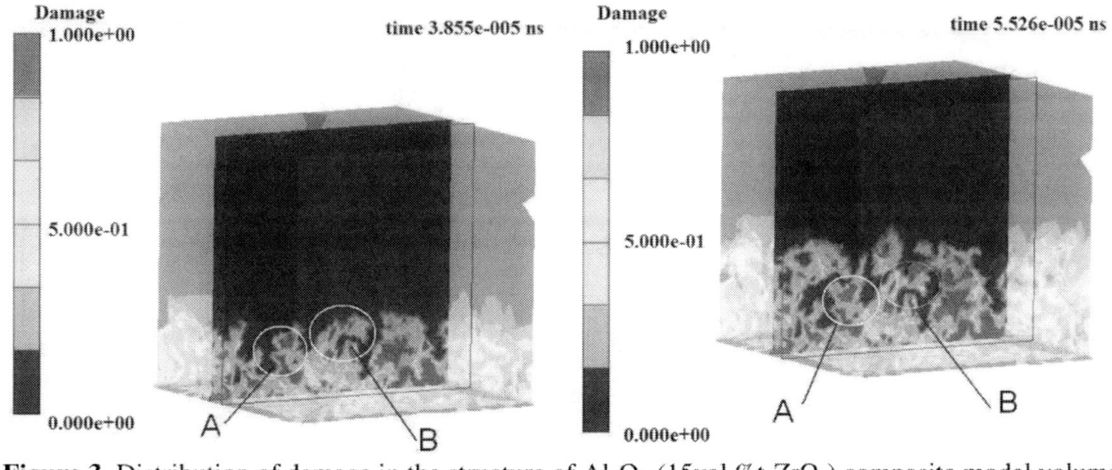

Figure 3. Distribution of damage in the structure of Al_2O_3-(15vol.%t-ZrO_2) composite model volume behind the shockwave front at consecutive moments of time of 0.0385 and 0.0553 ns

The results shown in Figure 3 for consecutive moments of time testify the formation of local damages of the ceramic matrix in the zone of shock transition. There was no considerable alteration detected to the size and configuration of local damage areas, for example, to area A and B in Figure 3, after the increase of time of hydrostatic compression by the shockwave front.

4. Conclusions

The modeling of compression wave distribution in a composite has demonstrated the formation of tensile strain areas. This leads to the emergence of microdamage at mesoscopic level in zones close to microvoids. At the amplitude of compression impulses not exceeding the limit of elasticity of a high density condensed matrix, the damage behind the front of shockwaves remains stationary and localized at the microscopic level. Possible formation of microcracks behind the front of a shockwave is connected with the presence of pore clusters. The zones of local damage get elongated; the orientation of axes of which is defined by the direction of the maximum shear stress.

The presence of submicron reinforcing particles near a pore cluster leads to the change in the shape of local damage area, while the increase of the relative number of damaged particles can be defined as the formation of a shear mesocracks in a model volume of the composite.

Thus, the modeling has shown that the value of effective shearing resistance is higher for the composite with higher concentration of particles. The obtained data agree well with the measurements of static strength characteristics of the considered class of composites [8].

Acknowledgments

The work was performed within the implementation of Tomsk State University Competitiveness Improvement Program and with partial financial support of the Ministry of Education and Science of the Russian Federation within State task # 16.2004.2014 / K.

References

[1] Kulkov S N, Buyakova S P 2007 *Nanotechnologies in Russia* **1-2** 119-32

[2] *Engineering Materials Handbook*. Desk Edition ASM International Society, 2001. Material Park Ohio USA, 44073-0002.

[3] Gogotsi G A 2003 *Ceramics International* **29** 777-84

[4] Cho H M , Yang P , Kattawar G W , et. all 2008 *Opt. Express* **16** 3931-48

[5] ANSYS AUTODYN Explicit software for nonlinear dynamics. Theory manual: *http://www.autodyn.org*.
[6] Skripnyak V A, Skripnyak E G, Kozulin A A, et. all 2009 *Russian Physics Journal* **12** 1300-08
[7] Skripnyak V A, Skripnyak E G, Kozulin A A, Pasko E G, Skripnyak V V, Korobenkov M V 2009 *Bulletin of the Tomsk politechnick university* **315 № 2** 113-7
[8] Buyakova S P, Kulkov S N 2010 *Inorganic materials* **46 № 10** 1277–80

IOP Conf. Series: Materials Science and Engineering **112** (2016) 012045 doi:10.1088/1757-899X/112/1/012045

Self-propagating high-temperature synthesis of the high-current emission lanthanum and niobium contained ceramics

O Yu Dolmatov[1], I V Shamanin and S S Chursin
National Research Tomsk Polytechnic University, Tomsk, Russia

[1]E-mail: dolmatov@tpu.ru

Abstract. The paper describes the production of metal-ceramic high-current emitters containing lanthanum hexaboride LaB_6 by self-propagating high-temperature synthesis. Tests of emitters as explosive-emission cathodes in an OMEGA-350 microsecond accelerator are presented. Experiments show that when a metal-ceramic cathode is used, the emitted energy was 12-17 % higher than that produced by tungsten and graphite cathodes and the beam "signature" is a circle with a small (~ 15 %) variation in radius.

1. Introduction

Accelerators of particle occupy a very important place in the industry. The most important element of devices generating flows of charged particles is cathode [1]. Cathodes made from lanthanum hexaboride (LaB_6) are widely used in accelerators to generate high-current electron beams in stationary and quasi-stationary modes. However, sintered pellets of LaB_6 are brittle and easily destroyed by heat shock [2,3]. Consequently cathodes with layers of LaB_6 plasma-sputtered onto substrates made from high melting- point metals are commonly used.

Here we present results of the synthesis of high-current emitters for explosion-emission cathodes containing LaB_6. The cathodes are produced by self-propagating high-temperature synthesis [4].

2. Experiment

The initial reagents are powders of La and Ti oxides, amorphous B and Si. The blend is additionally ballasted with a reactive stoichiometric mixture of Ni and Al powders taken in various proportions, thus the emitters are tougher due to the formation of a Ni-Al matrix containing inclusions of LaB_6 and TiB_2 compounds. The mixture is ground in a ball mill and they dried in a low vacuum. Emitter pellets are produced by moulding the mixture in press-molds of various diameters (up to 30 mm).

The explosive reaction is initiated at the edges of the pellet by heating them in a quartz tube furnace in air. Temperature diagrams generated by thermocouples placed on the sides and upper edges show that the process begins as surface combustion. After the combustion front has passed over the surface, a secondary heat explosion reaction begins inside the sample. The surface combustion acts as a "chemical furnace" that heats the center of the pellet.

The explosion consists of two reactions, viz.,

$$TiO_2 + \frac{10}{3}B \rightarrow TiB_2 + \frac{2}{3}B_2O_3, \quad (1)$$

$$\frac{1}{2}\mathrm{La_2O_3} + 7\mathrm{B} \rightarrow \mathrm{LaB_6} + \frac{1}{2}\mathrm{B_2O_3}, \quad (2)$$

with adiabatic combustion temperatures of 1400 and 1300 K, respectively [4], together with the synthesis of nickel aluminides.

Content from this work may be used under the terms of the Creative Commons Attribution 3.0 licence. Any further distribution of this work must maintain attribution to the author(s) and the title of the work, journal citation and DOI.
Published under licence by IOP Publishing Ltd

IOP Conf. Series: Materials Science and Engineering **112** (2016) 012045 doi:10.1088/1757-899X/112/1/012045

In order to test the emitters, several experiments were performed using the OMEGA-350 microsecond high-current electron beam accelerator. The accelerator OMEGA-350 allows generated electron beams with duration from 1 to 3 microseconds at a current density of 50 kA/cm^2 and energy of 350 keV electrons to ensuring full supply of energy of the beam more than 1 kJ generation efficiency to 80%. The complex includes a high-current vacuum diode, voltage pulse generator, and the system of collecting and processing information. Voltage pulse generator is a seven cascade generator, assembled in a classic case Arkadieva-Marksa. With voltage pulse generator may exercise the power vacuum diode or other physical equipment pulse voltage with a maximum peak value of 350 kV, negative or positive polarity in the manual (single) or automatic (multiple) modes.

3. Results and discussion

The temperature developed on the pellet surface during the surface phase is 1350-1450 K and then rises to 2200 K when the secondary heat explosion reaction starts. The synthesis is initiated when the surface temperature is over 1000 K.

The structure of the emitter surface is shown in figure 1. Several compositions were used to produce the emitter, viz. La_2O_3 (D 543 NTD OST 4S-194-81); TiO_2 (TU 6-09-2166-72); B (TU 113-12-132-83); Si (KR). The mixture is ballasted with powders of nickel (PNK-OT1) and aluminum (PA-4).

Figure 1. Electron-microscope photograph of the high-current emitter's surface (magnification 1:1000, E-modulation).

Many inhomogeneities (LaB_6 and TiB_2) can be seen on the emitter surface; some of the Si reacts with the TiO_2 to form the silicides TiSi, $TiSi_2$, and some Si forms inclusions 10-25 μm in size. These are uniformly distributed over the volume and surface.

Lanthanum hexaboride LaB_6 has a considerable field-emission current. The work function is considerably lower in the presence of La atoms in the surface layer of the emitter [5]. The LaB_6-Me(IV)B_2 compounds (where Me(IV) indicates a group IV metal, i.e., Ti, Zr, Hf) have relatively high thermal-emission characteristics.

The initial stages of the explosion emission involve thermal and field emissions. Field- and thermal-emission currents lead to the Joule heating of small points and to a subsequent explosion. Given the good field- and thermal-emission characteristics of materials, the time between the voltage pulse and beginning of the explosive emission can be shortened and, as a consequence, the time needed for the cathode plasma to form shortened. This, in turn, shortens the delay between the electron-current signal and the accelerating voltage pulse, and as a result, the energy transferred from the diode to the beam is higher.

The Si inclusions on the emitting surface increase the electric field at their boundaries. Thus, when there is no emission from some points, adjacent points stimulate explosive emission due to the potential differences that arise due to the currents passing through the points. As a result, the time needed for the plasma flares to merge on the emitting surface into a uniform plasma layer is shorter and the emitted electron beam is more uniform over its cross section.

Next step of experiments is test to emitters on the OMEGA-350 microsecond high-current electron beam accelerator.

The following figure shows oscillograms of accelerating voltage and beam current pulses outputted to the surface of the irradiated target.

Figure 2 and figure 3 show the results for the regime with current pulse duration of 2.4 microseconds.

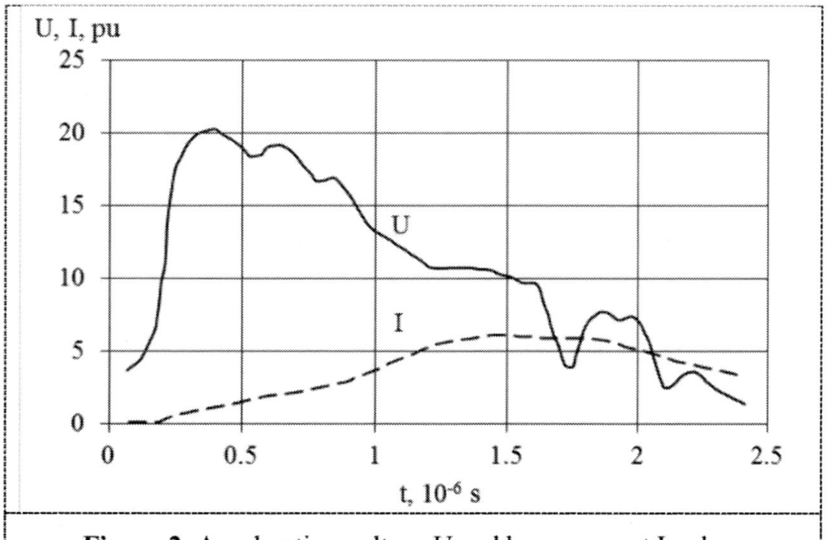

Figure 2. Accelerating voltage U and beam current I pulses.

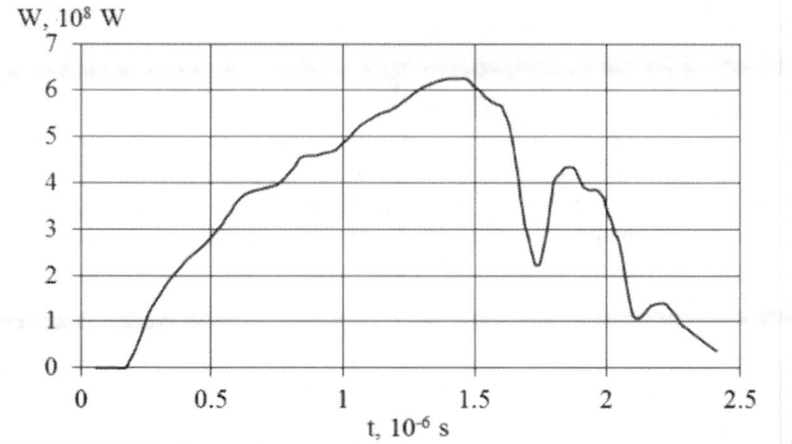

Figure 3. Beam power W within current pulse duration.

We obtained the following beam parameters:

- Beam energy storage 814 J.
- Beam generation efficiency 71 %.
- Energy density on irradiated surface 65…70 J/cm^2.

Figure 4 and figure 5 show the results for the regime with current pulse duration of 1.75 microseconds.

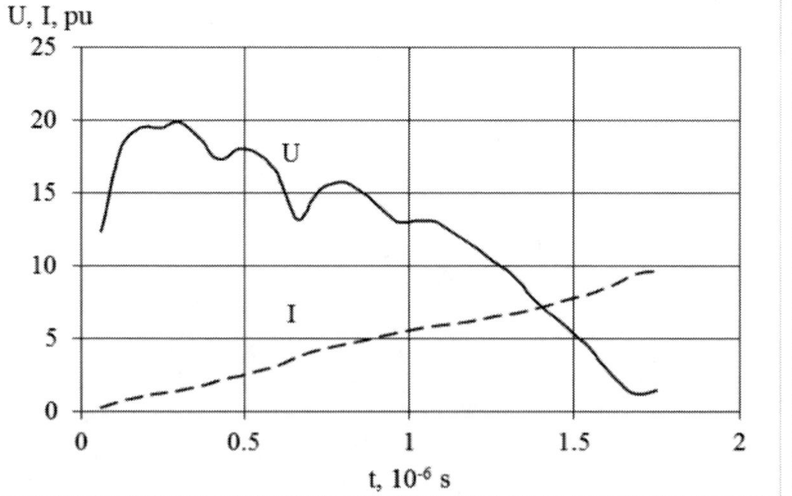

Figure 4. Accelerating voltage U and beam current I pulses.

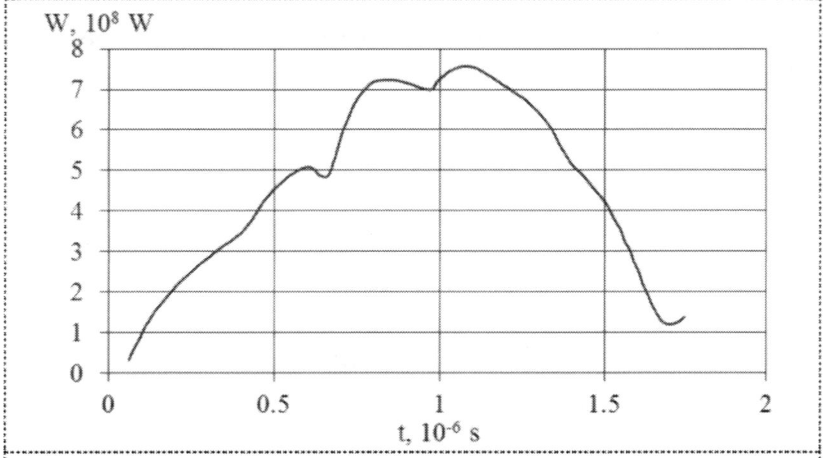

Figure 5. Beam power W within current pulse duration.

We obtained the following beam parameters:

- Beam energy storage 808 J.
- Beam generation efficiency 68 %.
- Energy density on irradiated surface 65…70 J/cm^2.

Also in experiments we used detectors of absorbed energy. The detector was a solid- state chemical dosimeter, viz., polycrystalline potassium nitrate in the form of pellets 16 mm in diameter and wide enough to absorb the beam fully. The density of the energy absorbed by the dosimeter in one pulse was calculated from the radiation-chemistry yield of nitrate ions $[NO_2^-]$. Experiments show that when a metal-ceramic cathode is used, the emitted energy was 12-17 % higher than that produced by tungsten and graphite cathodes.

4. Conclusions

The homogeneity of the cathode plasma was assessed from the "signature" of the electron beam output from the diode. The ES spectrographic photoplates and alkali-halide crystals were used as the sensitive elements. Experiments show that when a metal-ceramic cathode is used, the beam "signature" is a circle with a small (~ 15 %) variation in radius. The saturation of the crystal color is homogeneous within the circle. The negative picture of the "signature" is also homogeneous. The reproducibility is significantly worse when a tungsten cathode is used. The shape of the "signature" stochastically deviates from the circular to elliptical, and in some cases the beam fragments are observed. The explicit color inhomogeneity inside the "signature" is evidence of fragmentation.

References

[1] Abdullin E N, Ivanov N G, Losev V E, Morozov A N 2013 Generation of a large cross-section electron beam in the vacuum diode with rod current returns *Laser and Particle Beams* **31, 4** 7697-702

[2] Loginov L V, Naumkin A P, Fedotova L K and Shumilin A P 1988 *Abstracts of the 7th All-Union Symp. on High-Current Electronics part 1* (Tomsk) (in Russian) p. 149.

[3] Kresanov V S et al. 1987 Highly efficient electron emitter based on lanthanum hexaboride (M.: "Energoatomizdat") [in Russian]

[4] Merzhanov A G 1975 *Combustion Processes in Chemical Technology and Metallurgy* (Chernogolovka) (in Russian)

[5] Mesyats G A 1984 *Emission High-Current Electronics* (Novosibirsk) (in Russian)

Institute of Physics
Dirac House, Temple Back
Bristol BS1 6BE UK

ISSN: 1757-8981
ISBN 978-1-5108-2087-6

CURRAN ASSOCIATES INC.
proceedings.com